The Neutrophil: Cellular Biochemistry and Physiology

Editor

Maurice B. Hallett, Ph.D.
Senior Lecturer in Immunology
University Department of Surgery
University of Wales College of Medicine
Cardiff, U.K.

CRC Press, Inc.
Boca Raton, Florida

Library of Congress Cataloging-in-Publication Data

The Neutrophil: cellular biochemistry and physiology / editor,
Maurice B. Hallett.

p. cm.
Bibliography: p.
Includes index.
ISBN 0-8493-4808-0
1. Neutrophils. I. Hallett, Maurice B. (Maurice Bartlett), 1952-

QR185.8.N47N46 1989
612'. 112——dc19 88-2603

This book represents information obtained from authentic and highly regarded sources. Reprinted material is quoted with permission, and sources are indicated. A wide variety of references are listed. Every reasonable effort has been made to give reliable data and information, but the author and the publisher cannot assume responsibility for the validity of all materials or for the consequences of their use.

Direct all inquiries to CRC Press, Inc., 2000 Corporate Blvd., N.W., Boca Raton, Florida, 33431.

International Standard Book Number 0-8493-4808-0

Library of Congress Card Number 88-2603
Printed in the United States

Dedicated
to Jill, Jenny, and Tim.

PREFACE

The importance of neutrophils in combating infectious disease is dramatically demonstrated in chronic granulomatous disease. In this condition, neutrophils and macrophages fail to effectively kill infecting microbes. The consequences of this failure are catastrophic and fatal. Without effective neutrophils, the balance between the microbes and the animal swings greatly towards the microbe. Arguably, the evolution of man, therefore, has depended upon the efforts of these cells to keep the microbiological environment at bay. It is also being recognized that, in some situations, the "weapons" normally used in the fight against microbes can become directed against surrounding tissue. The resulting tissue damage may underlie many of the inflammatory diseases. Understanding of the mechanisms by which neutrophils perform their crucial antimicrobial role in health and the mechanisms that induce their aberrant behavior in disease is therefore of great importance.

As a consequence, there are a great many reviews on specific areas of interest within the cell. For example, some recent and important reviews have been given by Tauber,[1] Sklar and colleagues,[2] and Sha'afi and Molski.[3] An excellent and comprehensive review of the knowledge and state of understanding of the neutrophil (to 1977) has also been given in the book by Klebanoff and Clark.[4] However, since its publication, many great strides have been made towards understanding the mechanisms that control neutrophil activity.

Over the past 5 years, there has been a massive increase in our understanding of the ways in which stimuli evoke cellular responses. The expansion of knowledge is also reflected in the number of publications on the subject of neutrophils, rising from approximately 500 in 1977 to over 1000 in 1986. The aim of this book is to condense much of this important information concerning signal response coupling in neutrophils into a view of the cellular

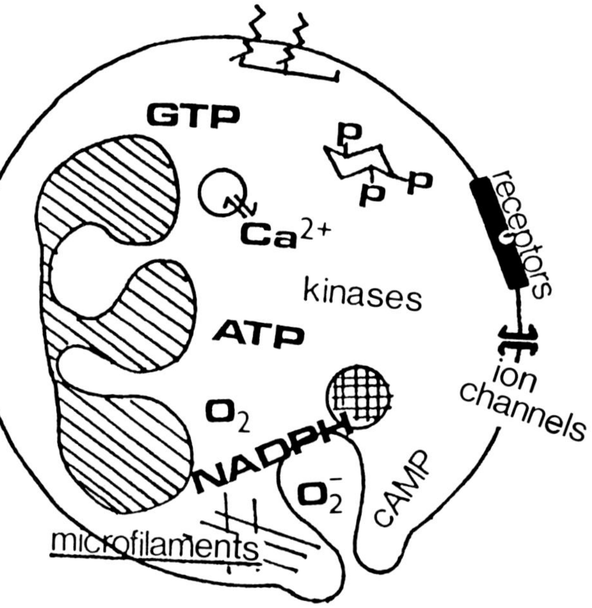

The neutrophil with some of the important intracellular constituents discussed in this volume shown. The reactions and relationships between them have been left deliberately ambiguous. After completion of this volume, the reader may perhaps feel confident about putting in a few definite arrows and even add a few speculative ones. (Space is also left for the inclusion of further components as their importance and identities becomes established.)

mechanisms that enables the cellular biochemistry and physiology to be understood within a conceptual framework. In areas where no such framework is clearly defined, it is hoped that the perspectives given will enable questions to be asked and answers sought. Despite the necessary process of condensation of the welter of data, each chapter has attempted to give a comprehensive review of its topic, over 1500 references being given in this volume. After the first chapter, in which some historical perspective to the study of neutrophils is given, together with an attempt to explore mechanisms for failure in pathological conditions for neutrophils, the first part of the text describes some of the cellular and intracellular events that result from the stimulation of neutrophils, i.e., chemotaxis, secretion, oxidase activation, arachidonate production, and cytoskeletal rearrangement during both phagocytosis and chemotaxis. The second part of the book concentrates on the mechanisms that couple stimulation to response, and includes discussion of intracellular Ca^{2+} and its relation to other intracellular changes, including kinase activation, phospholipid metabolism, and membrane electrophysiological changes.

REFERENCES

1. **Tauber, I. A.** Protein kinase C and the activation of the human neutrophil NADPH-oxidase, *Blood,* 69, 711, 1987.
2. **Omaann, G. M., Allen, R. A., Bokoch, G. M., Painter, R. G., Traynor, A. E., and Sklar, L. A.,** Signal transduction and cytoskeletal activation of the neutrophil, *Physiol. Rev.,* 67, 285, 1987.
3. **Sha'afi, R. I. and Molski, T. F. P.** Activation of the neutrophil, in *Progress in Allergy,* Vol. 39, Ishizaka, K., Kallos, P., and Waksman, B. H., Eds., S. Karger, Basel, 1987.
4. **Klebanoff, S. J. and Clark, R. A.,** *The Neutrophil: Function and Clinical Disorders,* North-Holland, Amsterdam, 1978.

THE EDITOR

Maurice B. Hallett is Senior Lecturer in Immunology at the University of Wales College of Medicine in Cardiff, U.K. He studied at University College London, receiving a B.Sc. in Pharmacology in 1974 and his Ph.D. in 1978. During this time, Dr. Hallett developed a strong interest in the intracellular mechanisms that control cell activation, particularly with the aim of understanding the manner in which alterations of cell signaling processes lead to allergic and inflammatory disease. This interest was strengthened at the Welsh National School of Medicine, where his research focused on the intracellular biochemistry and physiology of the neutrophil. He was appointed Lecturer at the University of Wales College of Medicine in 1984 and Senior Lecturer in 1987.

Dr. Hallett belongs to several scientific societies, including the Biochemical Society, the British Society for Cell Biology, the British Society for Immunology, and the Institute of Biology. His current research involves measurement of intracellular processes within living neutrophils taken from both normal and pathological tissues.

CONTRIBUTORS

Futwan A. Al-Mohanna
Research Officer
Department of Surgery
University of Wales
College of Medicine
Cardiff, Wales

Pierre Borgeat, Ph.D.
Professor
Unité de Recherche, Inflammation
et Immunologie-Rhumatologie
Centre Hospitalier de l'Universite Laval
Quebec, Canada

Shamshad Cockcroft, Ph.D.
Lecturer
Department of Experimental Pathology
School of Medicine
University College of London
London, England

Eryl Cooke, Ph.D.
Research Fellow
Department of Surgery
University of Wales
College of Medicine
Cardiff, Wales

Andrew Richard Cross, Ph.D.
Postdoctoral Research Associate
Department of Biochemistry
School of Medicine
University of Bristol
Bristol, England

Francesco Di Virgilio, M.D.
Associate Professor
Department of General Pathology
University of Padova
Padova, Italy

Elaine K. Gallin, Ph.D.
Head of Cellular Physiology Division
Department of Physiology
Armed Forces Radiobiology Research Institute
National Naval Medical Center
Bethesda, Maryland

Hans Gruler, Ph.D.
Professor of Physics
Department of Biophysics
University of Ulm
Ulm, West Germany

Maurice Bartlett Hallett, Ph.D.
Senior Lecturer in Immunology
Department of Surgery
University of Wales
College of Medicine
Cardiff, U.K.

Owen T. G. Jones, Ph.D.
Professor
Department of Biochemistry
School of Medicine
University of Bristol
Bristol, England

Daniel P. Lew, M.D.
Professor
Department of Infectious Diseases
Hopital Cantonal
Geneva, Switzerland

Leslie C. McKinney, Ph.D.
Research Physiologist
Department of Physiology
Armed Forces Radiobiology Research Institute
National Naval Medical Center
Bethesda, Maryland

Paul H. Naccache, Ph.D.
Associate Professor
Unité de Recherche, Inflammation et
Immunologie-Rhumatologie
Centre Hospitalier de l'Universite Laval
Quebec, Canada

Tullio Pozzan, M.D..
Professor
Department of General Pathology
University of Ferrara
Ferrara, Italy

Janet E. Rickard, Ph.D.
NATO Research Fellow
Department of Cell Biology
E. M. B. L.
Heidelberg, West Germany

P. Sheterline, Ph.D.
Senior Lecturer
Department of Medical Cell Biology
University of Liverpool
Liverpool, England

Ramadan I. Sha'afi, Ph.D.
Professor
Department of Physiology
University of Connecticut Health Center
Farmington, Connecticut

James E. Smolen, Ph.D.
Associate Professor
Departments of Pediatrics and Pathology
University of Michigan Hospitals
Ann Arbor, Michigan

Susan Treves, M.Sc.
Ph.D. candidate
Department of General Pathology
University of Padova
Padova, Italy

TABLE OF CONTENTS

Chapter 1
The Significance of Stimulus-Response Coupling in the Neutrophil for Physiology and Pathology 1
Maurice B. Hallett

Chapter 2
Characteristics and Mechanisms of Secretion by Neutrophils 23
James E. Smolen

Chapter 3
Biophysics of Leukocytes: Neutrophil Chemotaxis, Characteristics and Mechanisms.... 63
Hans Gruler

Chapter 4
The Molecular Mechanism of Oxygen Reduction by the Neutrophil Oxidase 97
Andrew Richard Cross and Owen T. G. Jones

Chapter 5
Mobilization, Metabolism, and Biological Effects of Eicosanoids in Polymorphonuclear Leukocytes 113
Paul H. Naccache, Ramadan I. Sha'afi, and Piêrre Borgeat

Chapter 6
The Cortical Actin Filament Network of Neutrophil Leukocytes during Phagocytosis and Chemotaxis 141
Peter Sheterline and Janet E. Rickard

Chapter 7
The Role of Inositol Phospholipid Metabolism and Diacylglycerol Production in Neutrophil Signal Transduction 167
Shamshad Cockcroft

Chapter 8
Role of Calcium in Neutrophil Activation 199
Francesco Di Virgilio, Susan Treves, Daniel Lew, and Tullio Pozzan

Chapter 9
Ca^{2+}-Dependent and Independent Mechanisms in Neutrophil Activation; Roles of Kinase C, Diacylglycerol, and Unidentified Intracellular Messengers 219
E. Cooke, F. A. Al-Mohanna, and M. B. Hallett

Chapter 10
Ion Transport in Phagocytes 243
Elaine Gallin and Leslie C. McKinney

Index 261

Chapter 1

THE SIGNIFICANCE OF STIMULUS-RESPONSE COUPLING IN THE NEUTROPHIL FOR PHYSIOLOGY AND PATHOLOGY

Maurice B. Hallett

TABLE OF CONTENTS

I. A Brief History of the Study of the Neutrophil 2
- A. Indentification of the Cell 2
- B. Early Speculation on the Role of the Neutrophil in Inflammation 2
- C. The Contributions of Metchnikoff 2
- D. The 19th Century Biochemistry of the Neutrophil 3
- E. Progress Since Metchnikoff 4
- F. More Recent Developments 4

II. The Involvement of Neutrophils in Pathology 5
- A. Inappropriate Activation of Neutrophils 5
- B. Why Do Neutrophils Go Wrong? 11
- C. Types of Cellular Responders 12
- D. Origins of Cellular Unreliability 12
- E. The Importance of Cell Size 14
- F. Probability and Cell Response 16
- G. Implications for Pathological Neutrophil Activation 17
- H. Consequences of Cellular Unpredictability for the Study of the Role of Neutrophils in Pathology 17
- I. Future Studies of Neutrophils 18

References 18

I. A BRIEF HISTORY OF THE STUDY OF THE NEUTROPHIL

A. Identification of the Cell

Although early light microscopists of the 18th century recognized that blood was composed of red spheres and larger white "corpuscles", it was not until the development of staining techniques, especially by Paul Ehrlich in the 1880s, that it was possible to determine that these "colourless" corpuscles or white corpuscles were of several types. The cells commonly called polymorphonuclear neutrophil leucocytes, PMNL, PMN, or simply neutrophils were first described by Ehrlich as "polynuclear cells".[1]

Several other researchers pointed out that the cell actually contained only one, but lobed, nucleus. Ehrlich consequently adapted the name in 1898 to produce the more accurate but clumsy term "cells with polymorphous nuclei".[2] This phrase was used by Metchnikoff to produce the more familiar name of polymorphonuclear leucocytes.[3] Metchnikoff also described the granules most commonly found within these cells in human blood as "neutrophile",[3] because they stained with neither basic aniline dyes (i.e., basophile) nor acidic aniline dyes, such as eosin (i.e., eosinophile).

However, despite the fact that this terminology has been widely adopted, Metchnikoff himself preferred the name "microphage". This name, he argued reflected the phagocytic function of these cells and distinguished them from the larger phagocytes that he called macrophages. While this latter term was universally adopted, the former was quickly abandoned. By 1907, Halliburton's *Handbook of Physiology*[4] was using the name "poly-morphonuclear cells". The term was subsequently copied by authors of other textbooks and thus quickly became established as the accepted scientific name for these cells.

B. Early Speculation on the Role of the Neutrophil in Inflammation

Light microscopical observations of pus revealed the presence of both "colourless corpuscles" and microorganisms. This led 19th century scientists to two confused speculations. The first was that the corpuscles were the result of "spontaneous generation". This idea was overturned by Waller[5] who demonstrated that these corpuscles originated from the blood. The corpuscles could also be seen leaving the circulation by a process described as diapedesis.[6,7] The second speculation was that the pus cells assisted the growth of the microorganisms. The evidence for such a proposal was that bacteria were rarely seen in tissues that were not also "infected" with white cells. The theory was further supported by the observation that bacteria were sometimes seen within the pus cells, presumably where they received essential sustinance. The theory was extended after the observations of Waller and Cronheim by the proposal that the microorganisms were carried to the site of infection by the white cells. This latter idea was widely held by many eminent pathologists, including Virchow, and was consequently hard to dislodge. Metchnikoff and others suggested that far from being assistants of infectious agents, these cells were their opponents. The resistance of accepted authority seemed to have spurred on Metchnikoff in particular. He needed convincing evidence for the rival theory because it could in no way accommodate the old one. In his quest for counter-argument he made the transition that occurred in many areas of biology at this time, from an observational approach to a true scientific experimental approach.

C. The Contributions of Metchnikoff

Elie (Ilya) Metchnikoff (1845—1916) approached the question of the function of phagocytic leucocytes from the solid ground of just two facts, namely, that these cells were able to phagocytose and also to move (chemotax). These two facts were easily verifiable and could not be disputed The thinking of biologists at the time at which Metchnikoff embarked on his research was strongly influenced by the publications of Wallace and Darwin,[8,9] and this led Metchnikoff to look for an evolutionary explanation for the presence of these cells in inflammatory sites. His approach was to search for organisms that displayed the same two key features of the phagocytic white cells. His attention was first drawn to amoeba and other unicellular organisms. It was obvious in these cells that their movement and phagocytosis were part of their feeding process, and were followed by killing and intracellular digestion of the captured microorganism. Comparing organisms from various phyla, Metchnikoff noted that as the complexity of the organism grew, digestion became the function of specific regions of the multicellular animal. For example, in metazoa, only the mesoderm retains the ability to phagocytose. These cells, he noted, also migrated to the site of injury, where they also exhibited phagocytosis. In animals with primitive vascular systems, Metchnikoff noted that phagocytic cells were carried to all parts of the body. He argued that the phagocytic cells in the mammalian circulation were the modern equivalent of these more ''primitive'' cells and thus were ''descended from'' the mesoderm phagocyte and, ultimately, from unicellular organisms similar to ameba. It was therefore reasonable to suppose that the modern phagocytic leucocyte retained its primitive function, namely, the killing and digestion of microorganisms.

Metchnikoff further extended this argument on the Darwinian principle that features without real benefit to the organism are lost through evolutionary processes. It was clear, he argued, that the presence of phagocytic cells was of considerable benefit to the organism, since all animals, from the most primitive to mammals, retained these cells. He therefore proposed that these phagocytic cells were a vital part of the natural immunity against infection by microorganisms.

This proposal was not met with wide approval. First, it contradicted established dogma, and second, Erhlich and others were at that time establishing a role for antibodies and complement in immunity. There were many observations about the lytic ability of sera that did not require the phagocytic theory. Baumgarten[10] published a detailed criticism of Metchnikoff's theory. The main philosophical argument seemed to be summed up by Baumgarten's remark that ''it matters little that his (Metchnikoff's) account of the development of the leukocyte . . . should be in conformity with the principles of evolution.'' Although there was a hint that Metchnikoff was flattered by the long essay of Baumgarten, especially since its title unequivocally credited the phagocytic theory to Metchnikoff alone (''Zur Kritik der Metschnikoffshcen Phagocytentheorie''), it without doubt stung Metchnikoff into experimental activity. He reported, ''I was able by numerous researches to refute point by point the objections.''[11] These experiments mainly involved injection of bacteria into various animals and collecting pus and other fluids containing ''microphages''. Within 10 years, these experiments had convinced the French scientific hierarchy, and Metchnikoff delivered his famous series of lectures to the Pasteur Institute on the comparative pathology of inflammation.[12]

Within the next decade Metchnikoff found many supporters, and experiments were performed that began to integrate natural immunity of phagocytosis with the antibody-complement theory. The new experiments and the comparative work were synthesized by Metchnikoff in his book of 1901 *(Immunity in Infective Diseases)*[3] which provided the basis for a unifying approach to immunity. The general acceptance of the phagocytic theory was acknowledged shortly after publication (and translation) by the award of the 1905 Nobel Prize for Physiology. It was shared by Metchnikoff and Erhlich and thus recognized the importance of both natural and acquired immunity.

D. The 19th Century Biochemistry of the Neutrophil

Following Metchnikoff's theory of the killing role of phagocytes, attempts were made to identify the biochemical mechanism. Although the microscopic observation of blue litmus powder within phagocytic vesicles clearly demonstrated that vacuoles became acidic,[12] it was clear that acidity alone could not provide an explanation for the killing. As early as 1864, Eichwald had demonstrated that pus contained products of proteolysis[13] and later proteases and lipases were isolated from pus.[14,15] These "digestive" enzymes seemed to provide the biochemical support for Metchnikoff's theory, although at about the same time inhibitors of these proteases were also being found in the serum.[16-18] The presence of these "antileukoproteases", it was later argued, was essential to prevent tissue damage at inflammatory sites.[19] However, there remained a puzzle. Since active antileukoprotease was taken up with the bacterium into the phagosome, this would inactivate the protease there and prevent killing. There seemed to be two conflicting evolutionary pressures, one driving the killing mechanism to greater efficiency by increasing the effectiveness of the protease, and the other driving the protective mechanism to increased efficiency by increasing the effectiveness of the antiproteases. Although this puzzle was to remain for the next 50 years, reports, which may have provided clues, appeared before the start of the 20th century. For example, Portier found that leucocytes contained "oxydase"[20] and Achalne found that "large quantities of oxydase" were in pus.[21]

E. Progress Since Metchnikoff

So much of the groundwork was laid between 1850 and the time of Metchnikoff's Nobel Prize, that the following 50 years seem to have added little but confirmation and detail. During this time, Comandon used time-lapse photography in 1917 to quantify chemotactic rate,[22,23] Fleming (1923) demonstrated the bacteriocidal activity of lyzosyme,[24] and Leishman described a method for quantifying the uptake of bacteria.[25] Probably the most significant progress was the discovery of the importance of oxygen to neutrophil killing. Although this was perhaps foreshadowed by the discovery of "oxydase activity", its relevance was not fully realized until the 1970s.

In 1933, two papers appeared describing the increase in oxygen consumption that accompanied phagocytotic stimulation of neutrophils.[26,27] Baldridge and Gerald used oxygen consumption to determine whether phagocytosis was a purely physical phenomenon or whether biological "energy" was required. Since oxygen was consumed, "energy" was clearly a requirement. However, as a greater understanding of ATP production was realized, the expected dependence on the mitochondrial oxidase was not found and oxygen consumption continued in the presence of cyanide.[28,29] A clue to the role of this nonmitochondrial oxidase had been provided by the isolation of the major lysosmal enzyme in the neutrophil, the peroxidase.[30] It was surprising that this enzyme was not identified earlier, since it is present in high concentration (>5% dry weight of the cell) and probably gives pus its greenish color. In fact, the enzyme was originally named verdoperoxidase because of its color, but this was later replaced by the more familiar name myeloperoxidase when the milk peroxidase, lactoperoxidase, was purified and also found to be green. Agner[31] showed that the neutrophil enzyme could effectively destroy the biological activity of toxins in the presence of hydrogen peroxide. If it was important in intracellular killing of phagocytosed microorganisms it was clear that a source of peroxide was necessary, perhaps from an oxidase. Later Selvaraj and Sbarra[32] showed that in the absence of oxygen neutrophils were able to phagocytose, but intracellular killing was weak or nonexistant. It is now widely accepted that the nonmitochondrial oxidase together with the myeloperoxidase system are responsible for the production of reactive species of oxygen in the phagosome, which are crucial for the killing process. This oxidase also has the property of being "switched on or off" as required to provide a transiently toxic environment within the phagosome. Such a system overcomes the problem associated with the conflicting evolutionary pressures of the destructive lysosomal enzymes and their extracellular neutralization.(See Table 1.)

F. More Recent Developments

Many quantitative details of the biochemistry and physiology of the neutrophil are con-

Table 1
SOME HIGHLIGHTS IN THE STUDY OF NEUTROPHILS (1880—1965)

Event	Author	Year	Ref.
Identification of the cell	Ehrlich	1886	2
Phagocyte theory	Metchnikoff	1895	3
Discovery of opsins	Wright and Douglas	1903	121
"Oxygen burst"	Ado; Baldrige and Gerard	1933	26, 27
Nonmitochondrial oxidase	Becker, Munder and Fischer Sbarra and	1958	28
	Karnovsky	1959	29
Incorporation of Pi into PI	Karnovsky	1960	123, 124
Importance of Ca^{2+} for responses	Woodin and Weineke	1963	125

tinuing to be added to our knowledge, some of which are summarized in Table 2. In the past fifteen years or so the most significant progress seems to have taken place in the area of signal-response coupling. Metchnikoff gives the impression that this area of knowledge was almost beyond science, the only possible connection between stimulus and response of the living "protoplasm" he cited was the mysterious psychophysiological law of Weber-Fechner, which states (accurately for the linear region of the semilogarithmic binding relationship) that "when the stimulus increases in geometric ratio, irritability increases in arithmetic ratio". The "protoplasm" of the 19th century has become a complex of intracellular signaling devices modulated by inorganic ions, such as calcium, and organic agents, such as nucleotides and phospholipids. It is the unfolding of the story of this transition to a molecular basis for stimulus-response coupling that occupies the majority of the remainder of this volume.

II. THE INVOLVEMENT OF NEUTROPHILS IN PATHOLOGY

A. Inappropriate Activation of Neutrophils

The interest in the signal transduction mechanisms of the neutrophil stems from a desire to understand not only general cell biology, but also the specific role of these mechanisms in pathological processes. As Metchnikoff had clearly demonstrated, the physiological function of these cells is the *prevention* of pathology by combatting infection. It is therefore surprising that these same cells are suspected of causing tissue damage in some diseases (e.g., in chronic inflammation). However, there is growing evidence that neutrophils can unleash their destructive power not only against the infecting microorganism but also against the surrounding extracellular matrix and neighboring cells.

There a two "weapons systems" used by neutrophils, sometimes characterized by their independence or dependence on oxygen. They are also characterized by the mechanisms that regulate and control them. The former group comprises the destructive enzymes, known for almost a century. The activities of these enzymes are limited only by access to their substrates, activity therefore being controlled by cellular location. Translocation of the enzymes to their substrates is achieved by fusion of the vesicle containing the enzymes (the granule) with the vesicle containing the endocytosed microorganism (the phagosome). The second system used by the neutrophil for the killing of microorganisms is the production of toxic oxygen metabolites. The difference between the mechanism that controls this system and the former is that whereas the former system is a collection of active enzymes awaiting access to their substrates, the activity of the latter system (the oxidase system) can be "switched on" and "switched off" when required. After phagocytosis, the oxidase is transiently activated to introduce toxic oxygen metabolites into the phagosome. The oxidase is then inactivated before egestion of the dead microorganism or death of the neutrophil.

Table 2
SOME PHYSICAL AND BIOCHEMICAL PROPERTIES OF THE NEUTROPHIL

Parameter	Value	Species	Ref.
	Physical		
Diameter	10 — 15 μm	Human	33,34
	10 μm	Rat	35
Volume	346 fℓ	Human	36
	555 fℓ	Rat	37
Organelle volume			
Cytoplasm	63%	Human	39
Nucleus	21%	Human	39
Granules	15.4%	Human	39
Mitochondria	0.6%	Human	39
Intracellular			
H_2O volume	274 fl	Human	38, 39
	268 fl	Rat	40
Surface area	290 μm^2	Human	38
Granule surface area	700 μm^2	Human	39
Water content	64 — 76%	Human	41 — 43
	80%	Rabbit	44 — 47
	85%	Horse	48
	78%	Dogfish	49
Wet weight	8.5 ng/cell	Rabbit	50
Dry weight	0.6 ng/cell	Rabbit	50
Protein content	44 pg/cell	Rabbit	51
	60 pg/cell	Rat	52
Density	1.08 — 1.09	Human	53
Granule density			
Specific	1.19	Human	54
Azurophil	1.23	Human	54
Coefficient of viscosity	ca. 130 poise	Human	55
	Ionic		
Na^+	ca. 29 m*M*	Human	41 — 43
	ca.80 m*M*	Rabbit	44 — 47
	89 m*M*	Dogfish	49
K^+	ca. 132 m*M*	Human	41 — 43
	ca. 104 m*M*	Rabbit	44 — 47
	134 m*M*	Dogfish	49
Na/K ratio	ca. 0.22	Human	41 — 43
	ca. 0.77	Rabbit	44 — 47
	0.66	Dogfish	49
Cl^-	80 m*M*	Human	38
Calcium			
Total	0.13 fmol/cell	Rabbit	51
Ionized			
(resting)	0.1 μ*M*	Human	57
	0.1 μ*M*	Rabbit	58
	0.1 — 0.3 μ*M*	Rat	59
(stimulated)	0.5 — 1.0 μ*M*	Human	57
	0.5 μ*M*	Rabbit	58
	0.8 — 1.0 μ*M*	Rat	59
Magnesium (total)	2.1 fmol/cell	Rabbit	50
(Cytosolic)	0.16 fmol/cell	Rabbit	50
pHi	7.25	Human	60
	7.25	Rabbit	61
	7.04	Rat	62

Table 2 (continued)
SOME PHYSICAL AND BIOCHEMICAL PROPERTIES OF THE NEUTROPHIL

Parameter	Value	Species	Ref.
Phosphates	1.3 fmol/cell	Rabbit	50
	Biochemical		
Nucleotides, etc.			
ATP	1.9 fmol/cell	Human	63
	0.8 fmol/cell	Rabbit	50
	2.7 fmol/cell	Guinea pig	64
	0.8 fmol/cell	Rat	40
ADP	81 amol/cell	Rat	40
AMP	16 amol/cell	Rat	40
Adenosine	5 amol/cell	Rat	40
Hypoxanthine	17 amol/cell	Rat	40
c-AMP	2.6 amol/cell	Human	65
	6.77 amol/cell	Rat	65
(Stimulated)	4.6 amol/cell	Human	65
	15.8 amol/cell	Rat	65
c-GMP	0.5 amol/cell	Human	66
(Stimulated)	50 amol/cell	Human	66
NADH	24 amol/cell	Human	67
	43 amol/cell	Guinea pig	68
NAD^+	73 amol/cell	Human	67
	154 amol/cell	Guinea pig	68
NAD/NADH	3.04	Human	67
	3.58	Guinea pig	68
NADPH	24 amol/cell	Human	67
	49 amol/cell	Guinea pig	68
$NADP^+$	19 amol/cell	Human	67
	25 amol/cell	Guinea pig	68
NADP/NADPH			
Resting	0.79	Human	67
	0.51	Guinea pig	68
Stimulated	1.06	Guinea pig	68
Enzymes			
Cytoplasmic			
LDH	99 nU/cell	Human	69
	180 nU/cell	Rat	40
Hexosekinase	1.7 nU/cell	Human	69
Pyruvate kinase	3.7 nU/cell	Human	69
Adenosine deaminase	c2.5 nU/cell	Rat	40
Adenosine kinase	c190 pU/cell	Rat	40
Nucleoside kinase	1.4 nU/cell	Rat	40
Xanthine oxidase	<500 pU/cell	Rat	40
Kinase C	38 μU/g	Human	70
	21 μU/g (1.2 fU/cell)	Rat	62
Kinase A	10.5 μU/g	Human	70
Kinase G	0.6 μU/g	Human	70
Granular			
Myeloperoxidase	132 nU/cell	Guinea pig	71
	39 nU/cell	Rabbit	71
	c.50 nU/cell	Rat	72

Table 2 (continued)
SOME PHYSICAL AND BIOCHEMICAL PROPERTIES OF THE NEUTROPHIL

Parameter	Value	Species	Ref.
Activity relative to human	1	Human	73
	0.12	Rabbit	73
	0.10	Guinea pig	73
	0.42	Rat	73
	0.16	Mouse	73
	0.36	Horse	73
Lysozyme	17.2 pg/cell	Human	73
	3 pg/cell	Rabbit	73
	3 pg/cell	Guinea pig	73
	2 pg/cell	Rat	73
	2 pg/cell	Mouse	73
	3.2 pg/cell	Horse	73
β-glucuronidase	154 nU/cell	Human	73
	14 nU/cell	Rabbit	73
	42 nU/cell	Guinea pig	73
	102 nU/cell	Rat	73
	14 nU/cell	Mouse	73
	20 nU/cell	Horse	73
Alk. phosphatase	22 nU/cell	Human	73
	198 nU/cell	Rabbit	73
	942 nU/cell	Guinea pig	73
	190 nU/cell	Rat	73
	<0.2 nU/cell	Mouse	73
	942 nU/cell	Horse	73
Collagenase	70 fg/cell	Human	74
Elastase	3 pg/cell	Human	74
Cathepsin G	1.5 pg/cell	Human	74
Other			
Calmodulin	4.2 amol/cell	Human	75
	0.7 amol/cell	Rabbit	76
Inositol trisphosphate	0.05 amol/cell	Rabbit	77
(Stimulated)	0.55 amol/cell	Rabbit	77
Diacylglycerol	50 amol/cell	Rabbit	78
Actin (total)	4.9 pg/cell	Pig	79
	5.6 pg/cell	Rabbit	80
	^{3}pg/cell	Rat	52
f-actin (% of total)	30%	Pig	79
	26%	Rabbit	80
	50%	Rat	52
Myosin	0.56 pg/cell	Rabbit	80
Phosphocreatine	<0.2 fmol/cell	Human	63
Low molecular weight SH	2.5 fmol/cell	Rat	81
Ferritin	4.6 fg/cell	Human	82
Lactoferrin	14 pg/cell	Human	83
DNA	6 pg/cell	Rabbit	84
RNA	1.5 pg/cell	Rabbit	84
Metabolic rate			
ATP formation	1.3 fmol/min/cell	Human	63
Lactate formation	1.3 fmol/min/cell	Human	63
	0.7 fmol/min/cell	Rabbit	84, 85
	1.9 fmol/min/cell	Rat	85

Table 2 (continued)
SOME PHYSICAL AND BIOCHEMICAL PROPERTIES OF THE NEUTROPHIL

Parameter	Value	Species	Ref.
Protein synthesis; lysine incorporation	0.013 amol/ min/cell	Rabbit	86
Glucose utilization	0.96 fmol/min/ cell	Human	87
(Stimulated)	1.13 fmol/min/ cell	Human	87
Glycogenolysis	10 amol/min/ cell	Human	87
(Stimulated)	16 amol/min/ cell	Human	87
Glycerol uptake	6.3 amol/min/ cell	Human	88
^{14}C-acetate incorporation	30 amol/min/ cell	Human	89
Oxygen consumption	(fmol/cell/min)		
Resting	0.35	Rat	90
	3 pg/cell	Rat	52
Stimulated	2.5	Human	91
	1.9	Rat	90
Superoxide production	3.0	Human	92
(Stimulated)	47	Human	92
H_2O_2 production (stim)	1.6	Human	93
Affinity for oxygen consumption;			
(Resting)	9.6 μ*M*	Rat	90
(Stimulated)	3.7 μ*M*	Rat	90
O_2 metabolite production			
(Resting)	45 μ*M*	Rat	94
(Stimulated)	27 μ*M*	Guinea pig	95
	22 μ*M*	Rat	94
Affinity for NADPH	8.6 μ*M*	Human	96
	26 μ*M*	Guinea pig	97
Affinity for NADH	0.8 m*M*	Human	96
	0.1 m*M*	Guinea pig	97
	0.6 — 1 m*M*	Rat	98
Membrane Properties			
Water permeability	0.65 Ωm/ min·atm	Human	36
	0.45 Ωm/ min·atm	Rabbit	36
(Heat of activation)	18.4 kcal/mol	Human	36
	14.0 kcal/mol	Rabbit	36
Electrophysiology			
Membrane potential	−53 mV	Human	99
	−67 mV	Bovine	100
Conductance	(μS/cm²)		
Membrane	5.3	Human	38
Na(g)	1.3 — 1.6	Human	38
K (g)	2.8	Human	38
C1 (g)	1	Human	38

Table 2 (continued)
SOME PHYSICAL AND BIOCHEMICAL PROPERTIES OF THE NEUTROPHIL

Parameter	Value	Species	Ref.
Single channel			
Type 1	18 — 25 pS	Human	111
Type 2	4 — 6 pS	Human	111
Membrane resistance	270kΩcm²	Human	38
Membrane permeability	(10^{-9} cm/sec)		
Sodium	4.3 — 5.6	Human	38
	52	Rabbit	56
Potassium	33 — 55	Human	38
	124	Rabbit	56
Chloride	4	Human	38
NH_4	38 — 57	Human	60
Calcium(stimulated)	36	Rabbit	56
Electrophoretic mobility	1.38μm/s/V/cm	Human	101
Mechanical			
Shear modulus, G			
Slow	135 dyn/cm²	Human	55
Fast	506 dyn/cm²	Human	55
Receptors (i, binding sites/cell; ii, dissociation const)			
f-Met-Leu-Phe	(i) 2×10^3	Human	102
	(ii) 10 n*M*	Human	102
	(i) 10^5	Rabbit	103
	(ii) 630	Horse	104
	(ii) 0.52 n*M*	Horse	104
f-Met-Leu-Phe	(i) 10^5	Rabbit	105
	(ii) 1 n*M*	Rabbit	105
C5a	(i) c3 $\times 10^5$	Human	106
	(ii) 1 n*M*	Human	106
Circulating Neutrophils (10^6/mℓ blood)			
	4.15	Human	107
	4.1	Rabbit	108
	3.1	Rat	108
	9.5	Cat	108
	8.2	Dog	108

The evolutionary pressures which armed these cells with their lethal weapons for killing infecting microorganisms must have been balanced by the requirement for the host tissue to be protected from their effects. The resultant forces would lead to mechanisms which maximized killing of the microorganism and minimized destruction of the host tissue. However, cellular control of these processes would seem to be imperfect and not totally reliable, since the extracellular environment contains both inhibitors of the destructive enzymes as a defense against phagosomal leakage and agents that reduce extracellular oxidative damage (Table 3). The question that arises is why the cellular control of such potentially hazardous events as activation of neutrophil oxidase and lysosomal enzyme release cannot be relied upon. In particular, why are the cellular control processes so unreliable that not only is a back-up system of neutralizing agents required, but these back-up systems can be overwhelmed, resulting in major tissue damage? Is the machinery of these cells intrinsically faulty? Are the mechanisms that control activation prone to fail if certain conditions prevail (i.e., a "hardware" or a "software" fault)? In the remainder of this chapter an attempt will be made to explore these questions. The discussion will be from a cell biological, rather

Table 3
EXTRACELLULAR PROTECTION AGAINST LYSOSOMAL AND OXIDATIVE DAMAGE

Agent	Concentration (plasma g/ℓ)	Ref.
Alpha-1 proteinase inhibitor	1.1 — 1.6	74
Alpha-2 macroglobulin	1.1 — 3.0	74
Alpha-1 antichymotrypsin	0.3 — 0.6	74
Anti-leukoprotease	0.125	74
Superoxide dismutase		
Plasma	26000 U/ℓ	126
Tissue	403 U/g	126

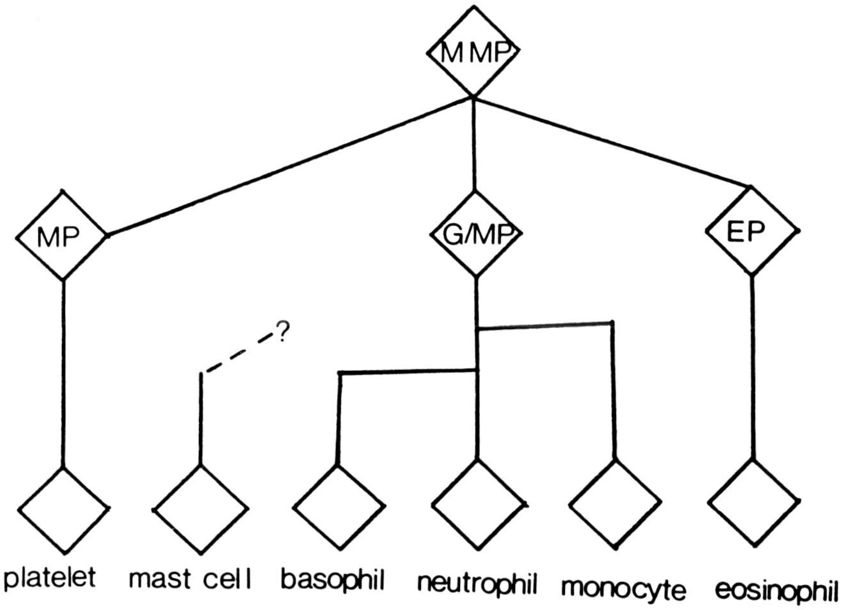

FIGURE 1. The relationship between neutrophil and other cells. MMP, mixed myeloid progenitor cell; MP, megakaryocyte progenitor; G/MP, granulocyte/monocyte progenitor; EP, eosinophil progenitor.

than clinical view point, and it is hoped that a novel perspective to these perplexing questions may emerge.

B. Why Do Neutrophils Go Wrong?

A first clue to this puzzle may be provided by looking at some of the cells that are related to the neutrophil (Figure 1). Close "relatives" of the neutrophil, the basophil and the eosinophil also overreact in inflammatory disease, while the other "relatives" also seem prone to overreact in allergic responses. The possibility exists that there may be a feature common to these cells which provides an explanation for their apparent fondness for overreacting to stimulation.

There are several features which cells of this group share. For example, the cells of this group, when responding to a physiological situation, rarely behave independently of each other. Signals generated by neutrophils arriving at the site of infection may recruit other cells which, in turn, generate more signals. Positive feedback loops such as this appear to operate with all cells of this group. There are obvious advantages to such a system, whereby amplification is inbuilt into the control of the response exerted by the population. However, a small proportion of aberrant cells perpetually emitting an inappropriate signal could in theory provoke a response from "well-behaved" neighboring cells that would escalate into an overreaction by the whole population. This may provide part of the answer to the mass disobedience that populations of neutrophils can exhibit. It cannot, however, provide the whole answer. In populations of other types of cells a similar strategy is used to amplify and coordinate the population response without problems. For example, electrically coupled cells in tissues such as muscle rarely behave unpredictably. Just as the reliability of muscle originates from the reliability of individual muscle cells, so the unreliability of neutrophil populations must originate from the unreliability of individual neutrophils. Is there a difference between the muscle cell and the neutrophil that makes one but not the other so unreliable?

C. Types of Cellular Responders

One important difference between the neutrophil and the muscle cell is in the type of cellular responses that are triggered. All the cell types belonging to the neutrophil family group can produce a number of cellular responses, which are often coordinated. For example, the neutrophil when stimulated has the ability to chemotax, phagocytose, secrete, and activate its oxidase. In order to function, the neutrophil must perform these responses appropriately and in a sequence that depends upon the precise conditions at the time of stimulation. The muscle cell, in contrast, produces a single response, namely, contraction. It is dedicated to this single response. This difference may be part of the underlying cause of the unreliability of the neutrophil.

Without considering the biochemical basis for stimulus-response coupling and merely using the "black-box" approach, cells fall naturally into four types of responder. These are illustrated in Figure 2. Skeletal muscle may be considered as an example of the first and most simple type. Cells of this type are capable of one response to stimulation acting through one intracellular mechanism. The second type of responder again is capable of one response, but the magnitude of the response can be influenced by agents which act on the cell through a second intracellular mechanism. The third type of responder has the ability to produce more than one response. Each response is controlled by a separate mechanism, and may thus be thought of as simply a number of type I response mechanisms within one cell. The fourth type is the most complex. Not only can the cell produce a number of separate responses, but the intracellular mechanisms that control them can interact to produce a synchronized and coordinated response.

The neutrophil in this classification would be a type IV responder. In vivo, the cell must chemotax towards the microorganism, phagocytose it, and then activate the oxidase system. Each response must be coordinated with the next to produce the required effect. Complex functions are achieved in tissues, such as kidney, by individual groups of cells performing more simple type I-II responses. In autonomous free living cells, like the neutrophils, each cell must perform the whole complex operation. The intracellular events controlling these responses are necessarily complex, involving a number of different intracellular "messengers" interacting to produce the final response. Could this difference help to explain the relative reliability of tissue compared with the neutrophil?

D. Origins of Cellular Unreliability

It is generally accepted that biological phenomena are controlled by physical and chemical processes. It may be considered that since physics (and hence also chemistry) is characterized

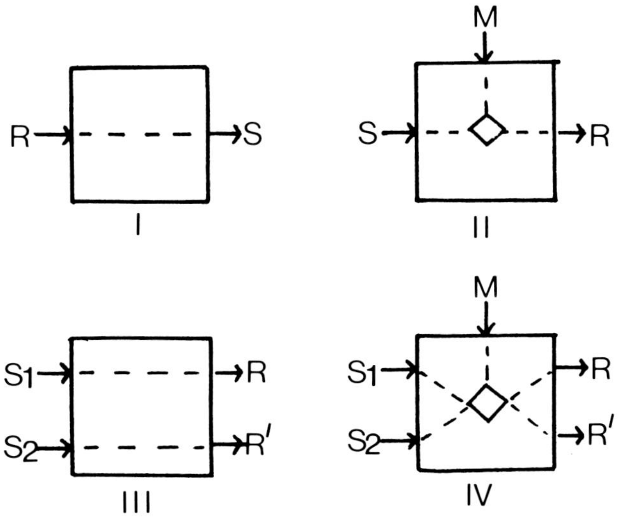

FIGURE 2. Types of cell responders. Type I: cells dedicated to a single response (R) induced by stimulus (S). Type II: cells dedicated to a single response induced by a single stimulus but with a second agent (M) which modifies the response. Type III: cells with two responses (R_1,R_2) induced separately by stimuli (S_1, S_2), without interaction. Type IV: integrated cells, with more than one stimulus and more than one response with interactions between each and with other modifiers.

by its unviolatible ''laws'' which predict phenomena quantitatively and reproducibly, these processes can play no role in the unpredictability of cell responses. However, as has been pointed out many times before, the events that underlie these physical ''laws'' are themselves random and hence unpredictable (see, for example, Schrodringer[109] and Eddington[110]). It may be argued that the very nature of these physical processes make cellular unreliability inevitable.

The ''laws'' that govern biochemical events within cells and elsewhere originate in the inevitability of physical processes to move towards and not away from an equilibrium. This tendency ''drives'' the reaction. An example of an approach to equilibrium is provided by the ''thought experiment'' given by Eddington[110] and illustrated in Figure 3. A container is divided into two compartments, one filled with water and the other with an ink solution. When the partition dividing the two fluids is removed, the ink will diffuse to fill the whole container with a uniform color (i.e., equilibrium). The ink will never spontaneously return to its original starting position. Furthermore, the time taken to reach this equilibrium can be accurately calculated from knowledge of the diffusion constant for the ink particles. The whole process, in fact, seems totally reliable and quantitatively predictable. How, then, can cellular unreliability originate here?

Perhaps the answer will become clearer when we adjust one detail of the experimental design, namely, the scale. Suppose the apparatus illustrated was just large enough to accommodate three ink particles. As the ink particles move with Brownian motion, there are sometimes three particles in one half, sometimes two, sometimes one, sometimes none. It would not be possible to predict when an equilibrium was reached or to be sure that the ink never returned to the starting position, with all the ink in just one side of the container. The smallness of the scale has undermined the concept of a ''reliable'' equilibrium and given a model of unreliability. This leads to the question of where the predictability of the first experiment came from? Individual particles of ink in both experiments show no preference

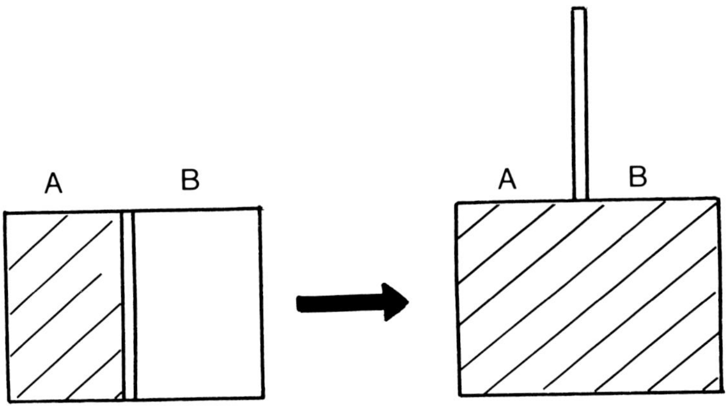

FIGURE 3. Thought experiment which demonstrates how cellular unpredictability may arise. I. Water-filled tank with two compartments, A and B. II. The same tank after removal of the dividing wall.

for either half of the container (once the partition is removed). So at any instant the probability of finding a particular ink particle in side "A" would be 1/2. The probability of finding all the particles in side "A" is therefore $1/2^n$, where n is the number of particles. It can therefore be seen that when the container is of "normal" (macro) proportions and the number of ink particles is probably more than 10^{10}, that the probability of all the particles being in the same half of the container would be so small that in practice we may be sure that it would *never* happen. This is not the case in the second experiment. We would expect all three ink particles to be in the same half of the container 1/8th (i.e., $1/2^3$) of the time. However, for the latter situation we could only predict probabilities and never be certain at any given time.

Similar arguments could be made for all natural processes that rely on an approach to equilibrium. The "law of mass action", for example, is only accurate when molecules act *en masse*. The certainty of the "law" again disappears as the number of interacting molecules decreases. These "laws" are therefore dependent on scale. The question then arises whether the "cellular scale" is such that the chemical reactions which occur within them are always reliable and hence can be predicted with certainty or are some reactions unreliable and "controlled" only by probabilty?

E. The Importance of Cell Size

For a random process such as diffusion of an intracellular messenger or binding of the messenger to its receptor, the accuracy of the "laws" that govern these phenomena is related to the total number of particles (N). Since the concentration of some important intracellular constituents is low (Table 2) and the cell size is small, an important question to be raised is how big must N be before the "law" is reliable? If we consider binding to a receptor, the law of mass action will predict from the K_d that a certain proportion of receptors would be occupied. However, if the proportion of occupied receptors could be counted from moment to moment it would be found that the proportion would flicker around the expected value as the random process of association and dissociation occurred. The scatter around the mean value is predicted for a random process by the Poisson distribution, the standard deviation being $\sqrt{N}$. The "accuracy" of the prediction can be calculated (Figure 4) and is seen to decrease rapidly as the number of interacting particles decreases. For thermodynamic processes to behave with reasonable precision, i.e., within $\pm 1\%$ of the expected in 99% of the cells, the cells must contain more than approximately 60,000 participating particles (ions, molecules, etc.).

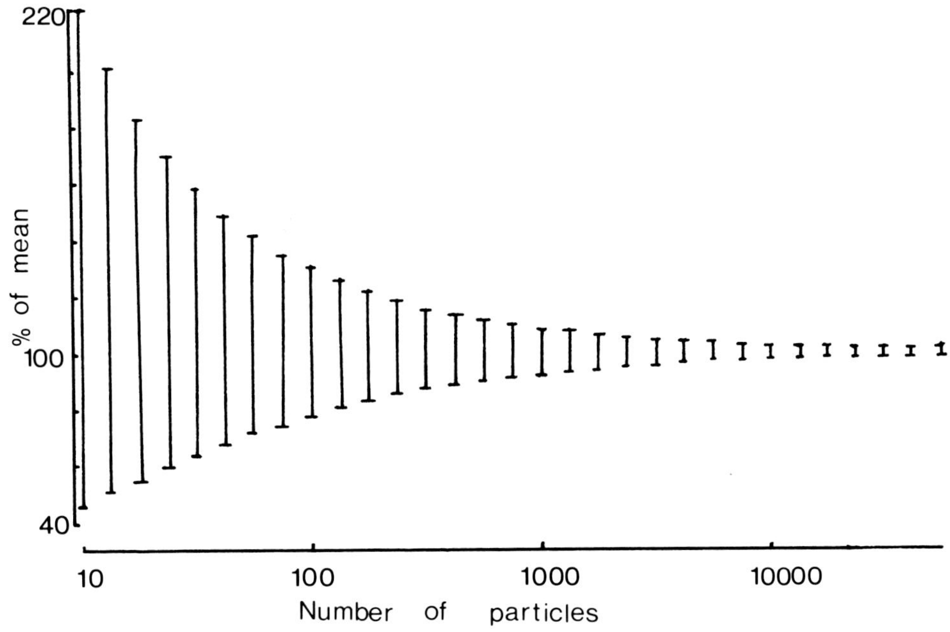

FIGURE 4. Effect of number of particles on the predictability of their behavior. The graph shows the 99% confidence limits (solid lines) around the predicted value (dotted line) as a function of the number of particles to be considered (N). The confidence limits as a fraction of the value N were calculated from the expression given by Wilson [126] i.e., $N + 3.32 \pm 2.576\sqrt{(N+1.7)}$.

The number of molecules of several important regulators within the cell is within the range at which the randomness of interaction will become evident. The number of formylated peptide receptors per cell, for example, has been reported to as low as 2000 [102] for humans and 600 for horses.[104] It can also be estimated from the number of PMA molecules per cell required for activation that less than 6000 PMA-receptors (presumably c-kinase molecules) are required for activation of the oxidase (see Chapter 9). In the resting neutrophil, there are only approximately 7500 to 15,000 free calcium ions in the whole cytoplasmic volume (Avogadro's number × volume (250 fℓ) × concentration (50 to 100 n*M*); the number of ions within the cytoplasm in the leading edge of a chemotactic cell or in the proximity of oxidase molecules probably being considerably less. The situation is similar for hydrogen ions, approximately 8000 in the whole cytoplasm at pH 7.25.[60,61] An extreme example of a low number of important ions is provided by considering that within an individual neutrophil granule (radius approximately 0.35μm) an intracellular pH of 7.4 represents only 4 hydrogen ions (i.e., Avogadro's number × volume (liters) × conc (*M*) $= 6.10^{23} \times 1.8.10^{-16} \times 4.10^{-5}$). At such a low number of hydrogen ions, the usual concepts of ionization of proteins becomes very imprecise. It can also be seen that whereas the average number of hydrogen ions per granule may be four, some granules would contain no protons, some one proton, some two, and so on. Of course, since the distribution is random the *population* of granules would behave in a manner predictable by the law of mass action. The behavior of processes within an individual granule would, however, be predicted only with probability.

As techniques become available, the randomness of the underlying physical processes is becoming more apparent and detectable. For example, the opening of individual ion channels in the plasma membrane when detected by patch clamping is seen to be a random event,[111] the action of an agent or membrane potential being to increase or decrease the probability of channel opening. It is not possible to predict in advance at any instant whether a particular channel will be open or closed. The question then arises whether randomness such as this can lead to unpredictabilty of cell response.

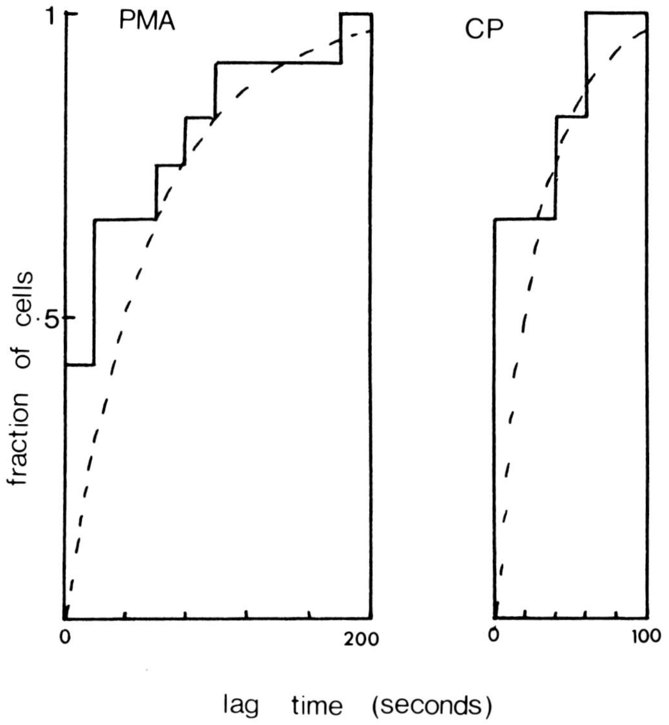

FIGURE 5. The time between stimulation and response measured from individual neutrophils. The fraction of cells with lag times less than that shown on the ordinate are shown by the histogram. The stimuli were PMA (0.1 μg/mℓ) and f-Met-Leu-Phe; CP(1 μM) and activation of the oxidase was detected microfluorometrically by the oxidation of dichlorofluorescin. The dotted line shows the expected distribution for a stochastic Brownian process from the equation $f(t) = 1 - e^{-\lambda t}$, where f(t) is the fraction of cells with lag time less than or equal to t, and λ is the reciprocal of the mean lag time, found to be 0.018/sec and 0.038/sec for PMA and CP, respectively.

F. Probability and Cell Response

It has been known for nearly 30 years that the release of acetylcholine from the nerve terminal provides an example of unreliability in the processes between stimulation and response. When stimulated under conditions that evoke the release of a few vesicles, the nerve will sometimes secrete none, sometimes one, sometimes two, and so on. It has been elegantly shown that although the number of quanta released for a given stimulation was unpredictable, in a long series of stimulations the frequency with which a given number of quanta were released was predicted by the Poisson distribution.[112-115] The release of each individual quantum was therefore controlled solely by probability. Similar experiments are more difficult to perform with individual neutrophils. However, secretion from individual neutrophils shows qualitatively the same characteristics. Release of 9-aminoacridine preloaded into the granules results in the relaxation of fluorescence quenching, an event detectable in individual neutrophils.[116] Weak stimulation resulted in some cells secreting a single unit of dye, some more, and some none. Similarly, the magnitude of activation of the neutrophil oxidase in individual cells as measured by oxidation of intracellular dichlorofluoresin was unpredictable.[117] Furthermore, the delay between stimulation and secretion or oxidase activation varied between cells, these processes being apparently random (Figure 5). There is therefore evidence that probabilistic phenomena also play a role in activation of the neutrophil.

G. Implications for Pathological Neutrophil Activation

If probabilistic phenomena underlie critical steps in the activation of neutrophils, the possibility exists that as with the neuron, the coupling of stimulation to response may be unreliable. There are, however, two features of stimulus-response coupling in the neutrophil that may have pathogenic significance. First, the intracellular mechanisms within the neutrophil are very elaborate with interactions at many stages, which must operate to provide the appropriate response upon stimulation (see following chapters). This may make the neutrophil particularly unreliable, because as the complexity of the intracellular mechanisms increase, the possibilities for error also increase. If each step in the control mechanism operates with an efficiency or probability of success, p, it can be seen that the efficiency or probability of success of the entire sequence will be p^n, where n is the number of steps. Obviously, as the number of steps increases, the probability of success for the pathway decreases. For example, if each step occurs with high probability, $p = 0.9$, the probability of success for a pathway with six such steps decreases to only 0.53. Second, in contrast to the neuron which is dedicated to a single response, i.e., a type I responder (Figure 2), the neutrophil can express a number of functions which are controlled by interacting intracellular mechanisms. Failure of the messenger to produce the expected intracellular chemical change merely results in failure to elicit the response in the type I responder, whereas in the neutrophil, and other cells of the type IV response group, failure of some of the intracellular mechanisms may result in outcomes other than nonresponse. A physiological stimulus may, for example, provoke a quantitatively or qualitatively different response. This inappropriate response may result in pathological damage.

A simple example in which unreliability in stimulus-response coupling could lead to an inappropriate and potentially pathogenic response is for activation of the neutrophil with *N*-formylated peptide. This results in a number of interacting intracellular events. At least two intracellular messengers are formed, Ca^{2+} which is essential for activation of the oxidase and for secretion of lysosomal contents (see Chapters 8 and 9) and also an unknown messenger which promotes actin polymerization (see Chapters 4 and 9). Agents, for example, cytochalasins and bacterial toxins, which inhibit this actin polymerization greatly enhance activation of the oxidase[52,118] to produce reactive oxygen metabolites extracellularly[119] and also enhance secretion of lysosomal contents.[120] Both these latter events could result in damage to neighboring cells and extracellular matrix. Polymerization of actin following activation by formylated peptides, therefore, exerts a necessary inhibitory influence on these potentially pathogenic events. If the control of the oxidase and actin polymerization were both controlled by separate probabilistic phenomena, then the possibility exists that under certain conditions, the signals for secretion and oxidase activation within an individual cell may ''succeed'' while the signals for actin polymerization within that cell may ''fail''. Such an occurrence would cause that particular cell to secrete lyzosomal enzymes and produce extracellular oxygen radicals. Hence, physiological stimulation of the cell would have provoked an inappropriate and potentially pathogenic response.

Of course, the above events would be insignificant if restricted to only one cell. However, if this inappropriate response was generated with a certain probability of occurrence, a proportion of cells at the inflammatory site exposed to the same conditions would also be expected to behave similarly. A critical number of cells behaving in a pathogenic manner may generate intercellular signals in sufficient concentration to provoke an inappropriate response from the whole population. Furthermore, since the behavior of neutrophil populations results from intercellular signaling with elaborate positive feedback, the response may be self perpetuating.

H. Consequences of Cellular Unpredictability for the Study of the Role of Neutrophils in Pathology

If probabilistic phenomena such as these play a role in the etiology of some inflammatory diseases, there are two predictions. The first is that since this is not a "hardware" fault, attempts to identify and establish the existence of a biochemical defect will fail. Cells isolated from the patient would be expected to behave normally (other than as a result of the inflammatory illness). The second is that since the trigger was local conditions, perhaps existing only transiently, attempts to identify the "initial cause" in chronically inflamed tissue may be difficult. The third prediction is that conditions in which the "probability for error" is maximum are those that are most likely to generate pathological activation of neutrophils and other cells, such as macrophages. The maximum probability of an inappropriate response of the type outlined above, where one intracellular messenger succeeds while another fails to transmit the appropriate intracellular signal, may be estimated. If the probability of success of each of the messengers individually is p then the probability of one messenger failing while the other succeeds will be $p(1\text{-}p)$. The maximum probability of this type of response would therefore be expected to occur when $p = 0.5$. It may, therefore, be expected that inappropriate activation is more likely to occur as a result of weak stimulation of the cells, when the probability of successful intracellular transmission was low, than as a result of strong stimulation.

I. Future Study of Neutrophils

As techniques become available for the study of individual neutrophils rather than of neutrophil populations, it would seem likely that the random and unpredictable nature of some aspects of the biochemistry and physiology of neutrophils will become apparent. Whether or not these phenomena are crucial for our understanding of the etiology of some inflammatory diseases, remains to be established. These phenomena must however be taken into account in any attempt to provide a complete understanding of the processes that trigger and control neutrophil responses.

REFERENCES

1. **Jones, W.,** The blood corpuscle considered in its different phases of development in the animal series, *Philos Trans. R. Soc. London,* 1, 63, 1846.
2. **Erhlich, P. and Lazarus E.,** *Die Anaemie. Wein,* 8, 49, 1898.
3. **Metchnikoff, E.,** *Immunity in Infective Deseases,* Cambridge University Press, London, 1905.
4. **Halliburton, W. D.,** *Handbook of Physiology,* John Murray, London, 1907.
5. **Waller, A.,** Microscopic observations on the perforation of the capillaries by the corpuscles of blood and on the origin of mucus and pus globules, *London Edinburgh Philos. Mag. J. Sci.,* 29, 271, 1846.
6. **Cohnheim, J.,** Uber Entezundung und Eitenung, *Virchow's Arch. Pathol. Anat.,* 40, 1, 1867.
7. **Cohnheim, J.,** Inflammation, in *Lectures in General Pathology,* Vol. 1, Syndenham Society of London, 242, 1807.
8. **Darwin, C. R. and Wallace, A. R.,** On the tendency of species to form varieties; and on the perpetuation of varieties and species by natural means of selection, *J. Proc. Linnean Soc.,* 3, 45, 1858.
9. **Darwin, C. R.,** *On the Origin of Species by Means of Natural Selection or the Preservation of Favoured Races in the Struggle for Life,* Cassel & Co., London, 1859.
10. **Baumgarten, L.,** Zur kritik der Metschnikoffshchen Phagocytentheorie, *Z. Klin Med. Bd.,* 15, 1888.
11. **Metchnikoff, E.,** *Lectures on the Comparative Physiology of Inflammation,* Kegan Paul, London, 1901.
12. **Mechnikoff, E.,** Leçon sur la pathologie comparée de inflammation, *Ann. l'Institute Pasteur,* 7, 348, 1893. 1893.
13. **Eichwald, E.,** Die Colloidentartung der Eierstocke, *Wurzb. Med. Z.,* 335, 1864.
14. **Muller, F.,** unpublished, cited by Kossel, *Z. Klin. Med.,* 13, 149, 1888.
15. **Leber, T.,** Über die Entstehung der Entzundung und die entzundungerregeden Scadliekeiten, *Fortschr. Med.,* 4, 460, 1888.
16. **Achalne, P.,** Researches sur le properties pathogenes de la trypsine et le pourvoir anti-trypsine du serum, *Ann. l'Inst. Pasteur,* 15, 737, 1901.

17. **Hildebrant, H.**, Weiters uber hydrolytische ferment, *Virchow's Arch.*, 131, 5, 1893.
18. **Opie, E. L.**, Enzymes and anti-enzymes of inflammatory exudate, *J. Exp. Med.*, 7, 316, 1905.
19. **Opie, E. L.**, Intracellular digestion; the enzymes and anti-enzymes concerned, *Physiol. Rev.*, 2, 552, 1922.
20. **Portier, P.**, L'oxydase du sang des mammiferes, sa location dans le leucocyte, *Soc. Biol. Mem.*, 50, 452, 1898.
21. **Achalne, P.**, Researches sur la presence de fermente solubles dans le pus, *C. R. Soc. Biol.*, 51, 568, 1899.
22. **Comandon, J.**, Phagocytose in vitro des hematozaires du calfat, *Compres R. Hebdomadaires Seances Memoires Soc. Biol.*, 80, 314, 1917.
23. **Comandon, J.**, Tactisme produit par l'amidon sur les leucocytes enrobement du charbon, *Compres R. Hebdomadaires Seances Memoires Soc. Biol.*, 82, 1171, 1919.
24. **Fleming, A.**, On a remarkable bacteriolytic element found in tissues and secretions, *Proc. R. Soc. B* 93, 306, 1923.
25. **Leishman, W. B.**, Note on a method of quantitatively estimating the phagocytoc power of the leucocytes of the blood, *Br. Med. J.*, 1, 73, 1902.
26. **Ado, A. D.**, Uber den Verlauf der oxydativen und glykolytischen Prozesse in der Leukocten des entzundeten Gewebes wahrend der Phagocytose, *Z. Ges. Exp. Med.*, 87, 473, 1933.
27. **Baldridge, C. W. and Gerard, R. W.**, The extra respiration of phagocytosis, *Am. J. Physiol.*, 103, 235, 1933.
28. **Becker, H., Munder, G., and Fischer, H.**, Uber den leukocytenstoff bei der phagocytose, *Hoppe-Seylers Z. Physiol. Chem.*, 313, 266, 1958.
29. **Sbarra, A. J. and Karnovsky, M. L.**, The biochemical basis of phagocytosis. I. Metabolic changes during the ingestion of particles by polymorphonuclear leukocytes, *J. Biol. Chem.*, 234, 1355, 1959.
30. **Agner, K.**, Verdoperoxidase. A ferment isolated from leukocytes, *Acta Physiol. Scand.*, 2(Suppl.8), 1, 1941.
31. **Agner, K.**, Detoxificating effect of verdoperoxidase on toxins, *Nature (London)*, 159, 271, 1947.
32. **Selvaraj, R. J. and Sbarra, A. J.**, Relationship of glycolytic and oxidative metabolism to particle entry and destruction in phagocytic cells, *Nature (London)*, 211, 1272, 1967.
33. **Ciba giegy**, Scientific Tables, 7th ed., Basel, 1975.
34. **Wintrobe, M. M.**, *Clinical Hematology*, Lea & Febiger, Philadelphia, 1967, 226.
35. **Blitstein, I. and Van den Berghe, L.**, Les dimensions des lymphocytes et des granulocytes a l'etat normal et au cours des affections heamatologiques, *Rev. Belge. Sci. Med.*, 16, 322, 1945.
36. **Hempling, H. G.**, Heats of activation for exosmotic flow of water across the membrane of leukocytes, *J. Cell. Physiol.*, 81, 1, 1973.
37. **Tivey, H., Li, J. G., and Osgood, E. E.**, Average volume of leakemic leukocytes, *Blood*, 6, 1013, 1951.
38. **Simchowitz, L. and De Weer, P.**, Chloride movements in human neutrophils, *J. Gen. Physiol.*, 88, 167, 1986.
39. **Scmid-Schonbein, G. W., Shih, Y. Y. and Chen, S.**, Morphometry of human leukocytes, *Blood*, 56, 866, 1980.
40. **Newby, A. C.**, The role of adenosine deaminase ecto(5′-nucleotidase) and ecto(non-specific phosphatase) in cyanide induced adenosine monophosphate catabolism in rat polymorphonuclear leucocytes, *Biochem. J.*, 186, 907, 1980.
41. **Lichtman, M. A. and Weed, R. I.**, The monovalent cation content and ATPase activity of human normal and leukemic granulocytes and lymphocytes; relationship to cell volume and morphological age, *Blood*, 43, 645, 1969.
42. **Baron, D. N. and Levin, G. E.**, Isolation of human leucocytes and analysis of their sodium, potassium, water and hydrogen ion concentration, *J. Physiol. (London)*, 246, 11P, 1974.
43. **Baron, D. N. and Ahmed, S. A.**, Intracellular concentration of water and principal electrolytes determined by analysis of isolated human leucocytes, *Clin. Sci.*, 37, 205, 1969.
44. **Wilson, D. L. and Mannery, J. F.**, Permeability of rabbit leukocytes to sodium, potassium and chloride, *J. Cell Comp. Physiol.*, 34, 493, 1947.
45. **Hempling, H. G.**, Potassium and sodium movements in rabbit polymorphonuclear leukocytes, *J. Cell Comp. Physiol.*, 44, 87, 1954.
46. **Elsbach, D. and Schwartz, I. L.**, Studies on sodium and potassium transport in rabbit polymorphonuclear leukocytes, *J. Gen. Physiol.*, 42, 883, 1959.
47. **Wilson, D. G.**, Direct effect of adrenal corticoid steroids on the electrolyte content of rabbit leukocytes, *Am. J. Physiol.*, 43, 493, 1957.
48. **Endres, G. and Herget, L.**, Mineralzusammensetzung der Blutpattchen und weissen Blutkorperchen, *Z. Biol.*, 88, 451, 1929.
49. **Hempling, H. G.**, K. and Na content in dogfish leukocytes, *Biol. Bull.*, 103, 303, 1952.
50. **Woodin, A. M. and Wieneke, A. A.**, The accumulation of calcium by the polymorphonuclear leucocyte treated with staphylococcal leucocidin and its significance in the extrusion of protein, *Biochem. J.*, 87, 481, 1963.

51. **Petroski, R. J., Naccache, P. H., Becker, E. L., and Sha'afi, R. I.,** Effect of the chemotactic peptide and cytochalasin B on the cellular level of calcium in rabbit neutrophils, *FEBS Lett.*, 100, 161, 1963.
52. **Al-Mohanna, F. A. and Hallett, M. B.,** Actin polymerization modifies activation of the neutrophil oxidase, *Biochem. Biophys. Acta*, 927, 366, 1987.
53. **Pertoft, H., Back, O., and Lindahl-Kiessling, K.,** Separation of various blood cells in colloidal silica-polyvinyl pyrrolidone gradients, *Exp. Cell Res.*, 50, 355, 1960.
54. **Bretz, U. and Baggionlini, M.,** *J. Cell Biol.*, G3, 251, 1974.
55. **Schmid-Schonbien, G. W., Sung, K-L. P., Tozeren, H., Skalak, R., and Chein, S.,** Passive mechanical properties of human leukocytes, *Biophys. J.*, 36, 234, 1981.
56. **Naccache, P. H., Showell, H., Becker, E. L., and Sha'afi, R. I.,** Changes in ionic movements across rabbit polymorphonuclear leukocytes membranes during lysosomal enzyme release, *J. Cell Biol.*, 75, 635, 1977.
57. **Pozzan, T., Lew, P. D., Wollheim, C. B., and Tsein, R. Y.,** Is cytosolic ionized calcium regulating neutrophil activation?, *Science*, 221, 1413, 1983.
58. **White, J. R., Naccache, P. H., Molski, T. F. P., Borgeat, P., and Sha'afi, R. I.,** Direct demonstration of increased intracellular concentration of free calcium in rabbit and human neutrophils, *Biochem. Biophys. Res. Commun.*, 113, 44, 1980.
59. **Hallett, M. B. and Campbell, A. K.,** Is intracellular Ca^{2+} the trigger for oxygen radical production by polymorphonculear leucocytes?, *Cell Calcium*, 5, 1, 1984.
60. **Simochowitz, L. and Roos, A.,** Regulation of intracellular pH in human neutrophils, *J. Gen. Physiol.*, 85, 443, 1985.
61. **Moliski, T. F. P., Naccache, P. H., Volpi, M., Wolpert, L. M., and Sha'afi, R. I.,** Specific modulation of the intracellular pH of rabbit neutrophils by chemotactic peptide, *Biochem. Biophys. Res. Commun.*, 94, 508, 1980.
62. **Cooke, E., Al-Mohanna, F. A., and Hallett, M. B.,** Ca^{2+} dependent and Ca^{2+}-independent mechanisms in neutrophils, in *The Neutrophil: Cellular Biochemistry and Physiology*, Hallett, M. B., Ed., CRC Press, Boca Raton, Fla., 1988.
63. **Borregaard, N. and Herlin, T.,** Energy metabolism neutrophil during phagocytosis, *J. Clin. Invest.*, 70, 550, 1982.
64. **Cohen, H. J. and Chovaniec, M. E.,** Superoxide generation in digitonin-stimulated guinea-pig granulocytes, *J. Clin. Invest.*, 61, 1088, 1970.
65. **Roberts, P. A., Patel, A., Barrow, S. C. L., Hallett, M. B., and Campbell, A. K.,** Cyclic AMP in PMN measured by chemiluminescence energy transfer assay, in *Analytical Applications of Biolumінesence and Chemiluminescence*, Kricka, L. J., Stanley, P. E., Thorpe, G. H. G., and Whitehead, T. P., Eds., Academic Press, New York, 1984, 175.
66. **Smith, R. J. and Ignarro, L. J.,** Bioregulation of lysosomal enzyme secretion from human neutrophils; roles of guanosine 3'5'-monophosphate and calcium in stimulus-secretion coupling, *Proc. Natl. Acad. Sci. U.S.A.*, 72, 108, 1975.
67. **Baehner, R. L., Johnson, R. B., and Nathan, D. G.,** Comparative studies of the metabolism and bacteriocidal characteristics of severely glucose-6-phosphate dehydrogenase deficient polymorphonuclear leukocyes and leukocytes from children with chronic granulomatous disease, *J. Reticulo-Endothelial Soc.*, 12, 150, 1972.
68. **Selvaraj, R. J. and Sbarra, A. J.,** Relationship of glycolytic and oxidative metabolism to particle entry and destruction in phagocytic cells, *Nature (London)*, 211, 1272, 1967.
69. **Diem, K. and Lentner, D. L., Eds.,** *Scientific Tables*, 7th ed., Documenta Giegy, Basel, 1975.
70. **Helfman, D. M., Barnes, K. C., Kinkade, J. M., Vogler, W. R., Shoji, M., and Kuo, F. F.,** Phospholipid-sensitive and Ca^{2+} -dependent protein phosphorylating system in various types of leukemic cells from human patients and from the leukemic cell lines HL60 and K562, *Cancer Res.*, 43, 2955, 1983.
71. **Romeo, D., Cramer, R., Marzi, T., Saranzo, M. R., Zubucchi, G., and Rossi, F.,** Peroxidase activity of alveolar and peritoneal macrophages, *J. Reticuloendothelial Soc.*, 13, 399, 1973.
72. **Hallett, M. B.,** unpublished.
73. **Rausch, D. G. and Moore, T. G.,** Granule enzymes in polymorphonuclear neutrophilic leukocytes; a phylogenetic comparison, *Blood*, 46, 913, 1975.
74. **Ohlsson, K.,** in *Inflammation: Basic Mechanisms, Tissue Injury Principles and Clinical Models*, Venge, P. and Lindbom, A. Eds., Almquist and Wiksell Int., Stockholm, 1985.
75. **Jones, H. P., Ghai, G., Petrone, W. F., and McCord, J. M.,** Calmodulin-dependent stimulation of the NADPH-oxidase in human neutrophils, *Biochim. Biophys. Acta*, 714, 152, 1982.
76. **Chafouleas, J. G., Dedman, J. R., Munjaal, R. P., and Means, A. R.,** Calmodulin: development and application of a sensitive radio-immunoassay, *J. Biol. Chem.*, 254, 10262, 1979.
77. **Bradford, P. G. and Rubin, R. P.,** Quantitative changes in inositol triphosphate in chemoattractant-stimulated neutrophils, *J. Biol. Chem.*, 261, 15644, 1986.

78. **Cockcroft, S., Barrowman, and Gomperts, B. D.**, Breakdown and synthesis of polyphosphoinositides in fmetLeuPhe- stimulated neutrophils, *FEBS Lett.*, 183, 259, 1985.
79. **Sheterline, P., Rickard, J. E., and Richards, R. C.**, Fc-receptor directed phagocytic stimuli induce transient actin assembly at an early stage of phagocytosis in neutrophilic leukocytes, *Eur. J. Cell Biol.*, 34, 8, 1984.
80. **White, J. R., Naccache, P. H., and Sha'afi, R. I.**, Stimulation by chemotactic peptide of actin association to cytoskeleton in rabbit neutrophils, *J. Biol. Chem.*, 258, 14544, 1983.
81. **Reed, P. W.**, Glutathione and hexose monophosphate shunt in phagocytosing and hydrogenperoxide stimulated rat leukocytes, *J. Biol. Chem.*, 244, 2459, 1969.
82. **Woodin, A. M.**, The effect of staphylococcal leucocidin on the leucocyte, *Biochem. J.*, 80, 562, 1961.
83. **Van der Weyden, M. B., Fong, H., Hallam, L., and Breidahl, M. I.**, A rapid and simple assay for human erythrocyte ferritin, *Clin. Chem. Acta*, 127, 397, 1983.
84. **Bennet, R. M. and Kokocinski, J.**, Lactoferrin content of peripheral blood cells, *Br. J. Haematol.*, 39, 509, 1979.
85. **Krebs, H. A. and Johnson, J.**, *Tab. Biol. (Amsterdam)*, 19, 100, 1940.
86. **Bennet, J. P., Cockcroft, S., Caswell, A. H., and Gomperts, B. D.**, Plasma membrane location of phosphatidyl inositol hydrolysis in rabbit neutrophils stimulated by formyl-methionyl-leucyl phenylalanine, *Biochem. J.*, 208, 801, 1982.
87. **Stjernholm, R. L. and Manak, R. C.**, Carbohyrate metabolism in leukocytes, *J. Reticuloendothelial Soc.*, 8, 550, 1970.
88. **Tibbling, G.**, Glycerol uptake in leukocytes and thombocytes, *Scand. J. Clin. Lab. Invest.*, 26, 185, 1970.
89. **Stjernholm, R. L., Dimitrov, N. V., and Pijanowski, L. J.**, Carbohyrate metabolism in leukocytes, *J. Reticuloendothelial Soc.*, 6, 194, 1969.
90. **Edwards, S. W., Hallett, M. B., Lloyd, D., and Campbell, A. K.**, Decrease in apparent km for oxygen after stimulation of respiration of rat polymorphonuclear leukocytes, *FEBS Lett.*, 161, 60, 1983.
91. **Weening, R. S., Roos, D., and Loos, J. A.**, Oxygen consumption of phagocytosing cells in human leukocyte and granulocyte preparations; a comparative study, *J. Lab. Clin. Med.*, 83, 570, 1974.
92. **Rosen, H. and Klebanoff, S. J.**, Chemiluminescence and superoxide production by myeloperoxidase-deficient leukocytes, *J. Clin. Invest.*, 58, 50, 1976.
93. **Homan-Muller, J. W. T., Weening, R. S., and Roos, D.**, Production of hydrogen peroxide by phagocytozing human granulocytes, *J. Lab. Clin. Med.*, 85, 198, 1975.
94. **Edwards, S. W., Hallett, M. B., and Campbell, A. K.**, Oxygen-radical production during inflammation may be limited by oxygen concentration, *Biochem. J.*, 217, 851, 1983.
95. **Kakinuma, K. and Kaneda, M.**, Apparent km of leukocyte O_2 and H_2O_2 forming enzyme for oxygen, *Adv. Exp. Med. Biol.*, 141, 351, 1982.
96. **Babior, B. M., Curnette, J. T., and Kipnes, R. S.**, Pyridine nucleotide-dependent superoxide production by a cell-free system from human granulocytes, *J. Clin. Invest.*, 56, 1035, 1975.
97. **Patriarca, P., Cramer, R., Moncalbo, S., Rossi, F., and Romeo, D.**, Enzymatic basis of the metabolic stimulation in leukocytes during phagocytosis; the role of the activated NADPH oxidase. *Arch. Biochem. Biophys.*, 145, 255, 1971.
98. **Reed, P. W. and Templeman, J.**, Phagocytosis-associated metabolism and enzymes in the rat polymorphonuclear leukocyte, *Am. J. Physiol.*, 216, 223, 1969.
99. **Seligmann, B. E. and Gallin, J. I.**, Comparison of indirect probes of membrane potential utilized in studies of human neutrophils, *J. Cell Physiol.*, 115, 105, 1983.
100. **Mottola, C. and Romeo, D.**, Calcium movement and membrane potential changes in the early phase of neutrophil activation by phorbol myristate acetate, *J. Cell Biol.*, 93, 129, 1982.
101. **Lichtman, M. A. and Weed, R. I.**, Electrophorectic mobility and *N*-acetyl neuraminic acid content of human normal and leukemic lymphocytes and granulocytes, *Blood*, 35, 12, 1970.
102. **Williams, L. T., Snyderman, R., Pike, M. C., and Lefkowitz, R. J.**, Specific receptor sites for chemotactic peptides on human polymorphonuclear leukocytes, *Proc. Natl. Acad. Sci. U.S.A.*, 74, 1204, 1977.
103. **Bennet, J. P. and Gomperts, B. D.** Use of cytochalasin B to distinguish between early and late events in neutrophil activation, *Biochim. Biophys. Acta*, 601, 584, 1980.
104. **Synderman, R. and Pike, M. C.**, *N*-formylated peptide receptors on equine leukocytes stimulate secretion but not chemotaxis, *Science*, 209, 493, 1980.
105. **Aswanikumar, S., Corcoran, B., and Schiffman, E.**, Demonstration of a receptor for rabbit neutrophils for chemotactic peptides, *Biochem. Biophys. Res. Commun.*, 74, 810-817, 1977.
106. **Chenoweth, D. J. and Hugli, T. E.**, Demonstration of specific C5a receptor sites on intact human polymorphonuclear leukocytes, *Proc. Natl. Acad. Sci. U.S.A.*, 75, 3943, 1977.
107. **Graham, H. J., Lowry, O. H., Wheelwright, F., Lenz, M. A., and Parish, H. H.**, Distribution of histamine among leukocytes and platelets, *Blood*, 10, 467, 1955.
108. **Albritton, E. C., Ed.** *Standard Values in Blood*, W. B. Saunders, Philadelphia, 1951, 57.

109. **Schrodinger, E.,** *What is Life? The Physical Aspect of Living Cells,* Cambridge University Press, London, 1944.
110. **Eddington, A. S.,** *The Nature of the Physical World,* J. M. Dent & Son, London, 1920.
111. **von Tscharner, V., Prod'hom, B., Baggiolini, M., and Reuter, H.,** Ion channels in human neutrophils are activated by a rise in the free cytosolic calcium concentration, *Nature (London),* 324, 369, 1986.
112. **del Castillo, J. and Katz, B.,** Quantal components of the end-plate potential, *J. Physiol. (London),* 124, 369, 1986.
113. **Boyd, I. A. and Martin, A. R.,** The end-plate potential in mammalian muscle, *J. Physiol. (London),* 132, 74, 1956.
114. **Liley, A. W.,** The quantal components of the mammalian end-plate potential, *J. Physiol. (London),* 133, 571, 1956.
115. **Katz, B.,** *Nerve, Muscle, and Synapse,* McGraw-Hill, New York, 1966.
116. **Hallett, M. B.,** Quantal secretion and response lag demonstrated in single rat neutrophils, *Biochim. Biophys. Acta,* 847, 15, 1985.
117. **Patel, A. K., Hallett, M. B., and Campbell, A. K.,** Threshold responses in individual human neutrophils measured by flow cytometry and microfluorometry, *Biochem. J.,* 248, 173, 1987.
118. **Al-Mohanna, F. A., Oshini, I., and Hallett, M. B.,** Botulinum C2 toxin enhances activation of the rat neutrophil oxidase; further evidence of a role for actin, *FEBS Lett.,* 219, 40, 1987.
119. **Hallett, M. B. and Campbell, A. K.,** Two distinct mechanisms for stimulation of oxygen-radical production by polymorphonuclear leucocytes, *Biochem. J.,* 216, 459, 1983.
120. **Davies, P., Fox, R. I., Polyzonis, M., and Allison, A. C.,** The inhibition of phagocytosis and the facilitation of exocytosis in rabbit polymorphonuclear leukocytes, *J. Lab. Invest.,* 28, 16, 1973.
121. **Wright, A. E. and Douglas, S. R.,** An experimental investigation of the role of blood fluids in connection with phagocytosis, *Proc. R. Soc. Lond. Ser. B,* 72, 364, 1903.
122. **Sbarra, A. J. and Karnovsky, M. L.,** The biochemical basis of phagocytosis, *J. Biol. Chem.,* 235, 2224, 1960.
123. **Karnovsky, M. L. and Wallach, D. F. H.,** The metabolic basis of phagocytosis: incorporation of inorganic phosphate into various classes of phosphatides during phagocytosis, *J. Biol. Chem.,* 236, 1895, 1961.
124. **Woodin, A. M. and Weineke, A. A.,** The accumulation of calcium by polymorphonuclear leucocytes treated by stapphylococcal, *Biochem. J.,* 87, 487, 1963.
125. **Maklund, S. L.,** Extracellular superoxide dismutase and other superoxide dismutase isoenzymes in tissues from nine species, *Biochem. J.,* 222, 699, 1984.
126. **Wilson, E. B.,** *An Introduction to Scientific Research,* McGraw-Hill, New York, 1952.

Chapter 2

CHARACTERISTICS AND MECHANISMS OF SECRETION BY NEUTROPHILS

James E. Smolen

TABLE OF CONTENTS

I. Introduction 24

II. Granule Descriptions 25
  A. Primary Granules 25
  B. Secondary Granules 25
  C. Tertiary Granules 25
  D. Assays of Granule Contents 26

III. Properties of Degranulation 26
  A. Secretagogues 26
  B. Kinetics of Degranulation 27
    1. End-Point Assays of Degranulation 27
    2. Continuous Assays of Degranulation 28

IV. Stimulus-Response Coupling 28
  A. Receptor-Ligand Interactions 29
  B. G-Proteins 29
  C. Phosphatidylinositol Metabolism 30
  D. Ions 30
    1. Calcium 30
    2. Other cations 32
    3. Anions 33
    4. pH 33
  E. Protein Kinases 33
  F. Lipids 34
    1. Phospholipases 34
    2. Arachidonic Acid Metabolism 35
      a. Cyclooxygenase Products 35
      b. Lipoxygenase Products 35
    3. Platelet Activating Factor 36
  G. Cytoskeleton 36
    1. Microfilaments 36
    2. Microtubules 37
  H. Cyclic Nucleotides 37

V. Fusion 38
  A. Synexins 39
  B. Calcium 39
  C. Lipids 39
  D. Osmolarity 39

VI. Disorders of Secretion 40
A. Chediak-Higashi Syndrome 40
B. Specific Granule Deficiency 40
C. Myeloperoxidase Deficiency 41

VII. Conclusion 41

References 42

I. INTRODUCTION

The primary role of neutrophils is host defense, to which end the contents of the secretory granules are tailored. However, some granule contents (particularly those enzymes designed for scavenging as well as microbicidal roles) are also capable of damaging host tissues. Under normal conditions, degranulation accompanies phagocytosis, wherein invading microorganisms are ingested. During degranulation, granules fuse with phagosomes containing the ingested particles, thereby discharging their contents in concentrated form in the immediate vicinity of their targets. Ideally, the extracellular milieu, host tissues, and the cytoplasm of the neutrophils are spared the deleterious effects of the granule contents. In practice, some of these contents are released outside the cell during particle ingestion, usually because of premature degranulation into those phagosomes that are still open at the cell surface.

Extensive extracellular degranulation (secretion) can be stimulated in some pathologic conditions or experimental systems. For example, a neutrophil might try to ingest an oversized particle, such as a large immune complex or opsonized surface; phagocytosis could not be accomplished but degranulation would be stimulated and extracellular discharge of granule contents would occur. These processes are felt to play important roles in initiating or exacerbating inflammatory reactions, such as arthritis and systemic lupus erythematosus. An experimental system for the study of these processes in vitro is provided by neutrophils whose ability to ingest particles is blocked by the presence of cytochalasin B, an alkaloid which disrupts microfilaments.[1,2] The surfaces of such cells can be stimulated by a variety of agents, such as particles, opsonins, lectins, and soluble chemotactic factors, leading to secretion of granule contents. Thus, cytochalasin B converts neutrophils from phagocytic cells into model secretory cells; consequently, it is possible to monitor extracellularly (after fusion of granules with surface membranes) processes that normally occur intracellularly (fusion of granules with phagosome membranes).

This review will examine the properties and mechanisms of secretion from neutrophils; other cell systems, including those as closely related as monocytes and macrophages, will not be discussed for the sake of brevity. The first topic will be the types of granules and their contents. A description of the agents used to induce degranulation will then be provided. The stimulus-response coupling pathway whereby cell surface signals are transduced into degranulation will be described. This will be followed by a discussion of the fusion process itself, wherein granule membranes and plasma membranes merge, thereby permitting the release of granule contents. Finally, this review will describe several disorders of neutrophil degranulation, which generally lead to impaired bactericidal capacity.

II. GRANULE DESCRIPTIONS

A. Primary Granules

One type of granule present in neutrophils is called "primary", on the basis of its early appearance in granulocytopoiesis, or "azurophil", on the basis of its histochemical properties. These granules contain all of the cellular myeloperoxidase (MPO; at least six isozymes have been reported[3]), low molecular weight cationic proteins (which can directly increase vascular permeability, cause histamine release, enhance phagocytosis, or exert permeant antimicrobial activity)[4-9], acid hydrolases (β- glucuronidase, acid phosphatase, α-mannosidase, *N*-acetyl-β- glucosaminidase), and neutral proteases. The last two categories appear to have digestive as well as bactericidal functions.[4,10-12]

With respect to the tissue destruction observed in inflammatory reactions, the most important azurophil granule contents are the proteases. There are three well-defined lysosomal proteases in azurophil granules of neutrophils. Elastase was the first of the proteases to be well characterized. This serine protease appears to have a molecular weight of 34,000 daltons.[13] It hydrolyzes typical pancreatic elastase substrates and a variety of elastin preparations from several sources. Cathepsin G is a chymotrypsin-like neutral protease with a molecular weight of 32,000 daltons. This enzyme hydrolyzes cartilage proteoglycans and insoluble collagen and seems to have considerable similarities with elastase.[14] Finally, cathepsin D has been shown to be present in granules[15] and has a molecular weight of 42,000 daltons. This enzyme can cleave leukokininogens to generate pharmacologically active leukokinins.[16] It is also active against proteoglycans. Working in concert, these neutral proteases can exacerbate inflammatory reactions by generating chemotactic factors from C5.[17]

At least two populations of azurophil granules have been resolved on the basis of velocity sedimentation[10,18] and isopycnic sucrose gradient sedimentation[19,20]: the slower population contains more myeloperoxidase, neutral protease, and β-glucuronidase while the faster one contains more lysozyme. Furthermore, the denser granule types release elastase during degranulation while the lighter ones do not.[20] These granule types also have different distributions of glycosidases.[21,22] Using morphologic criteria, three forms of azurophil granules have been observed using transmission and high-voltage electron microscopy.[23]

B. Secondary Granules

"Specific" or "secondary" granules appear at a later stage in granulocytopoiesis than do the "primary" variety. Specific granules contain lysozyme[10,24] (which hydrolyzes cell wall proteoglycan of some bacterial species and which is also found in azurophil granules), alkaline phosphatase (in rabbit and guinea pig neutrophils), collagenase,[25] vitamin B_{12} binding protein,[26,27] and lactoferrin.[10,28,29] The latter is an iron-binding protein with bacteriostatic properties. Collagenase is an important enzyme in inflammation since, when combined with a neutral protease, it can hydrolyze collagen fibrils. It is a metalloenzyme with a molecular weight of 70,000 to 76,000 daltons[13] which can also cleave Clq.[30] More recently, three collagenases have been isolated from neutrophils, having molecular weights of 75,000, 57,000, and 22,000 daltons.[31]

C. Tertiary Granules

A tertiary granule population, which cosediments with secondary granules in most preparative schemes, has also been reported.[32] This granule type has not yet made a substantial impact on studies of neutrophil secretion since it is difficult to resolve from specific granules and is currently only known to contain gelatinase,[33,34] the assay for which is tedious. This enzyme also has considerable biochemical similarities with neutrophil collagenase.[35] Tertiary granule membranes also appear to have a pool of glycoproteins which have been implicated in cell adhesion.[36,37] "Phosphasomes", which contain alkaline phosphatase, have also been isolated from neutrophils;[38] their relationship to gelatinase-containing tertiary granules is currently unknown.

D. Assays of Granule Contents

In spite of the wide variety of granule contents, most studies of neutrophil secretion employ only a handful of markers for which simple, convenient assays are available. For studies of azurophil granule secretion, either myeloperoxidase or β-glucuronidase is customarily measured. The former enzyme uses H_2O_2 to oxidize *o*-dianisidine to a colored product.[24] The latter enzyme splits the artificial substrate phenolphthalein-β-D-glucuronide, liberating phenolphthalein which can be measured colorimetrically in base.[39] The most reliable marker for specific granule contents is vitamin B_{12} binding protein.[27] The assay involves competition by the binding protein with bovine serum albumin for radiolabeled vitamin B_{12}. Radioimmunoassays for lactoferrin can also be used to measure secretion from specific granules; however, the use of this marker is complicated by the fact that lactoferrin adheres avidly to cell surfaces. The venerable assay for lysozyme, which is found in both specific and azurophil granules, is turbidometric; the rate of change in light scattering induced by degradation of cell walls of *Micrococcus lysodiekticus* is followed.[40] These markers are widely used by investigators because of their stability and ease of assay. The assays can all be completed the same day or within 8 hr of the experiment. Most of the markers are also stable to freezing.

In addition to endpoint determinations using the above-mentioned assays, a number of direct and indirect methods for continuously monitoring degranulation have been devised. Sklar et al.[41] employed methyl-succinyl-alanyl-alanyl-prolyl-valyl methylcoumarin amide as their elastase substrate. The release of aminomethylcoumarine, resulting from extracellular cleavage of the substrate by secreted elastase, can be continuously monitored fluorimetrically. Similarly, Korchak et al.[42] monitored the liberation of the chromophore *p*-nitrophenol from the substrate *N*-*t*-Boc-l-alanyl-*p*-nitrophenol as an indicator of elastase secretion.

Another continuous technique was developed using the cationic fluorophore 9-aminoacridine.[43] This lysosomotropic agent partitions into acidic spaces in the cells; in neutrophils, the primary acidic compartments are the azurophil granules. The dye is taken up into the granules to such a great extent that self-quenching occurs. Consequently, once it is released into the extracellular medium during degranulation, the total fluorescence of the cell suspension increases. The assay is quantitative since all of the granule-associated 9-aminoacridine can be competed out with NH_4Cl, permitting the percentage of degranulation to be measured.

Flow cytometry can also be used to indirectly monitor degranulation. Sklar et al[44] first showed that a decrease in right-angle light scatter correlates with elastase release. Both responses are similar in kinetics, dose-response characteristics, receptor-occupancy requirements for chemotactic peptide, and dependence upon cytochalasin B. In the absence of cytochalasin B, a different right angle light scatter response, potentially independent of secretion, is found. These results have been confirmed and extended by Fletcher and Seligmann,[45] who reported a correlation between right angle light scatter and secretion of lysozyme and β-glucuronidase as well as a decrease in ''granularity index'', the latter being a calculation dependent upon the extent of granule enzyme release.

III. PROPERTIES OF DEGRANULATION

A. Secretagogues

A variety of particulate and soluble stimuli can be used to provoke secretion from neutrophils. Most of these have immunologic import, such as immune complexes, opsonized particles, and chemotactic factors derived from serum factors or bacteria. The tumor promotor phorbol myristate acetate (PMA) and the lectin concanavalin A (Con A) are also widely employed. Three major points must be considered when a given stimulus is to be used. The first is the mode of action of the stimulus; does it engage cell surface receptors or does it bypass them? Examples of receptor-mediated stimuli include immune complexes, the complement-derived chemotactic peptide C5a, leukotriene B_4 (LTB_4), platelet-activating factor (PAF), and the synthetic chemotactic peptide *N*-formyl methionyl leucyl phenylalanine (FMLP). Stimuli which bypass cell surface receptors include PMA, 1-oleoyl 2-acetyl glycerol (OAG), and the calcium ionophores ionomycin and A23187. A second consideration is whether or not a given stimulus is a ''complete'' secretagogue. A complete secretagogue induces the release of both specific and azurophil granule contents.[46] An ''incomplete'' secretagogue, such as Con A or PMA, stimulates release from specific granules only.[47] A third consideration, related to the second, is which, if any, of the cytochalasins are employed to enhance secretion.[48] Most stimuli, such as FMLP, Con A, and C5a are virtually impotent as secretagogues without the alkaloid, although some discharge of specific granule contents might be observed. Limited secretion from specific granules can also be obtained with PMA in the absence of cytochalasin B. To induce large-scale release of specific granules or any discharge of azurophil granules, cytochalasin B must be present; this is especially true when particulate stimuli are employed. However, the investigator must realize that the results obtained with cytochalasin B may not be relevant to other neutrophil responses which are measured in the absence of the drug.

B. Kinetics of Degranulation

1. Endpoint Assays of Degranulation

Neutrophils respond very rapidly to surface stimulation. Degranulation begins with a lag period on the order of seconds and is completed within a few minutes, particularly if cytochalasin B is present. Perhaps the earliest study of the initial kinetics of degranulation was provided by the classic work of Bainton.[49] Granule discharge into phagosomes of rabbit neutrophils was observed by electron microscopy. Using histochemical stains, Bainton found that specific granule contents are released into the phagosome before the contents of azurophil granules, a process called ''sequential degranulation''.

Quantitative examinations of the kinetics of degranulation have recently appeared in the literature. The relative kinetics of degranulation into phagosomes of neutrophils were measured by Segal et al.[50] When the enzymatic contents of isolated phagosomes were assayed, it was found that lactoferrin and MPO appears in the phagosomes promptly, with no apparent lag periods. However, azurophil granule markers, such as β-glucuronidase and acid phosphatase, appear much later, with lag periods of 1 to 2 min. The authors suggested that the kinetics indicate that MPO is not contained in either specific or azurophil granules.

In order to obtain better kinetic measurements, a semicontinuous flow dialysis system was devised to monitor enzyme secretion from neutrophils treated with cytochalasin B.[51] Using a wide variety of stimuli, it was possible to show that lag periods do exist for lysosomal enzyme release and that the lag periods are comparable to those seen for superoxide anion (O_2^-) generation. The lag periods for both responses do not depend upon the dose of FMLP employed.

A rapid centrifugation technique, based upon the pelleting of cells through a layer of silicone oil, was also used to measure secretion of granule enzymes in response to a variety of stimuli.[52] This stopping method provides a time resolution on the order of seconds. Lag periods are stimulus dependent; as found by others,[51] FMLP provokes the most rapid secretion, with little or no apparent lag period. For all stimuli, the specific granule marker lactoferrin is secreted before release of two azurophil granule enzymes, thereby providing a quantitative basis for sequential degranulation.

2. Continuous Assays of Degranulation

The advent of the continuous monitoring techniques cited above permitted a clearer view of the rapidity of secretion. The 9-aminoacridine technique was first used by Smolen et al. to examine the kinetics of degranulation under a variety of conditions.[43] They reported that the rates of secretion and lag periods (approximately 5 sec or less with FMLP) are dependent upon the temperature and stimulus employed. A variety of calcium antagonists and anion channel blockers prove to modulate the rates, but not the lag periods, for secretion. By using a competitive inhibitor of FMLP binding, the authors also showed that only 5 sec of exposure to the chemotactic peptide are required for substantial degranulation.

The 9-aminoacridine technique was also used by Serhan et al.[53] in a kinetic analysis of the activation of neutrophils by leukotriene B_4. They reported that degranulation is very rapid (having a lag period of only a few seconds) and is one of the first responses of the cell to stimulation. Smolen[54] used both the rapid centrifugation and 9-aminoacridine methods for monitoring lag periods in response to FMLP and A23187. It was reported that compared to superoxide anion generation, the lag periods for degranulation are only slightly or not at all affected by calcium antagonists and anion channel blockers. Finally, Naccache et al.[55] used the 9-aminoacridine method to look at the effects of prior exposure to A23187 on degranulation induced by FMLP and leukotriene B_4. In essence, they found that the ionophore inhibited release induced by these latter stimuli.

Sklar et al.[41] reported that the release of elastase from human neutrophils, in the presence of cytochalasin B and stimulated with an analogue of FMLP, commences within 5 sec. The lag periods are relatively independent of stimulus concentration. Furthermore, only about 15 sec of continuous exposure to the stimulus are needed to provoke maximal release. Release of elastase was also compared to flow cytometry measurements by this group.[44]

Korchak et al.[42] reported a very short lag period for the release of elastase from neutrophils treated with cytochalasin B. In agreement with others,[41,43] they reported that after only 10 sec of receptor occupancy, the cells are committed to maximal degranulation. A lag period of approximately 6 sec was reported by the same workers for this response.[56] This work also compares the kinetics of degranulation with a number of other neutrophil responses measured by techniques featuring time resolution on the order of seconds.

Sklar and co-workers utilized right angle light scatter as an indicator of secretion from neutrophils in two recent reports. The kinetics, response time, and receptor occupancy requirements were compared for a variety of responses to chemotactic peptide.[57] Of interest is that right angle light scatter requires lower doses of stimulus, lower receptor occupancy, and less time to reach maximal response than elastase release. Furthermore, Sklar and Oades[490] reported that degranulation, as monitored by right angle light scatter in the presence of cytochalasin B, is strongly dependent upon the presence of intracellular Ca^{2+}. Transient changes in right angle light scatter in the absence of cytochalasin B, attributable to actin polymerization,[58] are insensitive to Ca^{2+}.

In contrast to these findings, McNeil et al.[59] were unable to demonstrate a relationship between right angle light scatter and degranulation. Rather, these authors report delayed (3 to 15 min) increases in right angle light scatter, which are attributed to changes in cell shape. Even under degranulating conditions (i.e., in the presence of cytochalasin B), no substantial decreases in scatter (which might be due to secretion) are observed. However, it is noteworthy that this report does not include actual measurements of degranulation, making a correlation (or lack thereof) between this response and decreases in right angle light scatter impossible to establish.

IV. STIMULUS-RESPONSE COUPLING

The transduction mechanisms leading from cell-surface interactions ultimately to secretion are currently under intense investigation. Because these topics are being covered in more

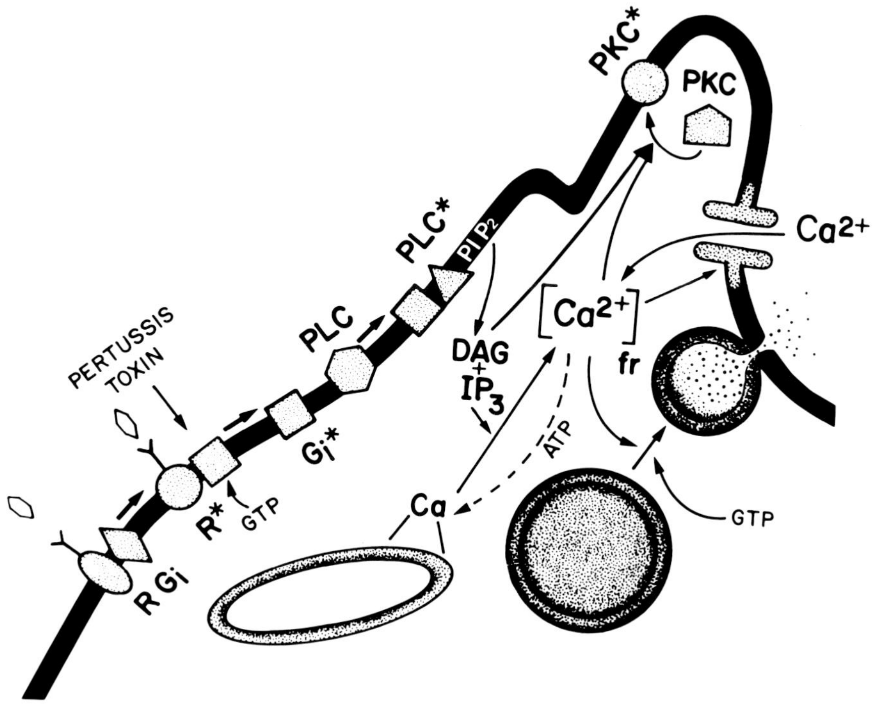

FIGURE 1. Transduction mechanism for secretion in neutrophils. The biochemical processes currently believed to be responsible for stimulus-secretion coupling are depicted. The abbreviations employed are as follows. R, R* — receptors; G_i, G_i* — guanine nucleotide binding proteins; PLC, PLC* — phospholipase C; PKC, PKC* — protein kinase C; DAG — diacylglycerol; IP_3 — inositol 1,4,5-trisphosphate. (From Smolen, J. E., Molecular mechanisms of leukocyte emigration, in *Leukocyte Emigration,* S. Karger AG, Basel, in press. With permission.)

detail elsewhere in this book, this review will, of necessity, be more cursory. Figure 1 schematically outlines the stimulus- response coupling pathway as it is currently understood. The succeeding discussion will approximately follow this scheme from left to right.

A. Receptor-Ligand Interactions

The initial step in neutrophil stimulation, a process leading ultimately to secretory events such as degranulation, generally involves either a ligand-receptor interaction or some other direct perturbation of the cell surface. The kinetics of receptor-ligand interactions appear to be rapid, when soluble stimuli are employed. Binding of the chemotactic peptide *N*-formyl-norleucyl-leucyl-phenylalanine (FNLP) to rabbit neutrophils or neutrophil plasma membranes is both time dependent and saturable.[60] FNLP, at saturating concentrations, occupies 50% of the available receptors on intact neutrophils within 30 sec. Similar kinetics were observed by Sklar et al.,[41] who employed a novel fluorimetric technique for measuring the binding of chemotactic peptide. They estimate that no more than 20% of the receptors need to be occupied to evoke an optimal secretory response. The kinetics of binding for other ligands have not been reported with such accuracy; however, this step is likely to be rapid with cell surface receptors.

B. G-Proteins

Recent results implicate guanine nucleotide-binding proteins (G-proteins) in the stimulus-response coupling pathway. In particular, neutrophil responses to f-Met-Leu-Phe and other stimuli which interact with defined receptors on the cell surface are blocked by pretreatment of the cells with pertussis toxin. Pertussis toxin covalently modifies an inhibitory G-protein, N_i; this protein was first identified as an inhibitory coupling species for the adenylate cyclase system in other cell types. Virtually all of the neutrophil responses to cell surface receptor-ligand interactions, such as elevation in intracellular Ca^{2+},[61-64] superoxide anion generation,[63,65-67] lysosomal enzyme release,[62,63,65,66,68-70] aggregation,[65,66] chemotaxis,[62,64-66,70] f-Met-Leu-Phe binding,[65] membrane receptor capping,[64] cytoskeletal changes,[71] phosphatidylinositol turnover,[63,68,69,72] arachidonic acid release,[73,74] and protein phosphorylation,[72] are blocked by this toxin. Stimuli which directly activate protein kinase C, such as PMA, are not affected by pertussis toxin.[67,70-72] When PAF is used as the stimulus, calcium mobilization appears to be unaffected,[68] while all other neutrophil responses are blunted.[66,68] Furthermore, guanine nucleotides can regulate the affinity of the f-Met-Leu-Phe receptor.[75,76] Treatment of intact neutrophils or isolated plasma membranes results in the ADP-ribosylation of a 41,000-dalton protein, which is very similar in its properties to N_i.[62,73,74,77] The fact that pretreatment with PMA results in increased ADP-ribosylation of this N_i-like protein suggests a role for protein kinase C in the regulation of its activity.[77]

An additional role for G-proteins is suggested by experiments using permeabilized neutrophils. In this system, guanine nucleotides were shown to induce secretion in their own right, as well as to enhance degranulation stimulated by Ca^{2+};[78,79] the effects of these agents are not blocked by pretreatment of the cells with pertussis toxin, suggesting (but not proving) the involvement of a different form of G-protein. Guanine nucleotides can act at more proximal steps in the stimulus-response coupling pathway of permeabilized neutrophils by stimulating polyphosphoinositide-specific phospholipase C activity.[80]

C. Phosphatidylinositol Metabolism

It currently appears that the next step in stimulus-response coupling involves the G-protein-induced activation of a polyphosphoinositide-specific phospholipase C. In particular, the relatively rare lipid phosphatidylinositol 4,5-bisphosphate is hydrolyzed to form two important second messengers, diacylglycerol (DAG) and inositol 1,4,5-trisphosphate (IP_3). DAG, of which class oleoyl acetyl glycerol (OAG) is a member, activates protein kinase C and will be discussed later in this review. IP_3 seems to release calcium from intracellular stores, thereby increasing intracellular free Ca^{2+} concentrations.

A number of researchers reported that typical secretagogues induce phosphatidylinositol turnover in neutrophils.[81-89] However, the kinetics of the response indicate that the hydrolysis of phosphatidylinositol *per se* is not involved in stimulus-response coupling.[85] It was later found that stimulation provokes major breakdowns of the polyphosphorylated species, phosphatidylinositol 4,5-bisphosphate (PIP_2) and phosphatidylinositol 4-monophosphate (PIP_1) [88,90,92-94] yielding IP_3 and IP_2 along with DAG.[90-93] Relatively high concentrations of calcium are required for this reaction in vitro.[90] Activation of these enzymes is not calcium dependent in the presence of guanine nucleotides.[80] In intact neutrophils, the turnover of these lipids commences within seconds.[91,92,94] Stimulation by A23187 or PMA, both of which bypass surface receptors, does not induce IP_3 formation.[92,93]

Once generated in response to receptor-ligand interactions, IP_3 itself then goes on to play a critical role in stimulus-response coupling; this compound binds to an intracellular receptor [95] and induces the release of calcium from intracellular stores, [96] probably located in the endoplasmic reticulum. As detailed below, the resulting increase in intracellular free Ca^{2+} is probably sufficient to induce degranulation on its own.

D. Ions

1. Calcium

Like other cell types, the intracellular Ca^{2+} concentration in neutrophils is maintained at submicromolar levels by a Ca^{2+}- ATPase pump.[97-99] Thus, elevations of Ca^{2+} above these low levels could serve in the capacity of a second messenger in stimulus-response coupling. The fact that degranulation can be induced by ionophores such as A23187 and ionomycin or by high concentrations of Ca^{2+} (presented to cells prepared without divalent cations[100]) has suggested that fluxes of divalent cations are required for stimulus-response coupling. Increased influxes of Ca^{2+} have been reported within the first few seconds following stimulation, accompanied by secretion and other cellular responses.[65,101] However, the fact that such influxes occur does not necessarily mean that they are crucial in stimulus-response coupling. Indeed, degranulation in response to PMA has been shown to be independent of extracellular Ca^{2+}.[102] This latter observation was extended to a wide variety of secretagogues, for which degranulation is not blocked by saturating concentrations of ethylene bis(oxyethylenenitrilo) tetraacetic acid (EGTA).[103] This strongly suggests that extracellular Ca^{2+} and a Ca^{2+} influx are not absolute requirements.

If a Ca^{2+} influx is not an absolute requirement for degranulation from neutrophils, then perhaps elevated and stimulatory levels of intracellular Ca^{2+} could be provided by their mobilization from intracellular stores. Evidence for such a mechanism was provided by the observation that TMB-8, an inhibitor of the mobilization of intracellular Ca^{2+}, completely blocks neutrophil responses.[102,103] TMB-8, Quin 2, and other intracellular calcium antagonists inhibit chemotaxis and oxidative metabolism.[104] Similarly the calmodulin inhibitors W-7 and trifluoperazine completely block degranulation,[103,105] without inhibiting stimulated changes in chlortetracycline (CTC) fluorescence,[105,106] a putative measure of membrane-bound calcium.

The rapid drop in fluorescence when CTC-loaded cells are exposed to stimulation indicates that membrane-bound deposits are being mobilized at a very early time in the stimulus-response coupling pathway.[105-107] The fact that these responses are not inhibited by EGTA emphasizes the intracellular nature of this process. Studies with the intracellular calcium probe Quin 2 indicate that following stimulation with FMLP, intracellular Ca^{2+} rises almost immediately from a resting level of 0.1 μM to approximately 1 μM.[101,108,109] A substantial rise is also observed in the presence of EGTA, suggesting that intracellular stores of Ca^{2+} can be mobilized. One of the few stimuli which apparently does not provoke an increase in intracellular free Ca^{2+} is PMA.[99,110] In fact, this stimulus enhances Ca^{2+} efflux from neutrophils and neutrophil cytoplasts.[111] Even in this case, intracellular free Ca^{2+} may underlie the activation process. PMA reduces the K_m for Ca^{2+} for its putative receptor, protein kinase C, into the submicromolar range.[112] With these altered kinetics, the kinase could be activated by the ''resting'' levels of intracellular Ca^{2+}.

Quin 2 has been used extensively in neutrophils to monitor changes in intracellular Ca^{2+} in response to other stimuli as well.[113-115] Unfortunately, this probe is also known to blunt the very intracellular Ca^{2+} response which it measures, by virtue of its chelating abilities, thereby underestimating the peak levels observed.[116-119] Indeed, quin 2 can be used as a potent calcium antagonist in neutrophils.[104,120,121] To obviate this difficulty, several other means have been developed for monitoring intracellular Ca^{2+} in these cells. One method employs the related fluorophore, Fura 2, which is less potent as a chelator.[122] In this study, it was found that spreading of neutrophils involves an increase in intracellular Ca^{2+}. Fused cell hybrids were used to introduce the calcium-sensitive photoprotein obelin into neutrophils;[123] these studies show that intracellular Ca^{2+} levels reach 6 μM following stimulation. Finally, Slonczewski et al. took advantage of the calcium-sensitivity of the intracellular enzyme phosphorylase a to show that free Ca^{2+} increases promptly during stimulation;[124] while this technique is totally nonperturbing, it is, unfortunately, not quantitative.

Thus, there is substantial evidence that intracellular free Ca^{2+} levels increase in stimulated neutrophils and that this Ca^{2+} can be derived from either intracellular or extracellular sources.[125] However, to substantiate a second messenger role for this cation, it is necessary to show that micromolar levels of free Ca^{2+} are sufficient to induce degranulation by the cells. For this purpose, permeabilized cell systems have been employed. Smolen and Stoehr[126] reported that the cholesterol-complexing agent saponin can be used to permeabilize human neutrophils in an EGTA- containing medium. These cells can be induced to secrete lysozyme in response to free Ca^{2+} in the micromolar range. However, these cells are quite fragile and do not secrete any azurophil granule constituents. These methodologic difficulties can be obviated by use of another cholesterol-complexing agent, digitonin.[96,127] Cells permeabilized with digitonin can be induced to secrete contents of both specific and azurophil granules following exposure to micromolar levels of free Ca^{2+}[127] alone, without any apparent cofactors such as Mg-ATP, cytochalasin B, or glucose. Secretion from specific granules is more sensitive than that from azurophil granules, perhaps providing a mechanistic basis for sequential degranulation. Release of granule constituents apparently involves fusion of the granule and plasma membranes; this is suggested both morphologically[127] and by the expression of a granule membrane antigen, Mo1, on the surface.[128] Exposure to degranulating concentrations of Ca^{2+} also results in resealing of the cells with respect to large molecular weight solutes.[127] Using elastase release, RLS, and 9-aminoacridine as continuous measures of secretion, it was shown that degranulation is very rapid, with a lag period of <10 sec and completion within approximately 2 min;[129] these results are very similar to those obtained in intact cells and suggest that the permeabilized neutrophil system is serving as an accurate model of degranulation.

In addition to digitonin, Sendai virus was used as a permeabilizing agent for neutrophils.[78,130] In this system, degranulation by both granule types is also stimulated by micromolar levels of free Ca^{2+}; however, somewhat lower concentrations are required than with the digitonin-permeabilized cells. Using both permeabilization systems, guanine nucleotides both serve as secretagogues in their own right and also augment release stimulated by Ca^{2+};[78,79,130] these data suggest a role for G-proteins in the fusion process. A potential role for products of phospholipase C, such as DAG or IP_3, was suggested by the observation that the inhibitor neomycin can inhibit Ca^{2+}-induced degranulation in digitonin-permeabilized neutrophils.[129]

While the data obtained with the permeabilized cell systems demonstrate that intracellular Ca^{2+} *can* alone serve as a second messenger, it is clear that the situation is not as simple in intact cells. Not only can PMA or DAG induce degranulation without any accompanying rise in intracellular Ca^{2+}, but also these agents can apparently work synergistically to augment secretion.[131-136] Furthermore, some other neutrophil responses, such as O_2^- generation and actin polymerization, appear to be independent of increases in intracellular Ca^{2+}.[137,138]

2. Other Cations

Surface interactions in secretory cells are usually accompanied by changes in membrane potential. To quantify these changes in neutrophils, which are too small to permit the insertion of microelectrodes, investigators resorted to a number of indirect techniques. Korchak and Weissmann first used the lipophilic cation triphenylmethylphosphonium ($TPMP^+$) as an indicator of stimulated changes in membrane electrochemical potential.[139,140] They found that the resting potential is negative and that initial hyperpolarizations, which are rapid and might be involved in stimulus-response coupling, follow exposure to agonists. These reports were followed by others using a variety of carbocyanine probes which confirm the negative resting potential, but, in contrast, find initial depolarizations following stimulation.[141-152] Unfortunately, both techniques are prone to experimental artifacts. The phosphonium probes can be toxic to cell function and take a very long time to equilibrate, limiting their usefulness with fresh cells and rapid changes in potential.[153] The carbocyanine dyes can also be toxic,[153] can be destroyed by MPO,[148] and may monitor mitochondrial, rather than plasma membrane, potential,[150] Nonetheless, in spite of the methodologic difficulties, experiments with both sets of probes indicate that changes in membrane potential occur within the first 5 sec following stimulation and may play some role in stimulus-response coupling.

These early changes in electrochemical potential suggest that prompt alterations in ion permeability are taking place. Simchowitz and co-workers[149] demonstrated the existence of an electrogenic Na^+-K^+ pump system which seems to generate the membrane potential in neutrophils. A (Na^+-K^+)-ATPase which can be activated by FMLP has been reported for rabbit cells.[154,155] A number of reports also demonstrated that enhanced Na^+ influxes and K^+ effluxes occur following stimulation.[156-162] Na^+ influx appears to be coupled to an amiloride-sensitive H^+ countertransport.[159,163-165] The potential importance of extracellular Na^+ in stimulus-response coupling is underscored by the fact that in its absence, degranulation is inhibited.[140,166] However, these same investigators showed that a Na^+ influx, induced by the specific ionophore monensin, is inadequate to stimulate secretion on its own.

3. Anions

Very little work has been done concerning the role of anions in stimulus-response coupling in neutrophils. Korchak et al.[167] first showed that several anion channel blockers inhibit extracellular degranulation without affecting superoxide anion generation; fusion of granules with phagosomes is apparently unaltered. They later showed that the effects of these inhibitors on SO_4^{2-} fluxes correlate closely with inhibition of secretion.[168] Smith et al.[169] confirmed these findings and extended them to a wide range of secretagogues. Smolen[54] also showed that anion channel blockers inhibit the rate of degranulation without affecting the lag periods of the responses.

4. pH

As with other cell types, the neutrophil contains several intracellular compartments with varying pH. As already mentioned, azurophil granules are acidic, as evidenced by their ability to take up 9-aminoacridine.[43] Indeed, the internal pH of human neutrophil granules is estimated as 5.6.[170] The acidity of these organelles may help contribute to a decline in phagolysosomal pH following degranulation into phagosomes;[171-174] the pH inside these vacuoles can reach as low as 6.

Activated neutrophils generate intracellular acids, which lead initially to a drop in pH; active extrusion of protons then leads to alkalinization of the cytoplasm.[175-177] The initial acidification seems to be generated by the metabolic burst following stimulation.[178-180] The subsequent rise in pH (approximately 0.6 units) requires extracellular Na^+ and is sensitive to amiloride, suggesting that a Na^+-H^+ antiporter is involved.[166,181-185] These biphasic changes in pH are also seen in neutrophil cytoplasts, which are devoid of granules and nuclei;[182] thus, degranulation is not the cause of these intracellular events. A Cl^-/HCO_3^- exchange may be involved in the recovery from intracellular alkalinization. When amiloride is used to block proton fluxes, chemotaxis to FMLP is inhibited;[185] the effect of this drug on degranulation has not yet been reported.

E. Protein Kinases

It has been known for some time that protein kinases are present in neutrophils[186-190] and that changes in protein phosphorylation occur following stimulation.[191] Phosphatases capable of dephosphorylating histone, a commonly used in vitro protein kinase substrate, are also present in neutrophils.[190,192]

Research has focused primarily upon two protein kinases: cyclic AMP-dependent protein kinase (PKA) and Ca^{2+}-activated, phospholipid-dependent protein kinase (protein kinase C or PKC). It might be expected that these enzymes are involved in stimulus-response coupling since their cofactors, cyclic AMP, Ca^{2+}, and DAG, are mobilized early following stimulation. Endogenous substrates for in vitro phosphorylation by both PKA and PKC were reported several years ago.[193,194] More often, investigators load cells with $^{32}P_i$ and look at endogenous proteins which are phosphorylated before and after stimulation of the intact neutrophils. While this technique does not identify which protein kinase is phosphorylating which substrates, it does provide a more general, overall view of kinase activity. It is not surprising that a wide variety of proteins are phosphorylated both before and after stimulation,[195-198] even in cytoplasts.[199,200] Phosphorylation or dephosphorylation of any particular target protein is dependent upon the timing and the stimulus employed; these changes often commence within a few seconds and generally correlate with the onset of degranulation. However, very few of the endogenous substrates have been identified in neutrophils; there is some evidence that these proteins include actin,[194] vimentin,[196] and a component of the NADPH oxidase system.[201]

The two protein kinases of neutrophils discussed above have been partially purified and characterized. In vitro assays of these enzymes can employ a variety of substrate proteins.[202] PKC can be activated by very low concentrations of Ca^{2+} in intact cells.[132] PKC also seems to be a receptor for phorbol esters in other cell types; this is also likely to be true for neutrophils.[203] There is some evidence that the binding site is shared by PMA, DAG (which may be the natural cofactor), bryostatin, and mezerein (which may be both an agonist or antagonist of PKC.[203-206] Arachidonate and other unsaturated fatty acids can also indirectly activate PKC in neutrophils.[207,208] Activation of PKC seems to result in "translocation" of PKC and its associated PMA-binding site from the cytosol to membrane fractions,[209] perhaps permitting the enzyme access to membrane-bound substrates. Binding of PKC and a specific Ca^{2+}-requiring protease to neutrophil membranes appears to require Ca^{2+};[210] proteolysis results in the formation of a soluble, Ca^{2+}-independent kinase.[211]

The close correlation between activation of PKC and neutrophil functions[198,203-205,208,212] and the synergy between PKC agonists and other stimuli [208,213,214] have suggested a role for this enzyme in stimulus-response coupling. The ability of PKA to stimulate phosphorylation of phosphatidylinositol[215] suggests that this enzyme may also intervene at more proximal steps in the transduction pathway. However, virtually all of these data were obtained indirectly and there is no evidence for an obligate role of PKC or PKA in stimulus-response coupling using stimuli other than protein kinase agonists. It has been reported that neutrophil-specific granules contain a potent inhibitor of PKC which may serve to dampen or at least regulate its activity in intact cells.[216] Furthermore, the PKC inhibitors H-7 and H-9 fail to inhibit neutrophil activation in response to several stimuli,[217,218] eliminating an obligate role for full activity of this enzyme. Finally, mezerein, which can completely block histone phosphorylation in the in vitro PKC assay, stimulates degranulation on its own.[206] Thus, PKC activity is not required for responses to physiologic stimuli, and the roles of this and other protein kinases in transduction are as yet unclear.

F. Lipids

1. Phospholipases

One of the first consequences of neutrophil stimulation to be studied was lipid turnover.[219-222] The fact that membrane fractions obtained before and after phagocytosis have substantially different fatty acid contents[223] suggests that phospholipases, particularly phospholipase A-type enzymes, are operative. Soluble stimuli also provoke phospholipid remodeling.[224] Later studies demonstrated the existence of these enzymes, particularly phospholipase A_2, in both the granules and membranes of neutrophils.[225-231] Phospholipase A_2 appears to require Ca^{2+} for activity.[232] The membrane-bound enzyme is activated during stimulation[233,234] and the granule-contained phospholipase A_2 is released during secretion.[228,235] The latter enzyme was purified away from a potent permeability-increasing protein, in conjunction with which it plays a bactericidal role.[236] The activity of neutrophil phospholipase A_2 can be reduced by glucocorticoids, which induce the generation of the cloned gene product lipomodulin[237] or macrocortin,[238] protein inhibitors of the enzyme; the activity of lipomodulin itself can be altered by protein phosphorylation.[240] The products of phospholipase A_2, namely, arachidonic acid and lysophosphatides, are both fusogenic and may play a role in degranulation. However, the main role of phospholipase A_2 appears to be the generation of arachidonic acid for subsequent metabolism (see below). The importance of this enzyme is suggested by the fact that one of its inhibitors, *p*-bromophenacyl bromide, is a potent antagonist of degranulation.[239]

2. Arachidonic Acid Metabolism

The stimulated release of arachidonic acid subsequent to activation of phospholipase A_2 may play a critical role in stimulus-response coupling. Not only is this and other unsaturated fatty acids released from activated neutrophils,[241-246] but also arachidonic acid itself can serve as a stimulus for these cells.[247-249] Direct activation of neutrophils may be a consequence of the ability of arachidonic acid and its metabolites to increase membrane permeability to Ca^{2+}.[250] In the next two sections, the main metabolites of arachidonic acid and their effects on neutrophil function will be discussed.

a. Cyclooxygenase Products

Stimulated neutrophils generate a number of stable prostaglandins of the E and F series, using the cyclooxygenase pathway of arachidonic acid metabolism.[251] These prostanoids generally inhibit cell functions, but their precise effects are dependent upon the particular prostaglandin and the type of response being examined.[252-257] The inhibition caused by the prostaglandins may be related to their ability to raise intracellular cyclic AMP levels.[257,258]

Perhaps the most potent cyclooxygenase product produced by activated neutrophils is thromboxane A_2, which is rapidly metabolized to the stable end product thromboxane B_2.[259-262] These compounds are powerful vasoconstrictors and may also enhance neutrophil chemotaxis,[263] aggregation,[261] and adhesiveness.[264] The strongest evidence that these and other cyclooxygenase products play crucial roles in degranulation is that a variety of non-steroidal antiinflammatory agents, which usually inhibit the lipoxygenase pathway as well as the cyclooxygenase pathway, blunt secretion in response to a variety of stimuli.[239,265-267]

b. Lipoxygenase Products

Borgeat and co-workers showed that arachidonic acid can also be metabolized by the lipoxygenase pathway to produce hydroxyeicosatetraenoic acids (HETEs), including 5-HETE, 12-HETE, and 5,12-diHETE.[268,269] These compounds and their hydroperoxy precursors (HPETEs) were shown to alter Ca^{2+} fluxes[250,270] and to induce a variety of neutrophil responses such as degranulation,[271-275] chemotaxis,[270,272,273,276-278] elevations of cyclic GMP levels,[273] and aggregation.[279] Production of HETEs correlates closely with degranulation.[280] The activity of the lipoxygenases is subject to product regulation.[281-283] In addition, HETEs may be reincorporated into cellular lipids[244,284] or exchanged with platelets.[285,286]

Stimulated neutrophils also produce a number of leukotrienes, which were formerly designated slow-reacting substance of anaphylaxis (SRS-A). One type of diHETE, leukotriene B_4 (LTB_4), is produced in the most copious amounts, although LTB_3[287] and LTB_5[288,289] can be obtained if fatty acids other than arachidonic acid are used as the substrate. LTB_4 and other leukotrienes can be released in response to a variety of stimuli,[290-296] even from neutrophil cytoplasts.[297] Receptors for LTB_4 were reported and partially purified,[298-302] although their specificity is disputed; [303] these receptors may become linked to the cytoskelton following occupancy.[304] Like the other HETEs, LTB_4 can alter Ca^{2+} fluxes[305,306] and can serve as a potent stimulus for degranulation,[307-310] chemotaxis,[311-315] and adherence;[312] LTA_4 and LTB_5 are somewhat less potent in these regards.[316,317] LTB_4 is catabolized by w-oxidation; the enzyme responsible for producing 20-OH- and 20-COOH-LTB_4 appears to be a cytochrome P-450 and has been partially purified.[318-320]

In addition to the fact that HETEs and LTB_4 serve as stimuli in their own right, the main evidence that these compounds are involved in stimulus- response coupling is pharmacologic. Inhibitors of lipoxygenase activity are potent antagonists of neutrophil function and are generally more powerful than are cyclooxygenase inhibitors.[239,248,260,277,310,321-325] The fact that these various lipids can be combined to either induce degranulation synergistically or to desensitize the neutrophil[326] strongly suggests a potential role in stimulus-response coupling. However, the observation that an inhibitor of leukotriene synthesis has little effect on aggregation induced by several stimuli casts some doubt on this hypothesis.[327]

3. Platelet Activating Factor

Another potent mediator of inflammation produced by stimulated neutrophils is 1-*O*-alkyl-2-acetyl-*sn*-glyceryl-3-phosphorylcholine, also known as platelet-activating factor (PAF).[328-331] Optimal synthesis requires extracellular Ca^{2+}.[331] PAF is heterogeneous with respect to the alkyl moiety,[332-335] although this has not been reported by all investigators;[336] acetate in the sn-2 position is required for activity. The synthesis and metabolism of PAF has also been extensively studied. 1-*O*-alkyl-2-acyl-*sn*-glyceryl-3-phosphorylcholines, particularly 1-*O*-alkyl-2-arachidonyl-*sn*-glyceryl-3-phosphorylcholine,[337] can be deacylated to form lyso-PAF;[338,339] arachidonic acid released from 1-*O*-alkyl-2-arachidonyl-*sn*-glyceryl-3-phosphorylcholine can be subsequently metabolized. The lyso-PAF is then acetylated with acetyl-CoA.[339-341] PAF is metabolized by an acetylhydrolase; the resulting lyso-PAF is often reacylated with arachidonic acid.[342,343] Not only is PAF synthesized by neutrophils, but it is also a potent stimulus for degranulation,[344–349,351,352] aggregation,[344,350,352,353] arachidonic acid metabolism,[348,350,353] glucose uptake,[352] superoxide generation,[351] and phospholipid turnover.[354] A suggestion that PAF may play a role in the stimulus-response coupling pathway is provided by the observation that this compound can exhibit cross-desensitization with other classic neutrophil secretagogues.[355,356]

G. Cytoskeleton

1. Microfilaments

The roles that microfilament proteins play in leukocyte function have recently been reviewed in detail.[357] In brief, the hyaline pseudopods of the neutrophil contain filament networks which are composed of filamentous actin and myosin. Actin comprises 10% of the protein of the neutrophil and exists in two states, as a globular monomer and as a double helical polymer of 0.1 to 10μm. The distribution of actin remains unknown in neutrophils, but many actin fibers, oriented in a random fashion, populate the peripheral cytoplasm and exist in a gel state.[358] Myosin represents about 1% of the cytoplasmic protein of neutrophils,[358] and, like striated muscle myosin, is a hexamer composed of two heavy polypeptide chains and four light chains. Actin in the pseudopods exists as a gel due to an actin- binding protein of high molecular weight (540,000 daltons) which cross-links actin filaments.[358] Actin-binding protein, an asymmetric dimer composed of two highly flexible chains,[359] comprises 1% of the total cell protein and is concentrated at the cell periphery along with myosin during phagocytosis.[360-362] Calcium concentrations greater than $2 \times 10^{-7}M$ dissolve the actin gel by activating another protein called gelsolin,[363] which shortens actin filaments between the cross-linking proteins.[364] When the calcium concentration falls below the critical level, gelsolin becomes inactive, permitting reannealing and gelation of the actin filaments to occur.[364,365] A number of other protein modulators of microfilament structure (profilin, acumentin) have been identified and are reviewed elsewhere.[357]

Myosin contracts the actin gel by hydrolysis of ATP which provides energy for contraction.[366,367] The interaction between actin and myosin can explain the force generation for movement and the dependence of phagocytosis on metabolic energy. Control of directional movement can be shown experimentally to arise from regulation of the network consistency of actin by gelsolin and Ca^{2+}.[368] Dysfunction of the contractile system can lead to impairment of chemotaxis and phagocytosis.

The role played by microfilaments in degranulation is suggested by the observation that in resting neutrophils the granules are separated from the plasma membrane by filament-rich hyaline ectoplasm. As already discussed, surface stimulation of the cell leads to an increased mobilization of Ca^{2+}, particularly in the periphery;[369] this increase in cytosolic Ca^{2+} would be expected to activate gelsolin, thereby shortening actin filaments and decreasing the viscosity in this area.[364] Indeed, it has been long observed that stimulation leads to such rearrangements of microfilaments and their regulatory proteins.[370-377] The decrease in viscosity of the hyaline ectoplasm could permit the granules to have access to the inner surface of the plasma membrane, thereby facilitating degranulation. The well-known ability of cytochalasins to enhance secretion[1,2] and block locomotion[378] is consonant with this model, since these alkaloids can split actin filaments.[379]

The enhanced turnover of microfilaments following stimulation of neutrophils is also manifest biochemically. The amount of actin associated with detergent-insoluble cytoskeleton fractions increases.[58,374,380-384] This cytoskeletal network has been visualized with high-voltage electron microscopy.[385] When chemotactic peptide is used as the stimulus, these changes in cytoskeletal actin may be associated with clearance of receptor-ligand complexes,[383] probably by endocytosis. Using techniques suitable for monitoring short-term kinetics, it was shown that polymerization commences within the first few seconds following stimulation and may thus play some role in stimulus-response coupling.[58,386]

2. *Microtubules*

Another important cytoskeletal structure in neutrophils is the microtubule, which is a hollow fiber, 24 nm in diameter. Microtubules assemble as cylindrical polymers of the 55,000-dalton protein tubulin, which is found as a heterodimer of α and β subunits. Soluble tubulin dimers in the cytoplasm appear to be in equilibrium with the assembled form, permitting ready "growth" or "dissolution" of these structures by shifting of the equilibrium. In many systems, microtubules can serve as "skeletal" support for the cell and perhaps for microfilaments and other contractile systems.

Microtubule assembly can be controlled in a variety of ways, both in vivo and in vitro. Agents which increase cGMP levels promote the assembly of tubulin. Ca^{2+} promotes dissolution of polymerized tubulin, as do agents which elevate cAMP levels; however, in neutrophils it appears that the state of microtubule assembly also controls cAMP levels.[387] Tubulin polymerization can also be controlled by redox conditions in vitro.[388] In neutrophils, redox conditions are a function of glutathione oxidation; indeed, the amount of oxidized glutathione, which increases as a consequence of oxidative metabolism in stimulated neutrophils, appears to regulate microtubule disassembly.[389,390] There is some evidence that oxidation increases the tyrosinolation of tubulin, which directly favors microtubule disassembly;[391,392] tubulin tyrosinolation is found largely in the plasma membrane fraction of stimulated neutrophils.[393]

Microtubules may play a role in stimulus-response coupling in neutrophils. The number of these structures increases following exposure to a number of stimuli and correlates closely with secretion.[370,394-396] No significant changes in microtubule numbers have been reported for chemotactic activation.[397] The fact that colchicine, which promotes microtubule disassembly, can inhibit phagocytosis, chemotaxis, surface adhesion, and degranulation suggests that these structures are important in all of these processes. Microtubules have been implicated in more subtle phenomena in that they can govern the planar disposition of plasma membrane components.[298-401]

H. Cyclic Nucleotides

Exposure of neutrophils to phagocytizable particles results in a rapid doubling of cyclic adenosine 3',5'-monophosphate (cAMP) levels.[402] This increment is prompt (maximal within

15 sec) and brief (returning to basal levels in 1 to 2 min), explaining why previous investigators, who had assayed cells after 2 to 5 min, were unable to detect this response.[403,404] Such a rapid response, which is comparable to the earliest changes in neutrophil biochemistry and physiology, could possibly play a role in stimulus-response coupling. Stimulated increments in cAMP levels were also reported in rabbit[405] and human[406-411] neutrophils. During the same time interval, no changes in cGMP levels are observed.[406,407] Stimulus- dependent lag periods are reported with some secretagogues.[406,411] The observations that changes in cAMP levels correlate closely with chemotaxis and degranulation,[411] display specific desensitization after repeated exposures to the same stimulus, and that this desensitization is accompanied by a parallel decrease in O_2^- generation[412] suggest that changes in cyclic nucleotide levels may be of mechanistic significance.

However, a number of lines of evidence suggest that changes in cAMP are not important in stimulus-response coupling in neutrophils. First, cAMP increases can be induced by low concentrations of FMLP, without accompanying degranulation or O_2^- generation,[406] suggesting that the former events are not sufficient for the latter. However, these dose-response data are disputed.[407] Second, generation of O_2^- and secretion of lysozyme, but not β-glucuronidase, can be induced by Con A and PMA, stimuli which do not provoke any changes in cAMP levels.[256] Similar observations were reported for other stimuli.[411] Thus, increments in cAMP do not appear to be necessary for discharge of specific granules or O_2^- generation. Finally, Simchowitz et al. showed that an inhibitor of adenylate cyclase, which blocks the stimulated increase in cAMP, does not impair neutrophil responses.[413] These findings seem to rule out cAMP in a second messenger role. However, redistribution of cAMP (without concomitant elevated levels) could still be of mechanistic significance, as immunochemical observations of cAMP show it to be localized near forming phagosomes.[414]

The adenylate cyclase of neutrophils appears to be coupled to microtubules, as interference with colchicine seems to increase enzyme activity and render the cells more sensitive to stimulation by histamine.[415] These histamine receptors have been examined and found to be of the H_2 class.[416] The adenylate cyclase of neutrophils was also found to be coupled to β-adrenergic receptors, which were closely examined;[417] uncoupling of the receptor appears to be the mechanism for β-adrenergic desensitization in these cells.[418] However, the adenergic receptor system does not appear to be involved in the increase in cAMP levels following stimulation with secretagogues; FMLP appears to increase cAMP levels indirectly through a Ca^{2+} dependent mechanism.[419] This could conceivably be accomplished by a calmodulin-sensitive phosphodiesterase,[419] which has been reported.[420] However, the degree to which neutrophil phosphodiesterase activities are sensitive to calmodulin has been disputed.[421,422]

One possible role for stimulated increments in cAMP is in feedback inhibition. Cyclic AMP-dependent protein kinase activity was detected in rabbit neutrophils [216,423] and the action of this enzyme could be either stimulatory or inhibitory. More to the point is the well-known observation that cAMP and agents which raise cellular cAMP levels inhibit neutrophil function.[256-258,403,424,425] Responses such as degranulation and O_2^- are inhibited differentially by these treatments.[426] However, unlike other cell types, inhibition of neutrophil metabolic responses by cAMP does not appear to be due to its effects on cytosolic Ca^{2+}.[427]

V. FUSION

The ultimate act in the secretory process, namely the fusion of the granule membranes with the phagosome or plasma membranes with the accompanying discharge of granule contents, has been virtually ignored in studies of neutrophils. It was shown early that actual degranulation, as assessed by cinemicrophotography, was very rapid, taking only a few seconds.[428] The fusion process is accompanied by the appearance of granule membrane markers on the cell surface,[36,37,429] even in permeabilized neutrophils.[128] Other than this, very little is known about the fusion process in neutrophils. This review, of necessity, will concentrate on fusion in better-studied systems, particularly of adrenal chromaffin granules, with the hope that similar mechanisms might be operative in neutrophils.

A. Synexins

Synexins belong to a class of Ca^{2+}-binding proteins found in adrenal chromaffin cells called chromobindins,[430] some of which are phosphorylated following stimulation.[431] At low micromolar concentrations of Ca^{2+}, synexin binds to chromaffin granules[432] and can undergo self-aggregation in the presence of *cis*-unsaturated fatty acids.[433] At high concentrations of Ca^{2+} (200 μM), it can induce the granules to aggregate and then fuse.[434,435] Aggregation can be sufficiently tight to promote resonance energy transfer[436,437] More than one type of synexin can be isolated from a given source material.[438]

B. Calcium

As discussed above, elevated levels of free Ca^{2+} are alone sufficient to induce degranulation from permeabilized neutrophils.[78,79,96,126-130] In vitro fusion events, not yet reported in neutrophils, have been studied in the chromaffin cell model and appear to require Ca^{2+}. The release of chromaffin granule contents can be stimulated by low micromolar concentrations of Ca^{2+} in the presence of isolated membranes,[439,440] perhaps providing a model system for exocytosis. Similar results were reported with sea urchin eggs.[441] Thus, as with the permeabilized neutrophils, Ca^{2+} alone appears to be sufficient for degranulation.

C. Lipids

Many of the lipids produced by stimulated neutrophils, particularly through the action of phospholipase A_2, are known to be fusogenic. Both arachidonic acid and lysophosphatides can promote fusion in vitro. Indeed, phospholipase A_2-treated plasma membranes can induce the release of chromaffin granule contents in a Ca^{2+}-dependent manner.[442] *cis*-Unsaturated fatty acids have been shown to induce the fusion of chromaffin granules aggregated by synexin and Ca^{2+}.[443] As examples of interactions between lipid structures and biological membranes, a Ca^{2+}-independent, trypsin-sensitive fusion of chromaffin granule membranes with liposomes was reported.[444] Similarly, a model system for degranulation was provided by the release of lysosomal enzymes from neutrophil granules in the presence of negatively charged liposomes.[445]

D. Osmolarity

There is substantial evidence that osmolarity can be important in the fusion process. Osmotic forces can control fusion of black lipid membranes,[446] and secretory granules of sea urchin eggs have been observed to swell irreversibly during Ca^{2+}-induced degranulation.[447] Like neutrophils, exocytosis by chromaffin cells is inhibited by hyperosmolarity.[435] Furthermore, chromaffin granules can be induced to lyse in the presence of Mg-ATP and permeant anions like Cl^-.[434] These results suggest that the action of a proton pump, compelling countertransport of permeant anions, may increase the osmolarity within the chromaffin granule; this would promote water transport, swelling, and eventual fusion with apposed plasma membranes.[435] In view of the acidity of azurophil granules,[170] a similar system may be operating in neutrophils. However, this model has been questioned by Holz[448] on the basis of experiments with digitonin-permeabilized chromaffin cells; in this system, degranulation is not highly sensitive to osmotic forces.

VI. DISORDERS OF SECRETION

A. Chediak-Higashi Syndrome

Patients with the rare autosomal recessive disease known as Chediak-Higashi syndrome have giant cytoplasmic granules in their neutrophils, monocytes, and lymphocytes.[449-452] Their susceptibility to infection can be explained by defective degranulation of abnormally large neutrophil lysosomes, delayed bactericidal responses, defects in oxidative metabolism, and impaired chemotaxis.[452-460] Beige mice, which have Chediak-Higashi syndrome, also lack lysosomal elastase and cathepsin G.[461] The abnormality in Chediak-Higashi syndrome appears to involve membrane function.[462-464] These neutrophils behave as if they were unable to regulate membrane activation. They spontaneously aggregate surface molecules into caps, exhibit high resting rates of oxidative metabolism, and demonstrate impaired adherence to surfaces with reduced chemotactic responses.[452,457] Beginning early in neutrophil development, there is ongoing fusion of giant azurophil granules with each other and with cytoplasmic membrane components;[464] this converts the azurophil granules into huge, abnormal[451] primary lysosomes which contain a reduced content of hydrolytic enzymes.[467] In spite of a normal ability to ingest particles, the neutrophils kill microorganisms relatively slowly due to inconsistent delivery of diluted hydrolytic enzymes from the giant granules into the phagosomes.[454,455,460] Molecular movements within the membrane are probably critical to cell function. Thus, altered membrane structure could lead to defective regulation of membrane activation by neutrophils in patients with Chediak-Higashi syndrome. Investigation by electron spin resonance spectrometry of neutrophil membranes from both humans and mice with Chediak-Higashi syndrome indicates increased fluidity of the cell membranes.[462,463] Little is known about the specific effect of membrane fluidity of cell function, although the ratio of polyunsaturated to saturated fatty acids is diminished. Surface receptor expression is affected, which can possibly lead to elevation of cAMP, disordered assembly of microtubules, and defective interaction of microtubules with lysosomal membranes, all of which are reported in this disorder.[258,391,468-470,] Treatment of the Chediak-Higashi syndrome neutrophils in vitro with ascorbic acid reduces membrane fluidity to near normal values and enhances the bactericidal activity of the neutrophils. When ascorbic acid is administered to patients in vivo, functional and clinical improvements are seen in some cases.[454,471] Functional improvement was also reported in some cases with lithium[472] and with colchicine,[473] which directly affects microtubules.

B. Specific Granule Deficiency

A small number of patients were identified whose neutrophils are deficient or lacking specific granules.[474-476] In many ways, these neutrophils share the functional abnormalities observed in Chediak-Higashi syndrome. Both sets of neutrophils have impaired adherence and chemotaxis and demonstrate an increased respiratory burst, although this is in dispute.[477] Some granule proteins are reduced in quantity and granule formation is morpholigically defective in developing myelocytes. Unlike the neutrophils of Chediak-Higashi syndrome, specific granule-deficient cells have normal membrane fluidity as monitored by electron spin resonance spectroscopy. In contrast to normal neutrophils, specific granule-deficient cells are unable to fully reduce their surface charge upon stimulation with chemotactic factors. These results imply that specific granule membranes and contents are normally recruited to the surface of the cells following stimulation, increasing surface "roughness" and cell size and diminishing negative surface charge.[478] These changes would thereby promote the attachment of neutrophils to other cell surfaces and to each other. The inability of specific granule-deficient cells to release lactoferrin, expand the surface membrane, and reduce the negative surface charge during stimulation could underlie the observed defects of adherence and chemotaxis following cellular activation with chemotactic factors.

The binding of iron to lactoferrin was shown to result in enhanced hydroxyl radical generation in neutrophils[479] and increased cytotoxicity.[480] It is of interest that specific granule-deficient neutrophils generate diminished amounts of hydroxyl radical upon stimulation, thereby implicating cell-associated lactoferrin in facilitating generation of this reactive oxygen species from O_2^- and H_2O_2.

More recently, it was shown that the neutrophils from one patient with specific granule deficiency have altered kinetics of O_2^- generation and patterns of endogenous protein phosphorylation. These changes may be related to an increase in resting cytosolic free Ca^{2+}, measured by using Quin 2, which could therefore increase the susceptibility of the granules, particularly specific granules, to release following suboptimal stimulation.

C. Myeloperoxidase Deficiency

The myeloperoxidase-H_2O_2-halide system is arguably the most potent bactericidal weapon employed by neutrophils. It is therefore not surprising that some patients with complete MPO deficiency suffer from opportunistic infections. What is more surprising is that MPO-deficiency is relatively widespread and that the number of patients with clinically significant infections is low;[482] this is likely due to the fact that the bactericidal mechanisms available to the neutrophils are so numerous and redundant that few bacteria can survive even in the absence of MPO.

A survey of the enzymology and clinical significance of MPO is beyond the scope of this review. However, this enzyme does seem to play a modulating role in degranulation. MPO can inactivate chemotactic factors[483] and thereby blunt neutrophil responsiveness. MPO can also inhibit the mechanisms underlying the respiratory burst[484] and phagocytosis, [485,486] perhaps by autooxidation of the neutrophils; the cells from patients with MPO deficiencies are metabolically more active than normal neutrophils. [485,486] Another commonly measured oxidative response of neutrophils, namely, chemiluminescence, is dependent upon MPO[487] and is absent from MPO-deficient patients,[488] thereby providing a diagnostic tool for identifying these individuals.

VII. CONCLUSION

The intent of this review was to cover degranulation by neutrophils; due to the scope of this topic, it was necessary to limit discussion of the literature on other neutrophil functions and on other closely related cell types. In many respects, this was unfortunate since these topics are closely related. However, other aspects of neutrophil function are being comprehensively reviewed in this volume and elsewhere. Reviews of secretion by other cell types, particularly adrenal chromaffin cells,[445,448,449] are also available.

It is apparent that degranulation by neutrophils has great significance, both clinically and scientifically. Clinically, secretion by neutrophils is involved in phagocytosis and subsequent bactericidal activity and in inflammation (arthritis, gout, systemic lupus erythematosus, and adult respiratory distress syndrome). Scientifically, the studies on receptor-ligand interactions, stimulus-response coupling, and membrane fusion constitute a burgeoning and exciting field of investigation. These studies are not only interesting in their own right but are also highly relevant to other secretory cell systems, with which they share many mechanistic similarities. Thus, such investigations may eventually lead to rationally designed therapies for secretory disorders in neutrophils and in other cells.

REFERENCES

1. **Zurier, R. B., Hoffstein, S., and Weissmann, G.,** Cytochalasin B: effect on lysosomal enzyme release from human leukocytes, *Proc. Natl. Acad. Sci. U.S.A.*, 70, 844, 1973.
2. **Zurier, R. B., Hoffstein, S., and Weissmann, G.,** Mechanisms of lysosomal enzyme release from human leukocytes. I. Effect of cyclic nucleotides and colchicine, *J. Cell Biol.*, 58, 27, 1973.
3. **Miyasaki, K. T., Wilson, M. E., Cohen, E., Jones, P. C., and Genco, R. J.,** Evidence for and partial characterization of three major and three minor forms of human neutrophil myeloperoxidase, *Arch. Biochem. Biophys.*, 246, 751, 1986.
4. **Zeya, H. I. and Spitznagel, J. K.,** Cationic protein-bearing granules of polymorphonuclear leukocytes: separation from enzyme-rich granules, *Science*, 163, 1069, 1972.
5. **Hallgren, R. and Venge, P.,** Cationic proteins of human granulocytes: enhancement of phagocytosis of Staphylococcus protein A-IgG complexes, *Inflammation*, 1, 237, 1976.
6. **Hodinka, R. L. and Modrzakowski, M. C.,** Bactericidal effect of granule contents from rat polymorphonuclear leukocytes, *Infect. Immun.*, 40, 139, 1983.
7. **Gennaro, R., Dolzani, L., and Romeo, D.,** Potency of bactericidal proteins purified from the large granules of bovine neutrophils, *Infect. Immun.*, 40, 684, 1983.
8. **Ganz, T., Selsted, M. E., Szklarek, D., Harwig, S. S. L., Daher, K., Bainton, D. F., and Lehrer, R. I.,** Defensins. Natural peptide antibiotics of human neutrophils, *J. Clin. Invest.*, 76, 1427, 1985.
9. **Selsted, M. E., Harwig, S. S. L., Ganz, T., Schilling, J. W., and Lehrer, R. I.,** Primary structures of three human neutrophil defensins, *J. Clin. Invest.*, 76, 1436, 1985.
10. **Spitznagel, J. K., Dalldorf, F. G., Leffel, M. S., Folds, J. D., Welsh, I. R. H., Cooney, M. H., and Martin, L. E.,** Character of azurophil and specific granules purified from human polymorphonuclear leukocytes, *Lab. Invest.*, 30, 744, 1974.
11. **Korn, E. D. and Weisman, R. A.,** Phagocytosis of latex beads by Acanthamoeba. II. Electron microscope study of the initial events, *J. Cell Biol.*, 34, 219, 1967.
12. **West, B. C., Rosenthal, A. S., Gelb, N. A., and Kimball, H. R.,** Separation and characterization of human neutrophil granules, *Am. J. Pathol.*, 77, 41, 1974.
13. **Ohlsson, K.,** Granulocyte collagenase and elastase and their interactions with a1-antitrypsin and a2-macroglobulin, in *Proteases and Biological Control*, Reich, E. and Rifkin, D. B., Eds., Cold Spring Harbor Laboratory, Cold Spring Harbor, New York, 1975, 591.
14. **Virca, G. D., Metz, G., and Schnebli, H. P.,** Similarities between human and rat leukocyte elastase and cathepsin G, *Eur. J. Biochem.*, 144, 1, 1984.
15. **Michell, R. H., Karnovsky, M. J., and Karnovsky, M. L.,** The distributions of some granule-associated enzymes in guinea-pig polymorphonuclear leucocytes, *Biochem. J.*, 116, 207, 1970.
16. **Movat, H. Z., Steinberg, S. G., Flavio, M., Habal, F. M., and Ranadive, N. S.,** Demonstration of a kinin-generating enzyme in the lysosomes of human polymorphonuclear leukocytes, *Lab. Invest.*, 29, 669, 1973.
17. **Ward, P. A. and Hill, J. H.,** C5 chemotactic fragments produced by an enzyme in lysosomal granules of neutrophils, *J. Immunol.*, 104, 535, 1970.
18. **Calamai, E. G. and Spitznagel, J. K.,** Characterization of rat polymorphonuclear leukocyte subcellular granules, *Lab. Invest.*, 46, 597, 1982.
19. **West, B. C., Rosenthal, A. S., Gelb, N. A., and Kimball, H. R.,** Separation and characterization of human neutrophil granules, *Am. J. Pathol.*, 77, 41, 1974.
20. **Garcia, R. C., Peterson, C. G., Segal, A. W., and Venge, P.,** Elastase in the different primary granules of the human neutrophil, *Biochem. Biophys. Res. Comm.*, 132, 1130, 1985.
21. **Rest, R. F., Cooney, M. H., and Spitznagel, J. K.,** Subcellular distribution of glycosidases in human polymorphonuclear leukocytes, *Biochem. J.*, 174, 53, 1978.
22. **West, B. C., Dunphy, C. H., and Moore, C. A.,** Human neutrophil *N*-acetyl-β-D-glucosaminidase: granule localization, *J. Lab. Clin. Med.*, 104, 60, 1984.
23. **Pryzwansky, K. B. and Brenton-Gorius, J.,** Identification of a subpopulation of primary granules in human neutrophils based upon maturation and distribution. Study by transmission electron microscopy cytochemistry and high voltage electron microscopy of whole cell preparations, *Lab. Invest.*, 53, 664, 1985.
24. **Bretz, U. and Baggiolini, M.,** Biochemical and morphological characterization of azurophil and specific granules of human neutrophilic polymorphonuclear leukocytes, *J. Cell Biol.*, 63, 251, 1974.
25. **Murphy, G., Reynolds, J. J., Bretz, U., and Baggiolini, M.,** Collagenase as a component of the specific granules of human neutrophil leukocytes, *Biochem. J.*, 162, 195, 1977.
26. **Kane, S. P. and Peters, T. J.,** Analytical subcellular fractionation of human granulocytes with reference to the localization of vitamin B12-binding proteins, *Clin. Sci. Mol. Med.*, 49, 171, 1975.
27. **Smith, G. P. and Peters, T. J.,** The release of granule components from human polymorphonuclear leukocytes in response to both phagocytic and chemical stimuli, *Biochim. Biophys. Acta*, 719, 304, 1982.

28. **Leffell, M. S. and Spitznagel, J. K.**, Association of lactoferrin with lysozyme in granules of human polymorphonuclear leukocytes, *Infect. Immun.*, 6, 761, 1972.
29. **Parmley, R. T., Takagi, M., Barton, J. C., Boxer, L. A., and Austin, R. L.**, Ultrastructural localization of lactoferrin and iron-binding protein in human neutrophils and rabbit heterophils, *Am. J. Pathol.*, 109, 343, 1982.
30. **Fletcher, D. S., Williams, H. R., and Lin, T. Y.**, Effects of human polymorphonuclear leukocyte collagenase on sub- component C1q of the first component of human complement, *Biochim. Biophys. Acta*, 540, 270, 1978.
31. **Hasty, K. A., Hibbs, M. S., Kang, A. H., and Mainardi, C. L.**, Secreted forms of human neutrophil collagenase, *J. Biol. Chem.*, 261, 5645, 986.
32. **Dewald, B., Bretz, U., and Baggiolini, M.**, Release of gelatinase from a novel secretory compartment of human neutrophils, *J. Clin. Invest.*, 70, 518, 1982.
33. **Sopata, I. and Dancewicz, A. M.**, Presence of a gelatin-specific proteinase and its latent form in human leucocytes, *Biochim. Biophys. Acta*, 370, 510, 1974.
34. **Murphy, G., Bretz, U., Baggiolini, M., and Reynolds, J. J.**, The latent collagenase and gelatinase of human polymorphonuclear neutrophil leucocytes, *Biochem. J.*, 192, 517, 1980.
35. **Williams, H. R. and Lin, T. Y.**, Human polymorphonuclear leukocyte collagenase and gelatinase. Comparison of certain enzymatic properties, *Int. J. Biochem.*, 16, 1321, 1984.
36. **Todd, R. F., III, Arnaout, M. A., Rosin, R. E., Crowley, C. A., Peters, W. A., and Babior, B. M.**, Subcellular localization of the large subunit of Mo1, a surface glycoprotein associated with neutrophil adhesion, *J. Clin. Invest.*, 74, 1280, 1984.
37. **Arnaout, M. A., Spits, H., Terhorst, C., Pitt, J., and Todd, R.F., III,** Deficiency of a leukocyte surface glycoprotein (LFA-1) in two patients with Mo-1 deficiency, *J. Clin. Invest.*, 74, 1291, 1984.
38. **Smith, G. P., Sharp, G., and Peters, T. J.**, Isolation and characterization of alkaline phosphatase-containing granules (phosphasomes) from human polymorphonuclear leucocytes, *J. Cell Sci.*, 76, 167, 1985.
39. **Brittinger, G. R., Hirschhorn, R., Douglas, S. D., and Weissmann, G.**, Studies on lysosomes. XI. Characterization of hydrolase-rich fraction from human lymphocytes, *J. Cell Biol.*, 37, 394, 1968.
40. In Worthington Enzyme Manual, Worthington Biochemical Co., Freehold, N. J., 1972, 100.
41. **Sklar, L. A., McNeil, V. M., Jesaitis, A. J., Painter, R. G., and Cochrane, C. G.**, A continuous, spectroscopic analysis of the kinetics of elastase secretion by neutrophils. The dependence of secretion upon receptor occupancy, *J. Biol. Chem.*, 257, 5471, 1982.
42. **Korchak, H. M., Wilkenfeld, C., Rich, A. M., Radin, A. R., Vienne, K., and Rutherford, L. E.**, Stimulus-response coupling in the human neutrophil. Differential requirements for receptor occupancy in neutrophil responses to a chemoattractant, *J. Biol. Chem.*, 259, 7439, 1984.
43. **Smolen, J. E., Korchak, H. M., and Weissmann, G.**, The kinetics of lysosomal degranulation of human neutrophils as measured by 9-aminoacridine quenching, *Biochim. Biophys. Acta*, 762, 145, 1983.
44. **Sklar, L. A., Oades, Z. G., and Finney, D. A.**, Neutrophil degranulation detected by right angle light scattering: spectroscopic methods suitable for simultaneous analyses of degranulation or shape change, elastase release, and cell aggregation, *J. Immunol.*, 133, 1483, 1984.
45. **Fletcher, M. P. and Seligmann, B. E.**, Monitoring human neutrophil granule secretion by flow cytometry: secretion and membrane potential changes assessed by light scatter and a fluorescent probe of membrane potential, *J. Leuk. Biol.*, 37, 431, 1985.
46. **Smolen, J. E. and Weissmann, G.**, Stimuli which provoke secretion of azurophil granules induce increments in adenosine cyclic 3′,5′-monophosphate, *Biochim. Biophys. Acta*, 672, 197, 1981.
47. **Estensen, R. D., White, J. G., and Holmes, B.**, Specific degranulation of human polymorphonuclear leukocytes, *Nature (London)*, 248, 347, 1974.
48. **Bentley, J. K. and Reed, P. W.**, Activation of superoxide production and differential exocytosis in polymorphonuclear leukocytes by cytochalasins A,B,C,D, and E. Effects of various ions, *Biochim. Biophys. Acta*, 678, 238, 1981.
49. **Bainton, D. F.**, Sequential degranulation of the two types of polymorphonuclear leukocyte granules during phagocytosis of microorganisms, *J. Cell Biol.*, 58, 249, 1973.
50. **Segal, A. W., Dorling, J., and Coade, S.**, Kinetics of fusion of the cytoplasmic granules with phagocytic vesicles in human polymorphonuclear leukocytes, *J. Cell Biol.*, 85, 42, 1980.
51. **Smolen, J. E., Korchak, H. M., and Weissmann, G.**, Initial kinetics of lysosomal enzyme secretion and superoxide anion generation by human polymorphonuclear leukocytes, *Inflammation*, 4, 145, 1980.
52. **Bentwood, B. J. and Henson, P. M.**, The sequential release of granule constituents from human neutrophils, *J. Immunol.*, 124, 885, 1980.
53. **Serhan, C. N., Radin, A., Smolen, J. E., Korchak, H., Samuelsson, B., and Weissmann, G.**, Leukotriene B_4 is a complete secretagogue in human neutrophils: a kinetic analysis, *Biochem. Biophys. Res. Comm.*, 107, 1006, 1982.

54. **Smolen, J. E.,** Lag period for superoxide anion generation and lysosomal enzyme release from human neutrophils: effects of calcium anatagonists and anion channel blockers, *J. Lab. Clin. Med.*, 104, 1, 1984.
55. **Naccache, P. H., Molski, T. F. P., Borgeat, P., and Sha'afi, R. I.,** Intracellular calcium redistribution and its relationship to fMet-Leu-Phe, leukotriene B_4, and phorbol ester induced rabbit neutrophil degranulation, *J. Cell. Physiol.*, 122, 273, 1985.
56. **Korchak, H. M., Vienne, K., Rutherford, L. E., Wilkenfeld, C., Finkelstein, M. C., and Weissmann, G.,** Stimulus response coupling in the human neutrophil. II. Temporal analysis of changes in cytosolic calcium and calcium influx, *J. Biol. Chem.*, 259, 4076, 1984.
57. **Sklar, L. A., Hyslop, P. A., Oades, Z. G., Omann, G. M., Jesaitis, A. J., Painter, R. G., and Cochrane, C. G.,** Signal transduction and ligand-receptor dynamics in the human neutrophil. Transient responses and occupancy-response relations at the formyl peptide receptor, *J. Biol. Chem.*, 260, 11461, 1985.
58. **Sklar, L. A., Omann, G. M., and Painter, R. G.,** Relationship of actin polymerization and depolymerization to light scattering in human neutrophils: dependence on receptor occupancy and intracellular Ca^{2+}, *J. Cell. Biol.*, 101, 1161, 1985.
59. **McNeil, P. L., Kennedy, A. L., Waggoner, A. S., Taylor, D. L., and Murphy, R. F.,** Light-scattering changes during chemotactic stimulation of human neutrophils: kinetics followed by flow cytometry, *Cytometry*, 6, 7, 1985.
60. **Sha'afi, R. I., Williams, K., Wacholtz, M. C., and Becker, E. L.,** Binding of the chemotactic synthetic peptide (^{3}H)formyl-norleu-leu-phe to plasma membrane of rabbit neutrophils, *FEBS Lett.*, 91, 305, 1978.
61. **Molski, T. F. P., Naccache, P. H., Marsh, M. L., Kermode, J., Becker, E. L., and Sha'afi, R. I.,** Pertussis toxin inhibits the rise in the intracellular concentration of free calcium that is induced by chemotactic factors in rabbit neutrophils: possible role of the "G proteins" in calcium mobilization, *Biochem. Biophys. Res. Comm.*, 124, 644, 1984.
62. **Goldman, D. W., Chang, F. H., Gifford, L. A., Goetzl, E. J., and Bourne, H. R.,** Pertussis toxin inhibition of chemotactic factor-induced calcium mobilization and function in human polymorphonuclear leukocytes, *J. Exp. Med.*, 162, 145, 1985.
63. **Krause, K. H., Schlegel, W., Wollheim, C. B., Andersson, T., Waldvogel, F. A., and Lew, P. D.,** Chemotactic peptide activation of human neutrophils and HL-60 cells. Pertussis toxin reveals correlation between inositol trisphosphate generation, calcium ion transients, and cellular activation, *J. Clin. Invest.*, 76, 1348, 1985.
64. **Lad, P. M., Olson, C. V., and Grewal, I. S.,** A step sensitive to pertussis toxin and phorbol ester in human neutrophils regulates chemotaxis and capping but not phagocytosis, *FEBS Lett.*, 200, 91, 1986.
65. **Lad, P. M., Olson, C. V., and Smiley, P. A.,** Association of the *N*-formyl-Met-Leu-Phe receptor in human neutrophils with a GTP-binding protein sensitive to pertussis toxin, *Proc. Natl. Acad. Sci. U.S.A.*, 82, 869, 1985.
66. **Lad, P. M., Olson, C. V., and Grewal, I. S.,** Platelet-activating factor mediated effects on human neutrophil function are mediated by pertussis toxin, *Biochem. Biophys. Res. Comm.*, 129, 632, 1985.
67. **Okamura, N., Uchida, M., Ohtsuka, T., Kawanishi, M., and Ishibashi, S.,** Diverse involvements of N_i protein in superoxide anion production in polymorphonuclear leukocytes depending on the type of membrane stimulants, *Biochem. Biophys. Res. Comm.*, 130, 939, 1985.
68. **Naccache, P. H., Molski, M. M., Volpi, M., Becker, E. L., and Sha'afi, R. I.,** Unique inhibitory profile of platelet activating factor induced calcium mobilization, polyphosphoinositide turnover and granule enzyme secretion in rabbit neutrophils towards pertussis toxin and phorbol ester, *Biochem. Biophys. Res. Comm.*, 130, 677, 1985.
69. **Bradford, P. G. and Rubin, R. P.,** Pertussis toxin inhibits chemotactic factor-induced phospholipase C stimulation and lysosomal enzyme secretion in rabbit neutrophils, *FEBS Lett.*, 183, 317, 1985.
70. **Shefcyk, J., Yassin, R., Volpi, M., Molski, T. F. P., Naccache, P. H., Munoz, J. J., Becker, E. L., Feinstein, M. B., and Sha'afi, R. I.,** Pertussis but not cholera toxin inhibits the stimulated increase in actin association with the cytoskeleton in rabbit neutrophils: role of the "G proteins" in stimulus-response coupling, *Biochem. Biophys. Res. Comm.*, 126, 1174, 1985.
71. **Becker, E. L., Kermode, J. C., Naccache, P. H., Yassin, R., Marsh, M. L., Munoz, J. J., and Sha'afi, R. I.,** The inhibition of neutrophil granule enzyme secretion and chemotaxis by pertussis toxin, *J. Cell. Biol.*, 100, 1641, 1985.
72. **Volpi, M., Naccache, P. H., Molski, T. F. P., Shefcyk, J., Huang, C. K., Marsh, M. L., Munoz, J., Becker, E. L., and Sha'afi, R. I.,** Pertussis toxin inhibits fMet-Leu-Phe- but not phorbol ester-stimulated changes in rabbit neutrophils: role of G proteins in excitation response coupling, *Proc. Natl. Acad. Sci. U.S.A.*, 82, 2708, 1985.
73. **Bokoch, G. M. and Gilman, A. G.,** Inhibition of receptor-mediated release of arachidonic acid by pertussis toxin, *Cell*, 39, 301, 1984.

74. **Okajima, F. and Ui, M.**, ADP-ribosylation of the specific membrane protein by islet-activating protein, pertussis toxin, associated with inhibition of a chemotactic peptide-induced arachidonate release in neutrophils. Possible role of the toxin substrate in Ca^{2+}-mobilizing biosignaling, *J. Biol. Chem.*, 259, 13863, 1984.
75. **Snyderman, R., Pike, M. C., Edge, S., and Lane, B.**, A chemoattractant receptor on macrophages exists in two affinity states regulated by guanine nucleotides, *J. Biol. Chem.*, 98, 444, 1984.
76. **Sklar, L. A., Bokoch, G. M., Button, D., and Smolen, J. E.**, Regulation of ligand-receptor dynamics by guanyl nucleotides: rapidly interconverting states for the neutrophil formyl peptide receptor, *Fed. Proc.*, 45, 5705a, 1986.
77. **Matsumoto, T., Molski, T. F. P., Volpi, M., Pelz, C., Kanaho, Y., Becker, E. L., Feinstein, M. B., Naccache, P. H., and Sha'afi, R. I.**, Treatment of rabbit neutrophils with phorbol esters results in increased ADP-ribosylation catalyzed by pertussis toxin and inhibition of the GTPase stimulated by fMet-Leu-Phe, *FEBS Lett.*, 198, 295, 1986.
78. **Barrowman, M. M., Cockcroft, S., and Gomperts, B. D.**, Two roles for guanine nucleotides in the stimulus-secretion sequence of neutrophils, *Nature (London)*, 319, 504, 1986.
79. **Smolen, J. E. and Stoehr, S. J.**, Guanine nucleotides reduce the free calcium requirement for secretion of granule constituents from permeabilized human neutrophils, *Biochim. Biophys. Acta*, 889, 171, 1986.
80. **Cockcroft, S. and Gomperts, B. D.**, Role of guanine nucleotide binding protein in the activation of polyphosphoinositide phosphodiesterase, *Nature (London)*, 314, 534, 1985.
81. **Rubin, R. P., Sink, L. E., Schrey, M. P., Day, A. P., Liao, C. S., and Freer, R. J.**, Secretagogues for lysosomal enzyme release as stimulants of arachidonyl phosphatidylinositol turnover in rabbit neutrophils, *Biochem. Biophys. Res. Comm.*, 90, 1364, 1979.
82. **Rubin, R. P., Sink, L. E., and Freer, R. J.**, Activation of (arachidonyl) phosphatidylinositol turnover in rabbit neutrophils by the calcium ionophore A23187, *Biochem. J.*, 194, 497, 1981.
83. **Cockcroft, S., Bennett, J. P., and Gomperts, R. D.**, The dependence on Ca^{2+} of phosphatidylinositol breakdown and enzyme secretion in rabbit neutrophils stimulated by formylmethionylleucylphenylalanine or ionomycin, *Biochem. J.*, 200, 501, 1981.
84. **Rubin, R. P., Sink, L. E., and Freer, R. J.**, On the relationship between formylmethionly-leucyl-phenylalanine stimulation and arachidonyl phosphatidylinositol turnover and lysosomol enzyme secretion by rabbit neutrophils, *Mol. Pharmacol.*, 19, 31, 1981.
85. **Cockcroft, S., Bennett, J. P., and Gomperts, B. D.**, Stimulus-secretion coupling in rabbit neutrophils is not mediated by phosphatidylinositol breakdown, *Nature (London)*, 288, 275, 1980.
86. **Yano, K., Hattori, H., Imai, A., and Nozawa, Y.**, Modification of positional distribution of fatty acids in phosphatidylinositol of rabbit neutrophils stimulated with formylmethionyl-leucyl-phenylalanine, *Biochim. Biophys. Acta*, 752, 137, 1983.
87. **Takanawa, T., Homma, Y., and Nagai, Y.**, Role of Ca^{2+} in phosphatidylinositol response and arachidonic acid release in formylated tripeptide- or Ca^{2+} ionophore A23187- stimulated guinea pig neutrophils, *J. Immunol.*, 130, 2849, 1983.
88. **Volpi, M., Yassin, R., Naccache, P. H., and Sha'afi, R. I.**, Chemotactic factor causes rapid decreases in phosphatidylinositol,4,5-bisphosphate and phosphatidylinositol 4-monophosphate in rabbit neutrophils, *Biochem. Biophys. Res. Comm.*, 112, 957, 1983.
89. **Hirayama, T. and Kato, I.**, A rapid stimulation of phosphatidylinositol metabolism in rabbit leukocytes by pseudomonal leukocidin, *FEBS Lett.*, 157, 46, 1983.
90. **Cockcroft, S., Baldwin, J. M., and Allan, D.**, The Ca^{2+}-activated polyphosphoinositide phosphodiesterase of human and rabbit neutrophil membranes, *Biochem. J.*, 221, 477, 1984.
91. **Dougherty, R. W., Godfrey, P. P., Hoyle, P. C., Putney, J. W., Jr., and Freer, R. J.**, Secretagogue-induced phosphoinositide metabolism in human leucocytes, *Biochem. J.*, 222, 307, 1984.
92. **Bradford, P. G. and Rubin, R. P.**, Characterization of formylmethionyl-leucyl-phenylalanine stimulation of inositol trisphosphate accumulation in rabbit neutrophils, *Mol. Pharmacol.*, 27, 74, 1985.
93. **Takenawa, T., Ishitoya, J., Homma, Y., Kato, M., and Nagai, Y.**, Role of enhanced inositol phospholipid metabolism in neutrophil activation, *Biochem. Pharmacol.*, 34, 1931, 1985.
94. **Cockcroft, S.**, Ca^{2+}-dependent conversion of phosphatidylinositol to phosphatidate in neutrophils stimulated with fMet-Leu-Phe or ionophore A23187, *Biochim. Biophys. Acta*, 795, 37, 1984.
95. **Spat, A., Bradford, P. G., McKinney, J. S., Rubin, R. P., and Putney, J. W., Jr.**, A saturable receptor for ^{32}P-inositol-1,4,5-trisphosphate in hepatocytes and neutrophils, *Nature (London)*, 319, 514, 1986.
96. **Prentki, M., Wollheim, C. B., and Lew, P. D.**, Ca^{2+} homeostasis in permeabilized human neutrophils. Characterization of Ca^{2+}-sequestering pools and the action of inositol 1,4,5-trisphosphate, *J. Biol. Chem.*, 259, 13777, 1984.
97. **Ochs, D. L. and Reed, P. W.**, ATP-dependent calcium transport in plasma membrane vesicles from neutrophil leukocytes, *J. Biol. Chem.*, 258, 10116, 1983.
98. **Ochs, D. L. and Reed, P. W.**, Ca^{2+}-stimulated, Mg^{2+}-dependent ATPase activity in neutrophil plasma membrane vesicles. Coupling to Ca^{2+} transport, *J. Biol. Chem.*, 259, 102, 1984.

99. **Lagast, H., Pozzan, T., Lew, P. D., and Waldvogel, F. A.,** Phorbol myristate acetate stimulated the plasma membrane calcium pump without raising cytoplasmic free calcium, $(Ca^{++})_i$, *Clin. Res.*, 31, 410a, 1983.
100. **Goldstein, I. M., Horn, J. K., Kaplan, H. B., and Weissmann, G.,** Calcium-induced lysozyme secretion from human polymorphonuclear leukocytes, *Biochem. Biophys. Res. Comm.*, 60, 807, 1974.
101. **Korchak, H. M., Rutherford, L. E., and Weissmann, G.,** Stimulus response coupling in the human neutrophil. I. Kinetic analysis of changes in calcium permeability, *J. Biol. Chem.*, 259, 4070, 1984.
102. **Smith, R. J. and Iden, S. S.,** Phorbol myristate acetate-induced release of granule enzymes from human neutrophils: inhibition by the calcium antagonist 8-(*N,N*-diethylamido)-octyl 3,4,5 trimethoxybenzoate hydrochloride, *Biochem. Biophys. Res. Comm.*, 91, 263, 1979.
103. **Smolen, J. E., Korchak, H. M., and Weissmann, G.,** The roles of extracellular and intracellular calcium in lysosomal enzyme release and superoxide anion generation by human polymorphonuclear leukocytes, *Biochim. Biophys. Acta*, 677, 512, 1981.
104. **Elferink, J. G. R. and Deierkauf, M.,** Involvement of intracellular Ca^{2+} in chemotaxis and metabolic burst by neutrophils: the use of antagonists of intracellular Ca^{2+}, *Res. Commun. Chem. Pathol. Pharmacol.*, 50, 67, 1985.
105. **Naccache, P. H., Molski, T. F. P., Alobaidi, T., Becker, E. L., Showell, H. J., and Sha'afi, R. I.,** Calmodulin inhibitors block neutrophil degranulation at a step distal from the mobilization of calcium, *Biochem. Biophys. Res. Comm.*, 97, 62, 1980.
106. **Smolen, J. E. and Weissmann, G.,** The effect of various stimuli and calcium antagonists on the fluorescence response of chlorotetracycline-labelled human neutrophils, *Biochim. Biophys. Acta*, 720, 172, 1982.
107. **Smolen, J. E., Eisenstat, B. A., and Weissmann, G.,** The fluorescence response of chlorotetracycline-loaded human neutrophils. Correlations with lysosomal enzyme release and evidence for a 'trigger' pool of calcium, *Biochim. Biophys. Acta*, 717, 422, 1982.
108. **White, J. R., Naccache, P. H., Molski, T. F. P., Borgeat, P., and Sha'afi, R. I.,** Direct demonstration of increased intracellular concentration of free calcium in rabbit and human neutrophils following stimulation with chemotactic factor, *Biochem. Biophys. Res. Commun.*, 113, 44, 1983.
109. **Pozzan, T., Lew, D. P., Wollheim, C. B., and Tsein, R. Y.,** Is cytosolic ionized calcium regulating neutrophil activation?, *Science*, 221, 1413, 1983.
110. **Sha'afi, R. I., White, J. R., Molski, T. F. P., Shefcyk, J., Volpi, M., Naccache, P. H., and Feinstein, M. B.,** Phorbol 12-myristate 13-acetate activates rabbit neutrophils without an apparent rise in the level of intracellular free calcium, *Biochem. Biophys. Res. Commun.*, 114, 638, 1983.
111. **Rickard, J. E. and Sheterline, P.,** Evidence that phorbol ester interferes with stimulated Ca^{2+} redistribution by activating Ca^{2+} efflux in neutrophil leucocytes, *Biochem. J.*, 231, 623, 1985.
112. **Niedel, J. E., Kuhn, L. J., and Vandenbark, G. R.,** Phorbol diester receptor copurifies with protein kinase C, *Proc. Natl. Acad. Sci. U.S.A.*, 80, 36, 1983.
113. **Lew, P. D., Dayer, J. M., Wollheim, C. B., and Pozzan, T.,** Effect of leukotriene B_4, prostaglandin E_2 and arachidonic acid on cytosolic-free calcium in human neutrophils, *FEBS Lett.*, 166, 44, 1984.
114. **Naccache, P. H., Molski, T. F. P., Borgeat, P., White, J. R., and Sha'afi, R. I.,** Phorbol esters inhibit the fMet-Leu-Phe- and leukotriene B_4-stimulated calcium mobilization and enzyme secretion in rabbit neutrophils, *J. Biol. Chem.*, 260, 2125, 1985.
115. **Strnad, C. F. and Wong, K.,** Calcium mobilization in fluoride activated human neutrophils, *Biochem. Biophys. Res. Commun.*, 133, 161, 1985.
116. **Lew, P. D., Wollheim, C. B., Waldvogel, F. A., and Pozzan, T.,** Modulation of cytosolic-free calcium transients by changes in intracellular calcium-buffering capacity: correlation with exocytosis and O_2^- production in human neutrophils, *J. Cell Biol.*, 99, 1212, 1984.
117. **Stickle, D. F., Daniele, R. P., and Holian, A.,** Cytosolic calcium, calcium fluxes, and regulation of alveolar macrophage superoxide anion production, *J. Cell. Physiol.*, 121, 458, 1984.
118. **Johnson, P. C., Ware, J. A., Cliveden, P. B., Smith, M., Dvorak, A. M., and Salzman, E. W.,** Measurement of ionized calcium in blood platelets with the photoprotein aequorin. Comparison with Quin 2, *J. Biol. Chem.*, 260, 2069, 1985.
119. **Haslam, R. J. and Davidson, M. M. L.,** Potentiation by thrombin of the secretion of serotonin from permeabilized platelets equilibrated with Ca^{2+} buffers. Relationship to protein phosphorylation and diacylglycerol formation, *Biochem. J.*, 222, 351, 1984.
120. **Elferink, J. G. R.,** Interference of quin2 with activation of rabbit neutrophils by phorbol myristate acetate, *IRCS Med. Sci.*, 13, 397, 1985.
121. **Elferink, J. G. R. and Deierkauf, M.,** The effect of quin2 on chemotaxis by polymorphonuclear leukocytes, *Biochim. Biophys. Acta*, 846, 364, 1985.
122. **Kruskal, B. A., Shak, S., and Maxfield, F. R.,** Spreading of human neutrophils is immediately preceded by a large increase in cytoplasmic free calcium, *Proc. Natl. Acad. Sci. U.S.A.*, 83, 2919, 1986.
123. **Hallett, M. B. and Campbell, A. K.,** Measurement of changes in cytoplasmic free Ca^{2+} in fused cell hybrids, *Nature (London)*, 295, 155, 1982.

124. **Slonczewski, J. L., Wilde, M. W., and Zigmond, S. H.,** Phosphorylase a activity as an indicator of neutrophil activation by chemotactic peptide, *J. Cell Biol.*, 101, 1191, 1985.
125. **Stapelton, C. L., Barnard, R. M., and Chandler, D. E.,** Intracellular or extracellular calcium can be used to trigger C5a-stimulated enzyme secretion from rabbit neutrophils, *Biochim. Biophys. Acta,* 763, 225, 1983.
126. **Smolen, J. E. and Stoehr, S. J.,** Micromolar concentrations of free calcium provoke secretion of lysozyme from human neutrophils permeabilized with saponin, *J. Immunol.*, 134, 1859, 1985.
127. **Smolen, J. E., Stoehr, S. J., and Boxer, L. A.,** Human neutrophils permeabilized with digitonin respond with lysosomal enzyme release when exposed to micromolar levels of free calcium, *Biochim. Biophys. Acta,* 886, 1, 1986.
128. **Smolen, J. E., Todd, R. F., III, and Boxer, L. A.,** Expression of a granule membrane marker on the surface of neutrophils permeabilized with digitonin. Correlations with Ca^{2+}-induced degranulation, *Am. J. Pathol.*, 124, 281, 1986.
129. **Smolen, J. E., Stoehr, S. J., Traynor, A. E., and Sklar, L. A.,** The kinetics of secretion from permeabilized human neutrophils: release of elastase and correlations with other granule constituents and right angle light scatter, *J. Leuk. Biol.*, 41, 8, 1987.
130. **Gomperts, B. D., Barrowman, M. M., and Cockcroft, S.,** Dual role for guanine nucleotides in stimulus-secretion coupling, *Fed. Proc.*, 45, 2156, 1986.
131. **Penfield, A. and Dale, M. M.,** Synergy between A23187 and 1-oleoyl-2-acetyl-glycerol in superoxide production by human neutrophils, *Biochem. Biophys. Res. Comm.*, 125, 332, 1984.
132. **Di Virgilio, F. D., Lew, D. P., and Pozzan, T.,** Protein Kinase C activation of physiological processes in human neutrophils at vanishingly small cytosolic Ca^{2+} levels, *Nature (London),* 310, 691, 1984.
133. **Hoult, J. R. S. and Nourshargh, S.,** Phorbol myristate acetate enhances human polymorphonuclear neutrophil release of granular enzymes but inhibits chemokinesis, *Br. J. Pharmacol.*, 86, 533, 1985.
134. **O'Flaherty, J. T., Schmitt, J. D., McCall, C. E., and Wykle, R. L.,** Diacylglycerols enhance human degranulation responses: relevancy to a multiple mediator hypothesis of cell function, *Biochem. Biophys. Res. Commun.*, 123, 64, 1984.
135. **Strnad, C. F. and Wong, K.,** Effect of the calcium ionophore A23187 on superoxide generation in phorbol ester stimulated human neutrophils, *Can. J. Physiol. Pharmacol.*, 63, 1543, 1985.
136. **Bender, J. G. and Van Epps, D. E.,** Stimulus interactions in release of superoxide anion (O_2^-) from human neutrophils. Further evidence for multiple pathways of activation, *Inflammation,* 9, 67, 1985.
137. **Apfeldorf, W. J., Melnick, D. A., Meshulam, T., Rasmussen, H., and Malech, H. L.,** A transient rise in intracellular free calcium is not a sufficient stimulus for respiratory burst activation in human polymorphonuclear leukocytes, *Biochem. Biophys. Res. Commun.*, 132, 674, 1985.
138. **Sha'afi, R. I., Shefcyk, J., Yassin, R., Molski, T. F. P., Volpi, M., Naccache, P. H., White, J. R., Feinstein, M. B., and Becker, E. L.,** Is a rise in intracellular concentration of free calcium necessary or sufficient for stimulated cytoskeletal-associated actin?, *J. Cell Biol.*, 102, 1459, 1986.
139. **Korchak, H. M. and Weissmann, G.,** Changes in membrane potential of human granulocytes antecede the metabolic responses to surface stimulation, *Proc. Natl. Acad. Sci. U.S.A.*, 75, 3818, 1978.
140. **Korchak, H. M. and Weissmann, G.,** Stimulus-response coupling in the human neutrophil: membrane potential changes and the role of extracellular Na^+, *Biochim. Biophys. Acta,* 601, 180, 1980.
141. **Jones, G. S., VanDyke, K., and Castranova, V.,** Purification of human neutrophils by centrifugal elutriation and measurement of transmembrane potential, *J. Cell. Physiol.*, 104, 425, 1980.
142. **Jones, G. S., VanDyke, K., and Castranova, V.,** Transmembrane potential changes associated with superoxide release from human granulocytes, *J. Cell. Physiol.*, 106, 75, 1981.
143. **Seligmann, B., Chusea, T. M., and Gallin, J. I.,** Human neutrophil heterogeneity identified using flow microfluorimetry to monitor membrane potential, *J. Clin. Invest.*, 68, 1125, 1981.
144. **Seligmann, B. E. and Gallin, J. I.,** Use of lipophilic probes of membrane potential to assess human neutrophil activation. Abnormality in chronic granulomatous disease, *J. Clin. Invest.*, 66, 493, 1980.
145. **Seligmann, B. and Gallin, J. I.,** Secretagogue modulation of the response of human neutrophils to chemoattractants: studies with a membrane potential sensitive cyanine dye, *Mol. Immunol.*, 17, 191, 1980.
146. **Tatham, P. E. R., Delves, P. J., Shen, L., and Roitt, I. M.,** Chemotactic factor-induced membrane potential changes in rabbit neutrophils monitored by the fluorescent dye 3,3'- dipropyldicarbocyanine iodide, *Biochim. Biophys. Acta,* 602, 285, 1980.
147. **Whitin, J. C., Chapman, C. E., Simons, E. R., Chovaniec, M. E., and Cohen, H. J.,** Correlation between membrane potential changes and superoxide production in human granulocytes by phorbol myristate acetate, *J. Biol. Chem.*, 255, 1874, 1980.
148. **Whitin, J. C., Clark, R. A., Simons, E. R., and Cohen, H. J.,** Effects of the myeloperoxidase system on fluorescent probes of granulocyte membrane potential, *J. Biol. Chem.*, 256, 8904, 1981.
149. **Simchowitz, L., Spilberg, I., and De Weer, P.,** Sodium and potassium fluxes and membrane potential of human neutrophils. Evidence for an electrogenic sodium pump, *J. Gen. Physiol.*, 79, 453, 1982.

150. **Korchak, H. M., Rich, A. M., Wilkenfeld, C., Rutherford, L. E., and Weissmann, G.,** A carbocyanine dye, $DiOC_6(3)$, acts as a mitochondrial probe in human neutrophils, *Biochem. Biophys. Res. Commun.*, 108, 1495, 1982.
151. **Kuroki, M., Kamo, N., Kobatake, Y., Okimasu, E., and Utsumi, K.,** Measurement of membrane potential in polymorphonuclear leukocytes and its changes during surface stimulation, *Biochim. Biophys. Acta*, 693, 326, 1982.
152. **Tauber, A. I. and Simons, E. R.,** Dissociation of human neutrophil membrane polarization, respiratory burst stimulation and phospholipid metabolism by quinacrine, *FEBS Lett.*, 156, 161, 1983.
153. **Seligmann, B. E. and Gallin, J. I.,** Comparison of indirect probes of membrane potential utilized in studies of human neutrophils, *J. Cell. Physiol.*, 115, 105, 1983.
154. **Sha'afi, R. I., Naccache, P., Raible, D., Krepcio, A., Showell, H., and Becker, E. L.,** Demonstration of ($Na^+ + K^+$)-sensitive ATPase activity in rabbit polymorphonuclear leukocyte membranes, *Biochim. Biophys. Acta*, 448, 638, 1976.
155. **Becker, E. L., Talley, J. V., Showell, H. J., Naccache, P. H., Sha'afi, R. I.,** Activation of the rabbit polymorphonuclear leukocyte membrane "Na^+,K^+"-ATPase by chemotactic factor, *J. Cell Biol.*, 77, 329, 1978.
156. **Elsbach, P. and Schwartz, I. L.,** Studies on the sodium and potassium transport in rabbit polymorphonuclear leukocytes, *J. Gen. Physiol.*, 42, 883, 1959.
157. **Naccache, P. H., Showell, H. J., Becker, E. L., and Sha'afi, R. I.,** Changes in ionic movements across rabbit polymorphonuclear leukocyte membranes during lysosomal enzyme release. Possible ionic basis for lysosomal enzyme release, *J. Cell Biol.*, 75, 635, 1977.
158. **Naccache, P. H., Showell, H. J., Becker, E. L., and Sha'afi, R. I.,** Sodium, potassium, and calcium transport across rabbit polymorphonuclear leukocyte membranes. Effect of chemotactic factor, *J. Cell Biol.*, 73, 428, 1977.
159. **Sha'afi, R. I., Molski, T. F. P., and Naccache, P. H.,** Chemotactic factors activate differentiable permeation pathways for sodium and calcium in rabbit neutrophils, *Biochem. Biophys. Res. Commun.*, 99, 1271, 1981.
160. **Showell, H. J., Naccache, P. H., Sha'afi, R. I., and Becker, E. L.,** The effects of extracellular K^+, Na^+, and Ca^{2+} on lysosomal enzyme secretion from polymorphonuclear leukocytes, *J. Immunol.*, 119, 804, 1977.
161. **Molski, T. F. P., Naccache, P. H., Borgeat, P., and Sha'afi, R. I.,** Similarities in the mechanisms by which formyl-methionyl-leucyl-phenylalanine, arachidonic acid and leukotriene B_4 increase calcium and sodium fluxes in rabbit neutrophils, *Biochem. Biophys. Res. Commun.*, 103, 227, 1981.
162. **Naccache, P. H., Showell, H. J., Becker, E. L., and Sha'afi, R. I.,** The chemotactic factors induced movement of calcium and sodium across rabbit neutrophil membranes: effect of desensitization to cytochalasin B, *J. Cell. Physiol.*, 100, 230, 1979.
163. **Grinstein, S. and Furuya, W.,** Amiloride-sensitive Na^+/H^+ exchange in human neutrophils: mechanism of activation by chemotactic factors, *Biochem. Biophys. Res. Commun.*, 122, 755, 1984.
164. **Grinstein, S., Goetz, J. D., Furuya, W., Rothstein, A., and Gelfand, E. W.,** Amiloride-sensitive Na^+-H^+ exchange in platelets and leukocytes: detection by electronic cell sizing, *Am. J. Physiol.*, 247, C293, 1984.
165. **Grinstein, S. and Furuya, W.,** Characterization of the amiloride-sensitive Na^+-H^+ antiport of human neutrophils, *Am. J. Physiol.*, 250, C283, 1986.
166. **Wright, J., Schwartz, J. H., Olson, R., Kosowsky, J. M., and Tauber, A. I.,** Proton secretion by the sodium/hydrogen ion antiporter in the human neutrophil, *J. Clin. Invest.*, 77, 782, 1986.
167. **Smith, R. J., Bowman, B. J., and Iden, S. S.,** Effects of an anion channel blocker, 4,4'-diisothiocyano-2,2'-disulfonic acid stilbene (DIDS), on human neutrophil function, *Biochem. Biophys. Res. Commun.*, 120, 964, 1984.
168. **Korchak, H. M., Eisenstat, B. A., Hoffstein, S. T., Dunham, P. B., and Weissmann, G.,** Anion channel blockers inhibit lysosomal enzyme secretion from human neutrophils without affecting generation of superoxide anion, *Proc. Natl. Acad. Sci. U.S.A.*, 77, 2721, 1980.
169. **Korchak, H. M., Eisenstat, B. A., Smolen, J. E., Rutherford, L. E., Dunham, P. B., and Weissmann, G.,** Stimulus-response coupling in the neutrophil: the role of anion fluxes in degranulation, *J. Biol. Chem.*, 257, 6916, 1982.
170. **Styrt, B. and Klempner, M. S.,** Internal pH of human neutrophil lysosomes, *FEBS Lett.*, 149, 113, 1982.
171. **Jacques, Y. V. and Bainton, D. F.,** Changes in pH within the phagocytic vacuoles of human neutrophils and monocytes, *Lab. Invest.*, 39, 179, 1978.
172. **Reinitz, D. M., Paape, M. J., and Mather, I. H.,** Effect of phagocytosed fat and casein on the intraphagosomal pH in bovine polymorphonuclear leukocytes, *Proc. Soc. Exp. Biol. Med.*, 170, 281, 1982.
173. **Bassoe, C. F., Laerum, O. D., Glette, J., Hopen, G., Haneberg, B., and Solberg, C. O.,** Simultaneous measurement of phagocytosis and phagosomal pH by flow cytometry: role of polymorphonuclear neutrophilic leukocyte granules in phagosome acidification, *Cytometry*, 4, 254, 1983.

174. **Cech, P. and Lehrer, R. I.**, Phagolysosomal pH of human neutrophils, *Blood*, 63, 88, 1984.
175. **Lynn, W. S. and Mohapatra, N.**, Control of leukocyte functions. The role of internal H^+ concentration and a membrane-bound esterase, *Inflammation*, 4, 329, 1980.
176. **Molski, T. F. P., Naccache, P. H., Volpi, M., Wolpert, L. M., and Sha'afi, R. I.**, Specific modulation of the intracellular pH of rabbit neutrophils by chemotactic factors, *Biochem. Biophys. Res. Commun.*, 94, 508, 1980.
177. **Sha'afi, R. I., Naccache, P. H., Molski, T. F. P., and Volpi, M.**, Chemotactic stimuli-induced changes in pHi of rabbit neutrophils, in *Intracellular pH: Its Measurement, Regulation, and Utilization in Cellular Functions*, Alan R. Liss, New York, 1982, 513.
178. **Lew, P. D., Dayer, J. M., Wollheim, C. B., and Pozzan, T.**, Effect of leukotriene B_4, prostaglandin E_2 and arachidonic acid on cytosolic-free calcium in human neutrophils, *FEBS Lett.*, 166, 44, 1984.
179. **Takanaka, K. and O'Brien, P. J.**, Proton release by polymorphonuclear leukocytes during phagocytosis or activation by digitonin or fluoride, *Biochem. Int.*, 11, 127, 1985.
180. **Grinstein, S., Furuya, W., and Biggar, W. D.**, Cytoplasmic pH regulation in normal and abnormal neutrophils. Role of superoxide generation and Na^+/H^+ exchange, *J. Biol. Chem.*, 261, 512, 1986.
181. **Simchowitz, L. and Roos, A.**, Regulation of intracellular pH in human neutrophils, *J. Gen. Physiol.*, 85, 443, 1985.
182. **Grinstein, S., Elder, B., and Furuya, W.**, Phorbol ester-induced changes of cytoplasmic pH in neutrophils: role of exocytosis in Na^+-H^+ exchange, *Am. J. Physiol.*, 248, C379, 1985.
183. **Simchowitz, L.**, Chemotactic factor- induced activation of Na^+/H^+ exchange in human neutrophils. I. Sodium fluxes, *J. Biol. Chem.*, 260, 13237, 1985.
184. **Simchowitz, L.**, Chemotactic factor-induced activation of Na^+/H^+ exchange in human neutrophils. II. Intracellular pH changes, *J. Biol. Chem.*, 260, 13248, 1985.
185. **Simchowitz, L. and Cragoe, E. J., Jr.**, Regulation of human neutrophil chemotaxis by intracellular pH, *J. Biol. Chem.*, 261, 6492, 1986.
186. **Tsung, P. K., Sakamoto, T., and Weissmann, G.**, Protein kinase and phosphatases from human polymorphonuclear leukocytes, *Biochem. J.*, 145, 437, 1975.
187. **Juhl, H. and Esmann, V.**, Phosphorylation of rabbit skeletal muscle glycogen synthase I by the cAMP dependent protein kinase, the cAMP independent synthase kinase and the phosvitin kinase from human polymorphonuclear leukocytes, *Mol. Cell. Biochem.*, 30, 131, 1980.
188. **Juhl, H.**, Phosphorylation of glycogen synthase i from human polymorphonuclear leukocytes, *Mol. Cell. Biochem.*, 35, 77, 1981.
189. **Juhl, H. and Esmann, V.**, Phosphorylation of glycogen synthase in a homogenate of human polymorphonuclear leukocytes, *Mol. Cell. Biochem.*, 35, 93, 1981.
190. **Nahas, N., Juhl, H., and Esmann, V.**, Chromatographic characteristics and subcellular localization of synthase phosphatase, phosphorylase phosphatase and histone phosphatase in human polymorphonuclear leukocytes, *Mol. Cell. Biochem.*, 58, 147, 1984.
191. **Schneider, C., Zanetti, M., and Romeo, D.**, Surface-reactive stimuli selectively increase protein phosphorylation in human neutrophils, *FEBS Lett.*, 127, 4, 1981.
192. **Smith, G. P. and Peters, T. J.**, Subcellular localisation and properties of histone phosphate phosphatase in human polymorphonuclear leukocytes: alterations in pregnancy and chronic granulocytic leukaemia and relationship to alkaline phosphatase, *Clin. Chim. Acta*, 111, 17, 1981.
193. **Helfman, D. M., Applebaum, B. D., Volger, W. R., and Kuo, J. F.**, Phospholipid-sensitive Ca^{2+}-dependent protein kinase and its substrates in human neutrophils, *Biochem. Biophys. Res. Commun.*, 111, 847, 1983.
194. **Huang, C. K., Hill, J. M., Jr., Bormann, B. J., Mackin, W. M., and Becker, E. L.**, Endogenous substrates for cyclic AMP-dependent and calcium-dependent protein phosphorylation in rabbit peritoneal neutrophils, *Biochim. Biophys. Acta*, 760, 126, 1983.
195. **Andrews, P. C. and Babior, B. M.**, Endogenous protein phosphorylation by resting and activated human neutrophils, *Blood*, 61, 333, 1983.
196. **Huang, C. K., Hill, J. M., Jr., Bormann, B. J., Mackin, W. M., and Becker, E. L.**, Chemotactic factors induced vimentin phosphorylation in rabbit peritoneal neutrophil, *J. Biol. Chem.*, 259, 1386, 1984.
197. **Okamura, N., Ohashi, S., Nagahisa, N., and Ishibashi, S.**, Changes in protein phosphorylation in guinea pig polymorphonuclear leukocytes by treatment with membrane-perturbing agents which stimulate superoxide anion production, *Arch. Biochem. Biophys.*, 228, 270, 1984.
198. **White, J. R., Huang, C. K., Hill, J. M., Jr., Naccache, P. H., Becker, E. L., and Sha'afi, R. I.**, Effect of phorbol 12-myristate 13-acetate and its analogue 4α-phorbol 12,13-didecanoate on protein phosphorylation and lysosomal enzyme release in rabbit neutrophils, *J. Biol. Chem.*, 259, 8605, 1984.
199. **Gennaro, R., Florio, C., and Romeo, D.**, Activation of protein kinase C in neutrophil cytoplasts. Localization of protein substrates and possible relationship with stimulus-response coupling, *FEBS Lett.*, 180, 185, 1985.

200. **Gennaro, R., Florio, C., and Romeo, D.,** Co-activation of protein kinase C and NADPH oxidase in the plasma membrane of neutrophil cytoplasts, *Biochem. Biophys. Res. Comm.*, 134, 305, 1986.
201. **Papini, E., Grzeskowiak, M., Bellavite, P., and Rossi, F.,** Protein kinase C phosphorylates a component of NADPH oxidase of neutrophils, *FEBS Lett.*, 190, 204, 1985.
202. **Tuy, F. P. D., Henry, J., Rosenfeld, C., and Kahn, A.,** Protein kinases in normal human blood cells, *Am. J. Hematol.*, 15, 105, 1983.
203. **Nishihira, J. and O'Flaherty, J. T.,** Phorbol myristate acetate receptors in human polymorphonuclear neutrophils, *J. Immunol.*, 135, 3439, 1985.
204. **O'Flaherty, J. T., Schmitt, J. D., Wykle, R. L., Redman, J. F., Jr., and McCall, C. E.,** Diacylglycerols and mezerein activate neutrophils by a phorbol myristate acetate-like mechanism, *J. Cell. Physiol.*, 125, 192, 1985.
205. **Berkow, R. L. and Kraft, A. S.,** Bryostatin, a non-phorbol macrocyclic lactone, activates intact human polymorphonuclear leukocytes and binds to the phorbol ester receptor, *Biochem. Biophys. Res. Commun.*, 131, 1109, 1985.
206. **Balazovich, K. J., Smolen, J. E., and Boxer, L. A.,** Ca^{++} and phospholipid-dependent protein kinase (protein kinase C) is not necessarily required for secretion by human neutrophils, *Blood*, 68, 810, 1986.
207. **McPhail, L. C., Clayton, C. C., and Snyderman, R.,** A potential second messenger role for unsaturated fatty acids: activation of Ca^{2+}-dependent protein kinase, *Science*, 224, 622, 1984.
208. **O'Flaherty, J. T., Schmitt, J. D., and Wykle, R. L.,** Interactions of arachidonate metabolism and protein kinase C in mediating neutrophil function, *Biochem. Biophys. Res. Commun.*, 127, 916, 1985.
209. **Nishihira, J., McPhail, L. C., and O'Flaherty, J. T.,** Stimulus-dependent mobilization of protein kinase C, *Biochem. Biophys. Res. Commun.*, 134, 587, 1986.
210. **Melloni, E., Pontremoli, S., Michetti, M., Sacco, O., Sparatore, B., Salamino, F., and Horecker, B. L.,** Binding of protein kinase C to neutrophil membranes in the presence of Ca^{2+} and its activation by a Ca^{2+}-requiring proteinase, *Proc. Natl. Acad. Sci. U.S.A.*, 82, 6435, 1985.
211. **Melloni, E., Pontremoli, S., Michetti, M., Sacco, O., Sparatore, B., and Horecker, B. L.,** The involvement of calpain in the activation of protein kinase C in neutrophils stimulated by phorbol myristic acid, *J. Biol. Chem.*, 261, 4101, 1986.
212. **Cooke, E. and Hallett, M. B.,** The role of C-kinase in the physiological activation of the neutrophil oxidase. Evidence from using pharmacological manipulation of C-kinase activity in intact cells, *Biochem. J.*, 232, 323, 1985.
213. **Penfield, A. and Dale, M. M.,** Synergy between A23187 and 1-oleoyl-2-acetyl-glycerol in superoxide production by human neutrophils, *Biochem. Biophys. Res. Commun.*, 125, 332, 1984.
214. **Dewald, B., Payne, T. G., and Baggiolini, M.,** Activation of NADPH oxidase of human neutrophils. Potentation of chemotactic peptide by a diacylglycerol, *Biochem. Biophys. Acta*, 125, 367, 1984.
215. **Farkas, G., Enyedi, A., Sarkadi, B., Gardos, G., Nagy, Z., and Farago, A.,** Cyclic AMP-dependent protein kinase stimulates the phosphorylation of phosphatidylinositol-4-monophosphate in a plasma membrane preparation from pig granulocytes, *Biochem. Biophys. Res. Commun.*, 124, 871, 1984.
216. **Balazovich, K. J., Smolen, J. E., and Boxer, L. A.,** Endogenous inhibitor of protein kinase C: association with human peripheral blood neutrophils but not with specific granule-deficient neutrophils or cytoplasts, *J. Immunol.*, 137, 1665, 1986.
217. **Wright, C. D. and Hoffman, M. D.,** The protein kinase C inhibitors H-7 and H-9 fail to inhibit human neutrophil activation, *Biochem. Biophys. Res. Commun.*, 135, 749, 1986.
218. **Gerard, C., McPhail, L. C., Marfat, A., Stimler-Garard, N. P., Bass, D. A., and McCall, C. E.,** Role of protein kinases in stimulation of human polymorphonuclear leukocyte oxidative metabolism by various agonists. Differential effects of a novel protein kinase inhibitor, *J. Clin. Invest.*, 77, 61, 1986.
219. **Berger, R. R. and Karnovsky, M. L.,** Biochemical basis of phagocytosis.V. Effect of phagocytosis on cellular uptake of extracellular fluid, and on the intracellular pool of L-α-glycerophosphate, *Fed. Proc.*, 25, 840, 1966.
220. **Elsbach, P. and Farrow, S.,** Cellular triglyceride as a source of fatty acid for lecithin synthesis during phagocytosis, *Biochim. Biophys. Acta*, 176, 438, 1969.
221. **Shohet, S. B.,** Changes in fatty acid metabolism in human leukemic granulocytes during phagocytosis, *J. Lab. Clin. Med.*, 75, 659, 1970.
222. **Elsbach, P.,** Lipid metabolism by phagocytes, *Sem. Hematol.*, 9, 227, 1972.
223. **Smolen, J. E. and Shohet, S. B.,** Remodeling of granulocyte membrane fatty acids during phagocytosis, *J. Clin. Invest.*, 53, 726, 1974.
224. **Phillips, W. A., Mossmann, H., and Ferber, E.,** Changes in the incorporation of free fatty acids upon the stimulation of human polymorphonuclear leukocytes, *J. Leuk. Biol.*, 39, 267, 1986.
225. **Franson, R., Patriarca, P., and Elsbach, P.,** Phospholipid metabolism by phagocytic cells. Phospholipases A_2 associated with rabbit polymorphonuclear leukocyte granules, *J. Lipid Res.*, 15, 380, 1974.
226. **Franson, R., Weiss, J., Martin, L., Spitznagel, J. K., and Elsbach, P.,** Phospholipase A activity associated with the membranes of human polymorphonuclear leukocytes, *Biochem. J.*, 167, 839, 1977.

227. **Kaplan, L., Weiss, J., and Elsbach, P.,** Low concentrations of indomethacin inhibit phospholipase A_2 of rabbit polymorphonuclear leukocytes, *Proc. Natl. Acad. Sci. U.S.A.*, 75, 2955, 1978.
228. **Traynor, J. R. and Authi, K. S.,** Phospholipase A_2 activity of lysosomal origin secreted by polymorphonuclear leukocytes during phagocytosis or treatment with calcium, *Biochim. Biophys. Acta*, 665, 571, 1981.
229. **Tou, J.-S.,** Activation of the metabolism of the fatty acyl group in granulocyte phospholipids by phorbol myristate acetate, *Biochim. Biophys. Acta*, 665, 491, 1981.
230. **Victor, M., Weiss, J., Klempner, M. S., and Elsbach, P.,** Phospholipase A_2 activity in the plasma membrane of human polymorphonuclear leukocytes, *FEBS Lett.*, 136, 298, 1981.
231. **Smith, D. M., Jr. and Waite, M.,** Phospholipid metabolism in human neutrophil subfractions, *Arch. Biochem. Biophys.*, 246, 263, 1986.
232. **Franson, R. and Waite, M.,** Relation between calcium requirement, substrate charge, and rabbit polymorphonuclear leukocyte phospholipase A_2 activity, *Biochemistry*, 17, 4029, 1978.
233. **Clancy, R. M., Dahinden, C. A., and Hugli, T. E.,** Arachidonate metabolism by human polymorphonuclear leukocytes stimulated by N-formyl-Met- Leu-Phe or complement component C5a is independent of phospholipase activation, *Proc. Natl. Acad. Sci. U.S.A.*, 80, 7200, 1983.
234. **Bormann, B. J., Huang, C. K., Mackin, W. M., and Becker, E. L.,** Receptor-mediated activation of a phospholipase A_2 in rabbit neutrophil plasma membrane, *Proc. Natl. Acad. Sci. U.S.A.*, 81, 767, 1984.
235. **Lanni, C. and Becker, E. L.,** Release of phospholipase A_2 activity from rabbit peritoneal neutrophils by f-Met-Leu-Phe, *Am. J. Pathol.*, 113, 90, 1983.
236. **Elsbach, P., Weiss, J., Franson, R. C., Beckerdite-Quagliata, S., Schneider, A., and Harris, L.,** Separation and purification of a potent bactericidal/permeability-increasing protein and a closely associated phospholipase A_2 from rabbit polymorphonuclear leukocytes. Observations of their relationship, *J. Biol. Chem.*, 254, 1000, 1979.
237. **Hirata, F., Schiffmann, E., Venkatasubramanian, K., Salomon, D., and Axelrod, J.,** A phospholipase A_2 inhibitory protein in rabbit neutrophils induced by glucocorticoids, *Proc. Natl. Acad. Sci. U.S.A.*, 77, 2533, 1980.
238. **Carnuccio, R., Di Rosaa, M., Flower, R. J., and Pinto, A.,** The inhibition by hydrocortisone of prostaglandin biosynthesis in rat peritoneal leucocytes is correlated with intracellular macrocortin levels, *Br. J. Pharmac.*, 74, 322, 1981.
239. **Smolen, J. E. and Weissmann, G.,** Effects of indomethacin, 5,8,11,14-eicosatetraynoic acid, and *p*-bromophenacyl bromide on lysosomal enzyme release and superoxide anion generation by human polymorphonuclear leukocytes, *Biochem. Pharmacol.*, 29, 533, 1979.
240. **Hirata, F.,** The regulation of lipomodulin, a phospholipase inhibitory protein, in rabbit neutrophils by phosphorylation, *J. Biol. Chem.*, 256, 7730, 1981.
241. **Waite, M., DeChatelet, L. R., King, L., and Shirley, P. S.,** Phagocytosis-induced release of arachidonic acid from human neutrophils, *Biochem. Biophys. Res. Commun.*, 90, 984, 1978.
242. **Walsh, C. E., DeChatelet, L. R., Thomas, M. J., O'Flaherty, J. T., and Waite, M.,** Effect of phagocytosis and ionophores on release and metabolism of arachidonic acid from human neutrophils, *Lipids*, 16, 120, 1981.
243. **Lapetina, E. G., Billah, M. M., and Cuatrecasas, P.,** Rapid acylation and deacylation of arachidonic acid into phosphatidic acid of horse neutrophils, *J. Biol. Chem.*, 255, 966, 1980.
244. **Walsh, C. E., Waite, B. M., Thomas, M. J., and DeChatelet, L. R.,** Release and metabolism of arachidonic acid in human neutrophils, *J. Biol. Chem.*, 256, 7228, 1981.
245. **Walsh, C. E., DeChatelet, L. R., Chilton, F. H., Wykle, R. L., and Waite, M.,** Mechanism of arachidonic acid release in human polymorphonuclear leukocytes, *Biochim. Biophys. Acta*, 750, 32, 1983.
246. **Volpi, M., Molski, T. F. P., Naccache, P. H., Feinstein, M. B., and Sha'afi, R. I.,** Phorbol 12-myristate, 13-acetate potentiates the action of the calcium ionophore in stimulating arachidonic acid release and production of phosphatidic acid in rabbit neutrophils, *Biochem. Biophys. Res. Commun.*, 128, 594, 1985.
247. **Naccache, P. H., Showell, H. J., Becker, E. L., and Sha'afi, R. I.,** Arachidonic acid induced degranulation of rabbit peritoneal neutrophils, *Biochem. Biophys. Res. Commun.*, 87, 292, 1979.
248. **Andersen, B. R., Amirault, H. J., and LeBreton, G. C.,** The effects of arachidonic acid and other fatty acids on canine neutrophil metabolism, *Prostaglandins*, 22, 469, 1981.
249. **Panus, P. C., Eddy, L. J., and Longenecker, G. L.,** Measurement of malonyldialdehyde production during sodium arachidonate-induced polymorphonuclear leukocyte aggregation, *J. Pharmacol. Meth.*, 13, 179, 1985.
250. **Volpi, M., Naccache, P. H., and Sha'afi, R. I.,** Arachidonate metabolite(s) increase the permeability of the plasma membrane of the neutrophils to calcium, *Biochem. Biophys. Res. Commun.*, 92, 1231, 1980.
251. **Zurier, R. B. and Sayadoff, D. H.,** Release of prostaglandins by human polymorphonuclear leukocytes, *Inflammation*, 1, 93, 1975.

252. **Boxer, L. A., Allen, J. M., Schmidt, M., Yoder, M., and Baehner, R. L.,** Inhibition of polymorphonuclear leukocyte adherence by prostacyclin, *J. Lab. Clin. Med.*, 95, 672, 1980.
253. **Till, G., Kownatzki, E., Seitz, M., and Gemsa, D.,** Chemokinetic and chemotactic activity of various prostaglandins for neutrophil granulocytes, *Clin. Immunol. Immunopathol.*, 12, 111, 1979.
254. **Wong, K. and Freund, K.,** Inhibition of the *N*-formylmethionyl-leucyl- phenylalanine induced respiratory burst in human neutrophils by adrenergic agonists and prostaglandins of the E series, *Can. J. Physiol. Pharmacol.*, 59, 915, 1981.
255. **Fantone, J. C., Kunkel, S.L., and Ward, P. A.,** Suppression of human polymorphonuclear function after intravenous infusion of prostaglandin E_1, *Prost. Med.*, 7, 195, 1981.
256. **Smolen, J. E. and Weissmann, G.,** Stimuli which provoke secretion of azurophil granules induce increments in adenosine cyclic 3',5'-monophosphate, *Biochim. Biophys. Acta,* 672, 197, 1981.
257. **Smolen, J. E.,** The fluorescence response of chlortetracycline-loaded human neutrophils is modulated by prostaglandin E_1, but not by cyclic nucleotides, *FEBS Lett.*, 163, 119, 1983.
258. **Zurier, R. B., Weissmann, G., Hoffstein, S., Kammerman, S., and Tai, H. H.,** Mechanisms of lysosomal enzyme release from human leukocytes. II. Effects of cAMP and cGMP, autonomic agonists, and agents which affect microtubule function, *J. Clin. Invest.*, 53, 297, 1974.
259. **Goldstein, I. M., Malmsten, C. L., Kindahl, H., Kaplan, H. B., Radmark, O., Samuelsson, B., and Weissmann, G.,** Thromboxane generation by human peripheral blood polymorphonuclear leukocytes, *J. Exp. Med.*, 148, 787, 1978.
260. **Goldstein, I. M., Malmsten, C. L., Samuelsson, B., and Weissmann, G.,** Prostaglandins, thromboxanes, and polymorphonuclear leukocytes. Mediation and modulation of inflammation, *Inflammation,* 2, 309, 1977.
261. **Tahamont, M. V., Gee, M. H., and Flynn, J. T.,** Aggregation and thromboxane synthesis and release in isolated sheep neutrophils and lymphocytes in response to complement stimulation, *Prost. Leukotr. Med.*, 16, 181, 1984.
262. **Puustinen, T. and Uotila, P.,** Thromboxane formation in human polymorphonuclear leukocytes is inhibited by prednisolone and stimulated by leukotrienes B_4, C_4, D_4 and histamine, *Prost. Leukotri. Med.*, 14, 161, 1984.
263. **Kitchen, E. A., Root, J. R., and Dawson, W.,** Chemotactic activity of thromboxane B_2, prostaglandins and their metabolites for polymorphonuclear leukocytes, *Prostaglandins,* 16, 239, 1978.
264. **Spagnuolo, P. J., Ellner, J. J., Hassid, A., and Dunn, M. J.,** Thromboxane A_2 mediates augmented polymorphonuclear leukocyte adhesiveness, *J. Clin., Invest.*, 66, 406, 1980.
265. **Nakanishi, M. and Goto, K.,** Inhibitory effects of anti-inflammatory drugs on enzyme release from rabbit polymorphonuclear leukocyte lysosomes, *Biochem. Pharmacol.*, 24, 421, 1975.
266. **Northover, B. J.,** Effect of indomethacin and related drugs on the calcium ion-dependent secretion of lysosomal and other enzymes by neutrophil polymorphonuclear leukocytes in vitro, *Br. J. Pharmacol.*, 59, 253, 1977.
267. **Smith, R. J.,** Nonsteroid anti-inflammatory agents: regulators of the phagocytic secretion of lysosomal enzymes from guinea pig neutrophils, *J. Pharmacol. Exp. Ther.*, 207, 618, 1978.
268. **Borgeat, P., Hamberg, M., and Samuelsson, B.,** Transformation of arachidonic acid and homo-gamma-linoleic acid by rabbit polymorphonuclear leukocytes. Monohydroxy acids from novel lipoxygenases, *J. Biol. Chem.*, 251, 7816, 1976.
269. **Borgeat, P. and Samuelsson, B.,** Transformation of arachidonic acid by rabbit polymorphonuclear leukocytes. Formation of a novel dihydroxyeicosatetraenoic acid, *J. Biol. Chem.*, 254, 2643, 1979.
270. **Naccache, P. H., Sha'afi, R. I., Borgeat, P., and Goetzl, E. J.,** Mono- and dihydroxyeicosatetraenoic acids alter calcium homeostasis in rabbit neutrophils, *J. Clin. Invest.*, 67, 1584, 1981.
271. **Stenson, W. F. and Parker, C. W.,** Monohydroxytetraenoic acids (HETEs) induce degranulation of human neutrophils, *J. Immunol.*, 124, 2100, 1980.
272. **Goetzl, E. J. and Pickett, W. C.,** The human PMN leukocyte chemotactic activity of complex hydroxy-eicosatetraenoic acids (HETEs), *J. Immunol.*, 125, 1789, 1980.
273. **Goetzl, E. J., Hill, H. R., and Gorman, R. R.,** Unique aspects of the modulation of human neutrophil function by 12-l-hydroperoxy-5,8,10,14-eicosatetraenoic acid, *Prostaglandins,* 19, 71, 1980.
274. **Bokoch, G. M. and Reed, R. W.,** Effect of various lipoxygenase metabolites of arachidonic acid on degranulation of polymorphonuclear leukocytes, *J. Biol. Chem.*, 256, 5317, 1981.
275. **O'Flaherty, J. T. and Thomas, M. J.,** Effect of 15-lipoxygenase-derived arachidonate metabolites on human neutrophil degranulation, *Prosta. Leuk. Med.*, 17, 199, 1985.
276. **Goetzl, E. J., Brash, A. R., Tauber, A. I., Oates, J. A., and Hubbard, W. C.,** Modulation of human neutrophil function by monohydroxyeicosatetraenoic acids, *Immunology,* 39, 491, 1980.
277. **Goetzl, E. J.,** A role for endogenous mono-hydroxy-eicosatetraenoic acids (HETEs) in the regulation of human neutrophil migration, *Immunology,* 40, 709, 1980.
278. **Shak, S., Perez, H. D., and Goldstein, I. M.,** A novel dioxygenation product of arachidonic acid possesses potent chemotactic activity for human polymorphonuclear leukocytes, *J. Biol. Chem.*, 258, 14948, 1983.

279. **O'Flaherty., J. T., Hammett, M. J., Shewmake, T. B., Wykle, R. L., Love, S. H., McCall, C. E., and Thomas, M. J.,** Evidence for 5,12-dihydroxy-6-8-10,14-eicosatetraenoate as a mediator of human neutrophil aggregation, *Biochem. Biophys. Res. Commun.*, 103, 552, 1981.
280. **Ward, P. A., Sulavik, M. C., and Johnson, K. J.,** Activated rat neutrophils. Correlation of arachidonate products with enzyme secretion but not with O_2^- generation, *Am. J. Pathol.*, 120, 112, 1985.
281. **Vanderhoek, J. Y., Bryant, R. W., and Bailey, J. M.,** Regulation of leukocyte and platelet lipoxygenases by hydroxyeicosanoids, *Biochem. Pharmacol.*, 31, 3463, 1982.
282. **Vanderhoek, J. Y., Karmin, M. T., and Ekborg, S. L.,** Endogenous hydroxyeicosatetraenoic acids stimulate the human polymorphonuclear leukocyte 15-lipoxygenase pathway, *J. Biol. Chem.*, 260, 15482, 1985.
283. **Chang, J., Skowronek, M. D., and Lewis, A. J.,** Differential effects of monoHETES (monohydroxyeicosatetraenoic acids) on arachidonic acid metabolism in glycogen-elicited rat polymorphonuclear leukocytes, *Inflammation*, 9, 395, 1985.
284. **Stenson, W. F. and Parker, C. W.,** Metabolism of arachidonic acid in ionophore-stimulated neutrophils. Esterification of a hydroxylated metabolite into phospholipids, *J. Clin. Invest.*, 64, 1457, 1979.
285. **Marcus, A. J., Safier, L. B., Broekman, M. J., Ullman, H. L., Islam, N., Sorrell, T. C., Serhan, C. N., Weissmann, G., Oglesby, T. D., and Gorman, R. R.,** Production of metabolic products of arachidonic acid during cell-cell interactions, *J. Allergy Clin. Immunol.*, 74, 338, 1984.
286. **Marcus, A. J., Broekman, M. J., Safier, L. B., Ullman, H. L., Islam, N., Serhan, C. N., and Weissmann, G.,** Production of arachidonic acid lipoxygenase products during platelet-neutrophil interactions, *Clin. Physiol. Biochem.*, 2, 78, 1984.
287. **Jubiz, W. and Nolan, G.,** Leukotrienes produced by incubation of dihomo-γ-linolenic acid with human polymorphonuclear leukocytes, *Biochem. Biophys. Res. Commun.*, 114, 855, 1983.
288. **Prescott, S. M.,** The effect of eicosapentaenoic acid on leukotriene B production by human neutrophils, *J. Biol. Chem.*, 259, 7615, 1984.
289. **Strasser, T., Fischer, S., and Weber, P. C.,** Leukotriene B_5 is formed in human neutrophils after dietary supplementation with icosapentaenoic acid, *Proc. Natl. Acad. Sci. U.S.A.*, 82, 1540, 1985.
290. **Radmark, O., Malmsten, C., Samuelsson, B., Goto, G., Marfat, A., and Corey, E. J.,** Leukotriene A. Isolation from human polymorphonuclear leukocytes, *J. Biol. Chem.*, 255, 11828, 1980.
291. **Radmark, O., Malmsten, C., Samuelsson, B., Clark, D. A., Goto, G., Marfat, A., and Corey, E. J.,** Leukotriene A: stereochemistry and enzymatic conversion to leukotriene B, *Biochem. Biophys. Res. Commun.*, 92, 954, 1980.
292. **Claesson, H. E., Lundberg, U., and Malmsten, C.,** Serum-coated zymosan stimulates the synthesis of leukotriene B_4 in human polymorphonuclear leukocytes. Inhibition by cyclic AMP, *Biochem. Biophys. Res. Commun.*, 99, 1230, 1981.
293. **Aehringhaus, U., Wolbling, R. H., Konig, W., Patrono, C., Peskar, B. M., and Peskar, B. A.,** Release of leukotriene C_4 from human polymorphonuclear leucocytes as determined by radioimmunoassay, *FEBS Lett.*, 146, 111, 1982.
294. **Palmer, R. M. J. and Salmon, J. A.,** Release of leukotriene B_4 from human neutrophils and its relationship to degranulation induced by *N*-formyl-methionyl-leucyl-phenylalanine, serum-treated zymosan and the ionophore A23187, *Immunology*, 50, 65, 1983.
295. **Salari, H., Braquet, P., Naccache, P., and Borgeat, P.,** Characterization of effect of *N*-formyl-methionyl-leucyl-phenylalanine on leukotriene synthesis in human polymorphonuclear leukocytes, *Inflammation*, 9, 127, 1985.
296. **Claesson, H. E., Lindgren, J. A., and Gustafsson, B.,** Opsonized bacteria stimulate leukotriene synthesis in human leukocytes, *Biochim. Biophys. Acta*, 836, 361, 1985.
297. **Haines, K. A., Giedd, K. N., and Weissmann, G.,** Leukotriene B_4 synthesis and metabolism by neutrophils and granule-free cytoplasts, *Biochem. J.*, 233, 583, 1986.
298. **Goldman, D. W. and Goetzl, E. J.,** Specific binding of leukotriene B_4 to receptors on human polymorphonuclear leukocytes, *J. Immunol.*, 129, 1600, 1982.
299. **Fleisch, J. H., Rinkema, L. E., and Marshall, W. S.,** Pharmacologic receptors for the leukotrienes, *Biochem. Pharmacol.*, 33, 3919, 1984.
300. **Lin, A. H., Ruppel, P. L., and Gorman, R. R.,** Leukotriene B_4 binding to human neutrophils, *Prostaglandins*, 28, 837, 1984.
301. **Goldman, D. W. and Goetzl, E. J.,** Heterogeneity of human polymorphonuclear leukocyte receptors for leukotriene B_4. Identification of a subset of high affinity receptors that transduce the chemotactic response, *J. Exp. Med.*, 159, 1027, 1984.
302. **Lin, A. H., Ruppel, P. L., and Gorman, R. R.,** Detergent solubilization of human neutrophil leukotriene B_4 receptors, *Biochem. Biophys. Res. Commun.*, 128, 878, 1985.
303. **Clancy, R. M., Dahinden, C. A., and Hugli, T. E.,** Oxidation of leukotrienes at the w end: demonstration of a receptor for the 20-hydroxy derivative of leukotriene B_4 on human neutrophils and implications for the analysis of leukotriene receptors, *Proc. Natl. Acad. Sci. U.S.A.*, 81, 5729, 1984.

304. **Naccache, P. H., Molski, T. F. P., Borgeat, P., and Sha'afi, R. I.,** Association of leukotriene B_4 with the cytoskeleton of rabbit neutrophils. Effect of chemotactic factor and phorbol esters, *Biochem. Biophys. Res. Commun.*, 124, 963, 1984.
305. **Sha'afi, R. I., Naccache, P. H., Molski, T. F. P., Borgeat, P., and Goetzl, E. J.,** Cellular regulatory role of leukotriene B_4: its effects on cation homeostasis in rabbit neutrophils, *J. Cell. Physiol.*, 108, 401, 1981.
306. **Naccache, P. H., Molski, T. F. P., Becker, E. L., Borgeat, P., Picard, S., Vallerand, P., and Sha'afi, R. I.,** Specificity of the effect of lipoxygenase metabolites of arachidonic acid on calcium homeostasis in neutrophils, *J. Biol. Chem.*, 257, 8608, 1982.
307. **Showell, H. J., Otterness, I. G., Marfat, A., and Corey, E. J.,** Inhibition of leukotriene B_4-induced neutrophil degranulation by leukotriene B_4-dimethylamide, *Biochem. Biophys. Res. Commun.*, 106, 741, 1982.
308. **Prescott, S. M., Zimmerman, G. A., and Seeger, A. R.,** Leukotriene B_4 is an incomplete agonist for the activation of human neutrophils, *Biochem. Biophys. Res. Commun.*, 122, 535, 1984.
309. **O'Flaherty, J. T., Wykle, R. L., Thomas, M. J., and McCall, C. E.,** Neutrophil degranulation responses to combinations of arachidonate metabolites and platelet-activating factor, *Res. Commun. Chem. Pathol. Pharmacol.*, 43, 3, 1984.
310. **Smith, R. J., Iden, S. S., and Bowman, B. J.,** Activation of the human neutrophil secretory process with 5(*S*),12(*R*)-dihydroxy-6,14*cis*-8,10-*trans*-eicosatetraenoic acid, *Inflammation*, 8, 365, 1984.
311. **Goetzl, E. J. and Pickett, W. C.,** Novel structural determinants of the human neutrophil chemotactic activity of leukotriene B, *J. Exp. Med.*, 153, 482, 1981.
312. **Palmblad, J., Malmsten, C. L., Uden, A. M., Radmark, O., Engstedt, L., and Samuelsson, B.,** Leukotriene B_4 is a potent and stereospecific stimulator of neutrophil chemotaxis and adherence, *Blood*, 58, 658, 1981.
313. **Palmblad, J., Uden, A. M., Lindgren, J. A., Radmark, O., Hansson, G., and Malmsten, C. L.,** Effects of novel leukotrienes on neutrophil migration, *FEBS Lett.*, 144, 81, 1982.
314. **Jubiz, W.,** A reevaluation of the chemotactic potency of leukotriene B_4 (LTB_4), *Biochem. Biophys. Res. Commun.*, 110, 842, 1983.
315. **Dahinden, C. A., Clancy, R. M., and Hugli, T. E.,** Stereospecificity of leukotriene B_4 and structure-function relationships for chemotaxis of human neutrophils, *J. Immunol.*, 133, 1477, 1984.
316. **Beckman, J. K., Gay, J. C., Brash, A. R., Lukens, J. N., and Oates, J. A.,** Investigations of the biological activity of leukotriene A_4 in human polymorphonuclear leukocytes, *Biochem. Biophys. Res. Commun.*, 133, 23, 1985.
317. **Goldman, D. W., Pickett, W. C., and Goetzl, E. J.,** Human neutrophil chemotactic and degranulating activities of leukotriene B_5 (LTB_5) derived from eicosapentaenoic acid, *Biochem. Biophys. Res. Commun.*, 117, 282, 1983.
318. **Shak, S. and Goldstein, I. M.,** w-Oxidation is the major pathway for the catabolism of leukotriene B_4 in human polymorphonuclear leukocytes, *J. Biol. Chem.*, 259, 10181, 1984.
319. **Shak, S. and Goldstein, I. M.,** Carbon monoxide inhibits w-oxidation of leukotriene B_4 by human polymorphonuclear leukocytes: evidence that catabolism of leukotriene B_4 is mediated by a cytochrome P-450 enzyme, *Biochem. Biophys. Res. Commun.*, 123, 475, 1984.
320. **Shak, S. and Goldstein, I. M.,** Leukotriene B_4 w-hydroxylase in human polymorphonuclear leukocytes. Partial purification and identification as a cytochrome P-450, *J. Clin. Invest.*, 76, 1218, 1985.
321. **Goetzl, E. J., Derian, C., and Valone, F. H.,** The extracellular and intracellular roles of hydroxy-eicosatetraenoic acids in the modulation of polymorphonuclear leukocyte and macrophage function, *J. Reticuloendothel. Soc.*, 28, 105s, 1980.
322. **Showell, H. J., Naccache, P. H., Sha'afi, R. I., and Becker, E. L.,** Inhibition of rabbit neutrophil lysosomal enzyme secretion, non-stimulated and chemotactic factor stimulated locomotion by nordihydro-guairetic acid, *Life Sci.*, 27, 421, 1980.
323. **Smith, R. J., Yein, F. S., Speziale, S. C., and Bowman, B. J.,** A possible requirement for arachidonic acid lipoxygenation in the mechanism of phagocytic degranulation by human neutrophils stimulated by aggregated immunoglobulin G, *Biochem. Biophys. Res. Commun.*, 136, 310, 1986.
324. **Palmer, R. M. J. and Salmon, J. A.,** Comparison of the effects of some compounds on human neutrophil degranulation and leukotriene B_4 and thromboxane B_2 synthesis, *Biochem. Pharmacol.*, 34, 1485, 1985.
325. **Vanderhoek, J. Y., Ekborg, S. L., and Bailey, J. M.,** Nonsteroidal anti-inflammatory drugs stimulate 15-lipoxygenase/leukotriene pathway in human polymorphonuclear leukocytes, *J. Allergy Clin. Immunol.*, 74, 412, 1984.
326. **O'Flaherty, J. T.,** Neutrophil degranulation: evidence pertaining to its mediation by the combined effects of leukotriene B_4, platelet-activating factor and 5-HETE, *J. Cell. Physiol.*, 122, 229, 1985.
327. **Dahl, M. L., Puustinen, T., and Uotila, P.,** U-60,257 has no effect on the metabolism of arachidonic acid in nonstimulated human polymorphonuclear leukocytes, *Res. Commun. Chem. Path. Pharmacol.*, 44, 355, 1984.

328. **Sugiura, T., Onuma, Y., Sekiguchi, N., and Waku, K.,** Ether phospholipids in guinea pig polymorphonuclear leukocytes and macrophages. Occurrence of high levels of 1-*O*-alkyl-2-acyl-*sn*-glycero-3-phosphocholine, *Biochim. Biophys. Acta,* 712, 515, 1982.
329. **Virella, G., Espinoza, A., Patrick, H., and Colwell, J. A.,** Polymorphonuclear leucocytes release a factor(s) that induces platelet aggregation and ATP release after interaction with insoluble and surface-fixed immune complexes, *Clin. Exp. Immunol.,* 49, 684, 1982.
330. **Virella, G., Lopes-Virella, M. F. L., Shuler, C., Sherwood, T., Espinoza, G. A., Winocour, P., and Colwell, J. A.,** Release of PAF by human polymorphonuclear leucocytes stimulated by immune complexes bound to Sepharose particles and human erythrocytes, *Immunology,* 50, 43, 1983.
331. **Ludwig, J. C., McManus, L. M., Clark, P. O., Hanahan, D. J., and Pinckard, R. N.,** Modulation of platelet-activating factor (PAF) synthesis and release from human polymorphonuclear leukocytes (PMN): role of extracellular Ca^{2+}, *Arch. Biochem. Biophys.,* 232, 102, 1984.
332. **Mueller, H. W., O'Flaherty, J. T., and Wykle, R. L.,** The molecular species distribution of platelet-activating factor synthesized by rabbit and human neutrophils, *J. Biol. Chem.,* 259, 14554, 1984.
333. **Pinckard, R. N., Jackson, E. M., Hoppens, C., Weintraub, S. T., Ludwig, J. C., McManus, L. M., and Mott, G. E.,** Molecular heterogeneity of platelet-activating factor produced by stimulated human polymorphonuclear leukocytes, *Biochem. Biophys. Res. Commun.,* 122, 325, 1984.
334. **Weintraub, S. T., Ludwig, J. C., Mott, G. E., McManus, L. M., Lear, C., and Pinckard, R. N.,** Fast atom bombardment-mass spectrometric identification of molecular species of platelet-activating factor produced by stimulated polymorphonuclear leukocytes, *Biochem. Biophys. Res. Commun.,* 129, 868, 1985.
335. **Oda, M., Satouchi, K., Yasunaga, K., and Saito, K.,** Molecular species of platelet-activating factor generated by human neutrophils challenged with ionophore A23187, *J. Immunol.,* 134, 1090, 1985.
336. **Clay, K. L. K., Murphy, R. C., Andres, J. L., Lynch, J., and Henson, P. M.,** Structure elucidation of platelet activating factor derived from human neutrophils, *Biochem. Biophys. Res. Commun.,* 121, 815, 1984.
337. **Chilton, F. H., Ellis, J. M., Olson, S. C., and Wykle, R. L.,** 1-*O*-alkyl-2-arachidonoyl-*sn*-glycero-3-phosphocholine. A common source of platelet-activating factor and arachidonate in human polymorphonuclear leukocytes, *J. Biol. Chem.,* 259, 12014, 1984.
338. **Swendsen, C. L., Ellis, J. M., Chilton, F. H., III, O'Flaherty, J. T., and Wykle, R. L.,** 1-*O*-2-*sn*-glycero-3-phosphocholine: a novel source of arachidonic acid in neutrophils stimulated by the calcium ionophore A23187, *Biochem. Biophys. Res. Commun.,* 113, 72, 1983.
339. **Jouvin-Marche, E., Ninio, E., Beaurain, G., Tence, M., Niaudet, P., and Benveniste, J.,** Biosynthesis of Paf-acether (platelet-activating factor). VII. Precursors of Paf-acether and acetyl-transferase activity in human leukocytes, *J. Immunol.,* 133, 892, 1984.
340. **Lee, T. C., Malone, B., Wasserman, S. I., Fitzgerald, V., and Snyder, F.,** Activities of enzymes that metabolize platelet-activating factor (1-alkyl-2-acetyl-*sn*-glycero-3-phosphocholine) in neutrophils and eosinophils from humans and the effect of a calcium ionophore, *Biochem. Biophys. Res. Commun.,* 105, 1303, 1982.
341. **Mueller, H. W., O'Flaherty, J. T., and Wykle, R. L.,** Biosynthesis of platelet activating factor in rabbit polymorphonuclear neutrophils, *J. Biol. Chem.,* 2, 6213, 1983.
342. **Chilton, F. H., O'Flaherty, J. T., Ellis, J. M., Swendsen, C. L., and Wykle, R. L.,** Metabolic fate of platelet-activating factor in neutrophils, *J. Biol. Chem.,* 2, 6357, 1983.
343. **Chilton, F. H., O'Flaherty, J. T., Ellis, J. M., Swendsen, C. L., and Wykle, R. L.,** Selective acylation of lyso platelet activating factor by arachidonate in human neutrophils, *J. Biol. Chem.,* 2, 7268, 1983.
344. **O'Flaherty, J. T., Wykle, R. L., Miller, C. H., Lewis, J. C., Waite, M., Bass, D. A., McCall, C. E., and DeChatelet, L. R.,** 1-*O*-alkyl-*sn*-glyceryl-3-phosphorylcholines. A novel class of neutrophil stimulants, *Am. J. Pathol.,* 103, 70, 1981.
345. **Wykle, R. L., Miller, C. H., Lewis, J. C., Schmitt, J. D., Smith, J. A., Surles, J. R., Piantadosi, C., and O'Flaherty, J. T.,** Stereospecific activity of 1-*O*-alkyl-2-*O*-acety-*sn*-glycero-3-phosphocholine and comparison of analogs in the degranulation of platelets and neutrophils, *Biochem. Biophys. Res. Commun.,* 100, 1651, 1981.
346. **O'Flaherty, J. T., Wykle, R. L., Lees, C. J., Shewmake, T., McCall, C. E., and Thomas, M. J.,** Neutrophil-degranulating action of 5,12-dihydroxy-6,8,10,14-eicosatetraen acid and 1-*O*-alkyl-2-*O*-acetyl-*sn*-glycero-3-phosphocholine. Comparison with other degranulating agents, *Am. J. Pathol.,* 105, 264, 1981.
347. **Smith, R. J. and Bowman, B. J.,** Stimulation of human neutrophil degranulation with 1-*O*-octadecyl-2-*O*-acetyl-*sn*-glyceryl-3-phosphorylcholine: modulation by inhibitors of arachidonic acid metabolism, *Biochem. Biophys. Res. Commun.,* 104, 1495, 1982.
348. **Chilton, F. H., O'Flaherty, J. T., Walsh, C. E., Thomas, M. J., Wykle, R. L., DeChatelet, L. R., and Waite, B. M.,** Platelet activating factor. Stimulation of the lipoxygenase pathway in polymorphonuclear leukocytes by 1-*O*-alkyl-2-*O*-acetyl-*sn*-glycero-3-phosphocholine, *J. Biol. Chem.,* 257, 5402, 1982.

349. **Smith, R. J., Bowman, B. J., and Iden, S. S.,** Characteristics of 1-*O*-hexadecyl- and 1-*O*-octadecyl-2-*O*-acetyl-*sn*-glyceryl-3-phosphorylcholine-stimulated granule enzyme release from human neutrophils, *Clin. Immunol. Immunopathol.,* 28, 13, 1983.
350. **Lin, A. H., Morton, D. R., and Gorman, R. R.,** Acetyl glyceryl ether phosphorylcholine stimulates leukotriene B_4 synthesis in human polymorphonuclear leukocytes, *J. Clin. Invest.,* 70, 1058, 1982.
351. **Jouvin-Marche, E., Poitevin, B., and Benveniste, J.,** Platelet-activating factor (PAF-acether), an activator on neutrophil functions, *Agents Actions,* 12, 716, 1982.
352. **O'Flaherty, J. T., Salzer, W. L., Cousart, S., McCall, C. E., Paintadosi, C., Surles, J. R., Hammett, M. J., and Wykle, R. L.,** Platelet-activating factor and analogues: comparative studies with human neutrophils and rabbit platelets, *Res. Commun. Chem. Pathol. Pharmacol.,* 39, 291, 1983.
353. **Dahl, M. L.,** Aggregating and prostanoid-releasing effects of platelet-activating factor and leukotrienes on human polymorphonuclear leukocytes and platelets, *Int. Arch. Allergy Appl. Immunol.,* 76, 145, 1985.
345. **Tou, J. S.,** Platelet-activating factor promotes arachidonate incorporation into phosphatidylinositol and phosphatidylcholine in neutrophils, *Biochem. Biophys. Res. Comm.,* 127, 1045, 1985.
355. **O'Flaherty, J. T., Lees, C. J., Miller, C. H., McCall, C. E., Lewis, J. C., Love, S. H., and Wykle, R. L.,** Selective desensitization of neutrophils: further studies with 1-*O*-alkyl-*sn*-glycero-3-phosphocholine analogues, *J. Immunol.,* 127, 731, 1981.
356. **O'Flaherty, J. T., Wykle, R. L., McCall, C. E., Shewmake, T. B., Lees, C. J., and Thomas, M.,** Desensitization of the human neutrophil degranulation response: studies with 5,12-dihydroxy-6,8,10,14-eicosatetraenoic acid, *Biochem. Biophys. Res. Commun.,* 101, 1290, 1981.
357. **Southwick, F. S. and Stossel, T. P.,** Contractile proteins in leukocyte function, *Sem. Hematol.,* 20, 305, 1983.
358. **Boxer, L. A. and Stossel, T. P.,** Interactions of actin, myosin, and an actin-binding protein in chronic myelogenous leukemia leukocytes, *J. Clin. Invest.,* 57, 964, 1976.
359. **Hartwig, J. H. and Stossel, T. P.,** Structure of macrophage actin-binding protein molecules in solution and interacting with actin filaments, *J. Mol. Biol.,* 145, 563, 1981.
360. **Boxer, L. A., Richardson, S., and Floyd, A.,** Identification of actin-binding protein in membrane of polymorphonuclear leukocytes, *Nature (London),* 263, 249, 1976.
361. **Davies, W. A. and Stossel, T. P.,** Peripheral hyaline blebs (podosomes) of macrophages, *J. Cell Biol.,* 75, 941, 1977.
362. **Stendahl, O. I., Hartwig, J. H., Brotschi, E. A., and Stossel, T. P.,** Distribution of actin-binding protein and myosin in macrophages during spreading and phagocytosis, *J. Cell Biol.,* 84, 215, 1980.
363. **Yin, H. L. and Stossel, T. P.,** Control of cytoplasmic actin gel-sol transformation by gelsolin, a calcium-dependent regulatory protein, *Nature (London),* 281, 583, 1979.
364. **Yin, H. L., Zaner, K. S., and Stossel, T. P.,** Ca^{2+} control of actin gelation. Interaction of gelsolin with actin filaments and regulation of actin gelation, *J. Biol. Chem.,* 255, 9494, 1980.
365. **Yin, H. L., Hartwig, J. H., Maruyama, K., and Stossel, T. P.,** Ca^{2+} control of actin filament length. Effects of macrophage gelsolin on actin polymerization, *J. Biol. Chem.,* 256, 9693, 1981.
366. **Shibata, N., Tatsumi, N., Tanaka, K., Okamura, Y., and Senda, N.,** A contractile protein possessing Ca^{2+} sensitivity (natural actomyosin) from leucocytes. Its extraction and some of its properties, *Biochim. Biophys. Acta,* 256, 565, 1972.
367. **Stossel, T. P. and Pollard, T. D.,** Myosin in polymorphonuclear leukocytes, *J. Biol. Chem.,* 248, 8288, 1973.
368. **Stendahl, O. I. and Stossel, T. P.,** Actin-binding protein amplifies actomyosin contraction, and gelsolin confers calcium control on the direction of contraction, *Biochem. Biophys. Res. Commun.,* 92, 675, 1980.
369. **Hoffstein, S. T.,** Ultrastructural demonstration of calcium loss from local regions of the plasma membrane of surface stimulated human granulocytes, *J. Immunol.,* 123, 1395, 1979.
370. **Hoffstein, S. and Weissmann, G.,** Microfilaments and microtubules in calcium ionophore-induced secretion of lysosomal enzymes from human polymorphonuclear leukocytes, *J. Cell Biol.,* 78, 769, 1978.
371. **Berlin, R. D. and Oliver, J. M.,** Analogous ultrastructure and surface properties during capping and phagocytosis in leukocytes, *J. Cell Biol.,* 77, 789, 1978.
372. **Boyles, J. and Bainton, D. F.,** Changing patterns of plasma membrane-associated filaments during the initial phases of polymorphonuclear leukocyte adherence, *J. Cell Biol.,* 82, 347, 1979.
373. **Valerius, N. H., Stendahl, O., Hartwig, J. H., and Stossel, T. P.,** Distribution of actin-binding protein and myosin in polymorphonuclear leukocytes during locomotion and phagocytosis, *Cell,* 24, 195, 1981.
374. **Fechheimer, M. and Zigmond, S. H.,** Changes in cytoskeletal proteins of polymorphonuclear leukocytes induced by chemotactic peptides, *Cell Motility,* 3, 349, 1983.
375. **Wolosewick, J. J.,** Distribution of actin in migrating leukocytes in vivo, *Cell Tissue Res.,* 236, 517, 1984.
376. **Stewart, D. I. H. and Crawford, N.,** Redistribution of membrane-bound and cytosolic actin in rabbit polymorphonuclear leucocytes during phagocytosis, *Biochem. J.,* 225, 807, 1985.
377. **Howard, T. H. and Oresaio, C. O.,** The kinetics of chemotactic peptide-induced change in F-actin content, F-actin distribution, and the shape of neutrophils, *J. Cell Biol.,* 101, 1078, 1985.

378. **Howard, T. H., Casella, J., and Lin, S.,** Correlation of the biologic effects and binding of cytochalasins to human polymorphonuclear leukocytes, *Blood,* 57, 399, 1981.
379. **Maruyama, K., Hartwig, J. H., and Stossel, T. P.,** Cytochalasin B and the structure of actin gels. II. Further evidence for the splitting of F-actin by cytochalasin B, *Biochim. Biophys. Acta,* 626, 494, 1980.
380. **White, J. R., Naccache, P. H., and Sha'afi, R. I.,** The synthetic chemotactic peptide formyl-methionyl-leucyl-phenylalanine causes an increase in actin associated with the cytoskeleton of rabbit neutrophils, *Biochem. Biophys. Res. Commun.,* 108, 1144, 1982.
381. **White, J. R., Naccache, P. H., and Sha'afi, R. I.,** Stimulation by chemotactic factor of actin association with the cytoskeleton in rabbit neutrophils. Effects of calcium and cytochalasin B, *J. Biol. Chem.,* 258, 14041, 1983.
382. **Wallace, P. J., Wersto, R. P., Packman, C. H., and Lichtmann, M. A.,** Chemotactic peptide-induced changes in neutrophil actin conformation, *J. Cell Biol.,* 99, 1060, 1984.
383. **Jesaitis, A. J., Tolley, J. O., Painter, R. G., Sklar, L. A., and Cochrane, C. G.,** Membrane-cytoskeleton interactions and the regulation of chemotactic peptide-induced activation of human granulocytes: the effects of dihydrocytochalasin B, *J. Cell. Biochem.,* 27, 241, 1985.
384. **Yassin, R., Shefcyk, J., White, J. R., Tao, W., Volpi, M., Molski, T. F. P., Naccache, P. H., Feinstein, M. B., and Sha'afi, R. I.,** Effects of chemotactic factors and other agents on the amounts of actin and a 65,000-mol-wt protein associated with the cytoskeleton of rabbit and human neutrophils, *J. Cell Biol.,* 101, 182, 1985.
385. **Pryzwansky, K. B., Schliwa, M., and Porter, K. R.,** Comparison of the three-dimensional organization of unextracted and Triton-extracted human neutrophilic polymorphonuclear leukocytes, *Eur. J. Cell Biol.,* 30, 112, 1983.
386. **Howard, T. H. and Oresajo, C. O.,** A method for quantifying F-actin in chemotactic peptide activated neutrophils: study of the effect of tBOC peptide, *Cell Mot.,* 5, 545, 1985.
387. **Malawista, S. E., Oliver, J. M., and Rudolph, S. A.,** Microtubules and cyclic AMP in human leukocytes: on the order of things, *J. Cell Biol.,* 77, 881, 1978.
388. **Mellon, M. G. and Rebhun, L. I.,** Sulfhydryls and the in vitro polymerization of tubulin, *J. Cell Biol.,* 70, 226, 1976.
389. **Burchill, B. R., Oliver, J. M., Pearson, C. B., Leinbach, E. D., and Berlin, R. D.,** Microtubule dynamics and glutathione metabolism in phagocytizing human polymorphonuclear leukocytes, *J. Cell Biol.,* 76, 439, 1978.
390. **Oliver, J. M., Spielberg, S. P., Pearson, C. B., and Schulman, J. D.,** Microtubule assembly and function in normal and glutathione synthetase-deficient polymorphonuclear leukocytes, *J. Immunol.,* 120, 1181, 1978.
391. **Nath, J., Flavin, M., and Gallin, J. I.,** Tubulin tyrosinolation in human polymorphonuclear leukocytes: studies in normal subjects and in patients with the Chediak-Higashi syndrome, *J. Cell Biol.,* 95, 519, 1982.
392. **Nath, J. and Gallin, J. I.,** Studies in normal and chronic granulomatous disease neutrophils indicate a correlation of tubulin tyrosinolation with the cellular redox state, *J. Clin. Invest.,* 71, 1273, 1983.
393. **Nath, J. and Gallin, J. I.,** Ionic requirements and subcellular localization of tubulin tyrosinolation in human polymorphonuclear leukocytes, *J. Immunol.,* 136, 628, 1986.
394. **Goldstein, I., Hoffstein, S., Gallin, J., and Weissmann, G.,** Mechanisms of lysosomal enzyme release from human leukocytes: microtubule assembly and membrane fusion induced by a component of complement, *Proc. Natl. Acad. Sci. U.S.A.,* 70, 2916, 1973.
395. **Hoffstein, S., Soberman, R., Goldstein, I., and Weissmann, G.,** Concanavalin A induces microtubule assembly and specific granule discharge in human polymorphonuclear leukocytes, *J. Cell Biol.,* 68, 781, 1976.
396. **Hoffstein, S., Goldstein, I. M., and Weissmann, G.,** Role of microtubule assembly in lysosomal enzyme secretion from human polymorphonuclear leukocytes, *J. Cell Biol.,* 73, 242, 1977.
397. **Anderson, D. G., Wible, L. J., Hughes, B. J., Smith, C. W., and Brinkley, B. R.,** Cytoplasmic microtubules in polymorphonuclear leukocytes: effects of chemotactic stimulation and colchicine, *Cell,* 31, 719, 1982.
398. **Ukena, T. E. and Berlin, R. D.,** Effect of colchicine and vinblastine on the topographical separation of membrane functions, *J. Exp. Med.,* 136, 1, 1972.
399. **Oliver, J. M., Ukena, T. E., and Berlin, R. D.,** Effects of phagocytosis and colchicine on the distribution of lectin-binding sites of the cell surface, *Proc. Natl. Acad. Sci. U.S.A.,* 71, 394, 1974.
400. **Smolen, J. E. and Karnovsky, M. L.,** The effect of phagocytosis on guinea pig granulocyte membrane markers, *J. Cell. Physiol.,* 102, 71, 1980.
401. **Oliver, J. M. and Berlin, R. D.,** Surface and cytoskeletal events regulating leukocyte membrane topography, *Sem. Hematol.,* 20, 282, 1983.
402. **Herlin, T., Petersen, C. S., and Esmann, V.,** The role of calcium and cyclic adenosine 3′,5′-monophosphate in the regulation of glycogen metabolism in phagocytozing human polymorphonuclear leukocytes, *Biochim. Biophys. Acta,* 542, 63, 1978.

451. **West, B. C.,** Chediak-Higashi syndrome neutrophils are characterized by the absence of both normal azurophil granules, *Am. J. Pathol.*, 122, 177, 1986.
452. **Root, R. K., Rosenthal, A. S., and Balestra, D. K.,** Abnormal bactericidal, metabolic, and lysosomal functions of Chediak-Higashi syndrome leukocytes, *J. Clin. Invest.*, 51, 649, 1972.
453. **Stossel, T. P., Root, R. K., and Vaughan, M.,** Phagocytosis in chronic granulomatous disease and the Chediak- Higashi syndrome, *N. Engl. J. Med.*, 286, 120, 1972.
454. **Boxer, L. A., Watanabe, A. M., and Rister, M.,** Correction of leukocyte function in Chediak-Higashi syndrome by ascorbate, *N. Engl. J. Med.*, 39, 1041, 1976.
455. **Stossel, T. P., Root, R. K., and Vaughan, M.,** Phagocytosis in chronic granulomatous disease and the Chediak-Higashi syndrome, *N. Engl. J. Med.*, 286, 120, 1972.
456. **White, J. G. and Clawson, C. C.,** Chediak-Higashi syndrome: variable cytochemical reactivity of giant inclusions in polymorphonuclear leukocytes, *Ultrastruct. Pathol.*, 1, 223, 1980.
457. **Keller, H. U., Morell, A., Schaffner, T., Zimmermann, A., and Schaad, U.,** Impaired neutrophil locomotion associated with hyperadhesiveness in a patient with Chediak-Higashi syndrome, *Blood Cells*, 10, 37, 1984.
458. **Newburger, P. E., Robinson, J. M., Pryzwansky, K. B., Rosoff, P. M., Greenberger, J. S., and Tauber, A. I.,** Human neutrophil dysfunction with giant granules and defective activation of the respiratory burst, *Blood*, 61, 1247, 1983.
459. **Goihman-Yahr, M., Merino, F., Duque, P. R., Apitz, R., San Martin, B., de Gomez, M. H., de Roman, A., and Pereira, J.,** Functions of neutrophils in endemic Chediak-Higashi syndrome, *J. Leuk. Biol.*, 36, 549, 1984.
460. **Hamers, M. N., Bot, A. A. M., Weening, R. S., Sips, H. J., and Roos, D.,** Kinetics and mechanism of the bactericidal action of human neutrophils against Escherichia coli, *Blood*, 64, 635, 1984.
461. **Takeuchi, K., Wood, H., and Swank, R. T.,** Lysosomal elastase and cathepsin G in beige mice. Neutrophils of beige (Chediak-Higashi) mice lack lysosomal elastase and cathepsin G, *J. Exp. Med.*, 163, 665, 1986.
462. **Haak, R. A., Ingraham, L. M., Baehner, R. L., and Boxer, L. A.,** Membrane fluidity in human and mouse Chediak-Higashi leukocytes, *J. Clin. Invest.*, 64, 138, 1979.
463. **Ingraham, L. M., Burns, C. P., Boxer, L. A., Baehner, R. L., and Haak, R. A.,** Fluidity properties and liquid composition of erythrocyte membranes in Chediak-Higashi syndrome, *J. Cell Biol.*, 89, 510, 1981.
464. **White, J. G. and Clawson, C. C.,** The Chediak-Higashi syndrome; the nature of the giant neutrophil granules and their interactions with cytoplasm and foreign particulates. I. Progressive enlargement of the massive inclusions in mature neutrophils. II. Manifestations of cytoplasmic injury and sequestration. III. Interactions between giant organelles and foreign particulates, *Am. J. Pathol.*, 98, 151, 1980.
465. **Oliver, J. M.,** Cell biology of leukocyte abnormalities — membrane and cytoskeletal function in normal and defective cells. A review, *Am. J. Pathol.*, 93, 221, 1978.
466. **Boxer, L. A., Allen, J. M., Watanabe, A. M., Besch, H. R., Jr., and Baehner, R. L.,** Role of microtubules in granulocyte adherence, *Blood*, 51, 1045, 1978.
467. **Boxer, L. A., Rister, M., Allen, J. M., and Baehner, R. L.,** Improvement of Chediak-Higashi leukocyte function by cyclic guanosine monophosphate, *Blood*, 49, 9, 1977.
468. **Boxer, L. A., Albertini, D. F., Baehner, R. L., and Oliver, J. M.,** Improved microtubule assembly and polymorphonuclear leucocyte function in the Chediak-Higashi syndrome correctable by ascorbic acid, *Br. J. Haematol.*, 43, 207, 1979.
469. **Ostland, R. E., Jr., Tucker, R. W., Leong, J. T., Okun, N., and Williamson, J. R.,** The cytoskeleton in Chediak-Higashi syndrome fibroblasts, *Blood*, 56, 806, 1980.
470. **Frankel, F. R., Tucker, R. W., Bruce, J., and Stenberg, R.,** Fibroblasts and macrophages of mice with the Chediak-Higashi-like syndrome have microtubules and actin cables, *J. Cell Biol.*, 79, 401, 1978.
471. **Gallin, J. I., Elin, R. J., Hubert, R. T., Fauci, A. S., Kaliner, M. A., and Wolff, S. M.,** Efficacy of ascorbic acid in Chediak-Higashi syndrome (CHS): studies in humans and mice, *Blood*, 53, 226, 1979.
472. **Oliver, J. M. and Zurier, R. B.,** Correction of characteristic abnormalities of microtubule function and granule morphology in Chediak-Higashi syndrome with cholinergic agents. Studies in vitro in man and in vivo in beige mouse, *J. Clin. Invest.*, 57, 1239, 1976.
473. **Kaplan, S. S., Joyce, R., Basford, R. E., and Zdziarski, U. E.,** Effect of lithium on Chediak-Higashi leukocytes: correction of impaired function, *J. Reticuloendothel. Soc.*, 30, 573, 1981.
474. **Breton-Gorius, J., Mason, D. Y., Buriot, D., Vilde, J. L., and Griscelli, C.,** Lactoferrin deficiency as a consequence of a lack of specific granules in neutrophils from a patient with recurrent infections. Detection by immunoperoxidase staining for lactoferrin and cytochemical electron microscopy, *Am. J. Pathol.*, 99, 413, 1980.
475. **Komiyama, A., Morosawa, H., Nakahata, T., Miyagawa, Y., and Akabane, T.,** Abnormal neutrophil maturation in a neutrophil defect with morphologic abnormality and impaired function, J. *Pediatr.*, 94, 19, 1979.

476. **Strauss, R. G., Bove, K. E., Jones, J. F., Mauer, A. M., and Fulginiti, V. A.**, An anomaly of neutrophil morphology with impaired function, *N. Engl. J. Med.*, 290, 478, 1974.
477. **Gallin, J. I., Fletcher, M. P., Seligmann, B. E., Hoffstein, S., Cehrs, K., and Mounessa, N.**, Human neutrophil-specific granule deficiency: a model to assess the role of neutrophil-specific granules in the evolution of the inflammatory response, *Blood*, 59, 1317, 1982.
478. **Boxer, L. A., Haak, R. A., Yang, H. H., Wolach, J. B., Whitcomb, J. A., Butterick, C. J., and Baehner, R. L.**, Membrane-bound lactoferrin alters the surface properties of polymorphonuclear leukocytes, *J. Clin. Invest.*, 70, 1049, 1982.
479. **Ambruso, D. R. and Johnston, R. B., Jr.**, Lactoferrin enhances hydroxyl radical production by human neutrophils, neutrophil particulate fractions, and an enzymatic generating system, *J. Clin. Invest.*, 67, 352, 1981.
480. **Vercellotti, G. M., van Asbeck, B. S., and Jacob, H. S.**, Oxygen radical-induced erythrocyte hemolysis by neutrophils. Critical role of iron and lactoferrin, *J. Clin. Invest.*, 76, 956, 1985.
481. **Curnutte, J. T., Mayo, L. A., Stoehr, S. J., Balazovich, K. J., Boxer, L. A., and Smolen, J. E.**, Enhanced superoxide production in a patient whose neutrophils have elevated levels of intracellular calcium and protein kinase C activity, *Blood*, 66, 77a, 1985.
482. **Parry, M. F., Root, R. K., Metcalf, J. A., Delaney, K. K., Kaplow, L. S., and Richar, W. J.**, Myeloperoxidase deficiency. Prevalence and clinical significance, *Ann. Int. Med.*, 95, 293, 1981.
483. **Clark, R. A.**, Chemotactic factors trigger their own oxidative inactivation by human neutrophils, *J. Immunol.*, 29, 2725, 1982.
484. **Kobayashi, M., Tanaka, T., and Usui, T.**, Inactivation of lysosomal enzymes by the respiratory burst of polymorphonuclear leukocytes. Possible involvement of myeloperoxidase-H_2O_2-halide system, *J. Lab. Clin. Med.*, 100, 896, 1982.
485. **Stendahl, O., Coble, B. I., Dahlgren, C., Hed, J., and Molin, L.**, Myeloperoxidase modulates the phagocytic activity of polymorphonuclear neutrophil leukocytes. Studies with cells from a myeloperoxidase-deficient patient, *J. Clin. Invest.*, 73, 366, 1984.
486. **Dri, P., Soranzo, M. R., Cramer, R., Menegazzi, R., Miotti, V., and Patriarca, P.**, Role of myeloperoxidase in respiratory burst of human polymorphonuclear leukocytes. Studies with myeloperoxidase-deficient subjects, *Inflammation*, 9, 21, 1985.
487. **Dahlgren, C. and Stendahl, O.**, Role of myeloperoxidase in luminol-dependent chemiluminescence of polymorphonuclear leukocytes, *Infect. Immun.*, 39, 736, 1983.
488. **Cohen, M. S., Shirley, P. S., and DeChatelet, L. R.**, Further evaluation of luminol-enhanced luminescence in the diagnosis of disorders of leukocyte oxidative metabolism: role of myeloperoxidase, *Clin. Chem.*, 29, 513, 1983.
489. **Burgoyne, R. D.**, Mechanisms of secretion from adrenal chromaffin cells, *Biochim. Biophys. Acta*, 779, 201, 1984.
490. **Sklar, L. A. and Oades, Z. G.**, Signal transduction and ligand-receptor dynamics in the neutrophil. Ca^{2+} modulation and restoration, *J. Biol. Chem.*, 260, 11,468, 1985.

Chapter 3

BIOPHYSICS OF LEUKOCYTES: NEUTROPHIL CHEMOTAXIS, CHARACTERISTICS, AND MECHANISMS

Hans Gruler

TABLE OF CONTENTS

I. Introduction 64

II. Characterization of Cellular Movement 64
A. Microscopical Observation and Data Reduction 64
B. Stationary Motion 64
C. Chemokinesis 64
D. Isotropic Random Walk 66
E. Anisotropic Random Walk 68
1. Nematic Contact Guidance 69
2. Smectic Contact Guidance 71
F. Directed Locomotion 72

III. Biophysics of Cell Movement 75
A. Phenomenology of Cell Locomotion 76
1. Cellular Environment and Type of Locomotion 76
2. Gain of Information 77
3. Directional Memory 78
4. Chemotaxis, Polar Order Parameter, and Cooperativity 78
5. Contact Guidance 82
6. Virtual Force and Torque 82
B. Intrinsic properties of the moving cell: Internal Programs of the Cell 84
1. Change of Direction of Locomotion 84
2. Locomotive Activity 85
3. Internal Clock 87
4. Directional Memory 87
5. Directional Sensing Mechanism 87
6. Concluding Remarks 89

IV. Biophysics of Disturbed Tactic Cellular Activity 90
A. Virus and Chemotaxis 90
B. Virus-Induced Order-Disorder Transition 91

Acknowledgments 93

References 93

I. INTRODUCTION

Cell movements are important in embryogenesis, wound healing, tumor invasion and metastasis, infection and inflammation.[1] The behavior of the cells is determined by the stimuli coming from the environment of the cell and by the past history of the cell (heredity and memory). Behavioral research is applied to single cells in order to obtain information about how the locomotory system works. It is possible to apply some basic physical principles to gain insight into the mechanism of cell motility by careful observation of cellular movement.[2,3] In order to determine the cellular mechanisms involved in such behavior it is often useful to describe the observed movement mathematically because mathematics may make it possible to obtain information that is not immediately obvious from the data.

This concept is not restricted to single cells; it can also be applied to a well-defined assembly of cells, e.g., as organs or even whole animals. Even more, the concept[4] is also valid for machines such as, e.g., robots, etc. In this study the method is applied to the movement of polymorphonuclear leukocytes.

II. CHARACTERIZATION OF CELLULAR MOVEMENT

The macroscopic picture of cell locomotion is as follows: the cell collects information from its environment and this information is used to guide the locomotory machinery of the cell. Thus, the symmetry of the cellular environment determines the symmetry of locomotion.

Cellular locomotion is the result of a sequence of biochemical reactions as well as biophysical events. A quantitative description of cell locomotion is necessary to learn more about the biophysics of this process. In this section we are only interested in the kinematics of the cells. In Section III the measured data are interpreted.

A. Microscopical Observation and Data Reduction

The best way to study cellular locomotion is to follow single cells over a long period of time. The kinematics of the moving cells can be determined if the center of gravity of the cell is calculated on the basis of consecutive frames. The microscopic image contains a large amount of information which is also a function of the microscopy technique used. The center of gravity which one should measure can be approximated by taking the center of the area defined by the contour line of the cell (Figure 1). The trajectory of the moving cell is obtained from consecutive frames (Figure 2).

B. Stationary Motion

A granulocyte adhering to a surface has the intrinsic possibility of motility. This is demonstrated by the fact that irregular cytoplasmic projections or protuberances are always formed without change in the original adhering position. This means that the mean displacement and the mean square displacement should only hover about zero, and the degree of fluctuation should be very small as compared with the size of the cell. Exposure of the cell to chemokinetic stimuli results in active movement accompanied by characteristic cell-shape changes and by change of the starting position.

Stationary motion can be quantitated by the average of the cell shape anisotropy.[3,5] The mean track velocity of a nonmoving cell should be zero. However, the measured mean track velocity is only about zero, since the center of gravity is approximated by the center of area and the cell shape has dynamic variations.

C. Chemokinesis

Exposing cells to chemokinetic stimuli[1] such as blood plasma, f-Met-Leu-Phe, etc. causes active cell locomotion, and this chemokinetic activity can be quantified as a function of cell displacement and velocity.

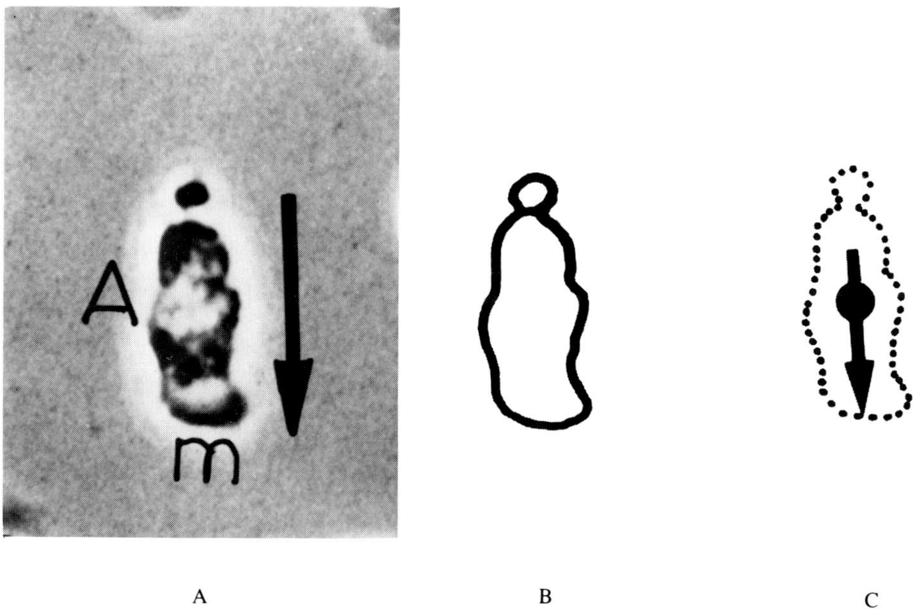

FIGURE 1. (A) Picture of a locomoting granulocyte photographed in a light microscope (phase contrast). The arrow indicates the direction of migration (leading front m). (B) Contour line of the cell. (C) The center of the cell and the cell orientation. The center of the cell and the elongation of the cell are determined from the contour line and by manually pointing to the leading front.

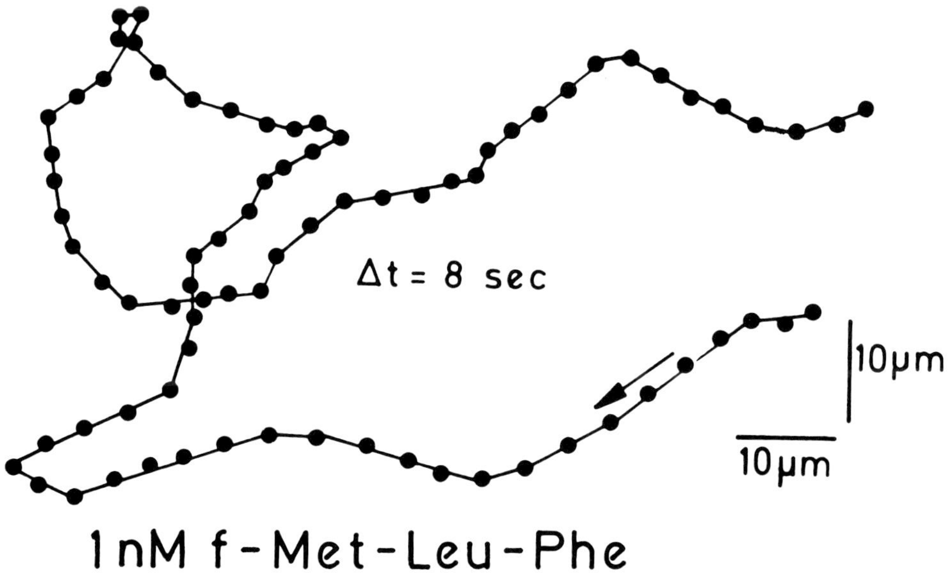

FIGURE 2. Trajectory of a single cell obtained from consecutively taken frames (20% calf serum and 1 n*M* f-Met-Leu-Phe).

The change in location along the track, Δs, of a moving cell is then

$$v_c = \frac{\Delta s}{\Delta t} \tag{1}$$

If the shape of the cell changes markedly, it is necessary to take the average center of area from two or more frames. The real path of an individual granulocyte, as shown in Figure 2, results in a snake-like coiled movement. The mean track velocity, $\langle V_c \rangle$, is 24 ± 4 μm/min at 37°C.[6] The velocity distribution function, $N(v_c)\Delta V_c$, determined from many cells is a single broadly distributed gaussian like curve [7,8] (Figure 3b). A similar broad gaussian-like curve is measured if a single cell is observed over a long period of time.[7] This indicates that the track velocity fluctuates around the mean value, but there is no indication that subpopultions of moving granulocytes are present. Later in the discussion, we will show that the track velocity distribution function is not established by a stochastic process or random process. An internal program of the cell is responsible for the chemokinetic activity.

The chemokinetic activity depends on the environmental conditions of the cell.[9-12] The response curve for the chemokinetic molecule f-Met-Leu-Phe is shown in Figure 3a. There is no locomotion for very low concentration, c. The track velocity is maximum at about 10 n*M* f-Met-Leu-Phe. This part of the response curve can be approximated by a hyperbolic curve.

$$\langle v_c \rangle = \frac{\langle v_{co} \rangle}{\left(1 + \frac{K_v}{c}\right)} \tag{2}$$

$\langle V_{co} \rangle$ is the maximum value of the mean track velocity. K_v/c is a dimensionless number which reflects the natural range of cellular sensitivity to the chemokinetic stimulus. If the concentration of the chemokinetic molecules equals K_v then the track velocity of the cell is half maximum. The chemokinetic constant, K_v, is for f-Met-Leu-Phe between 0.1 n*M* and 1 n*M*. The chemokinetic constant has about the same value as the receptor of high affinity to f-Met-Leu-Phe (0.53 n*M*).[13] The chemokinetic activity of locomoting granulocytes decreases for f-Met-Leu-Phe concentrations larger than 100 n*M*. This decreasing activity could be explained by the second binding site for the tripeptide (47 n*M*).[13]

The mean track velocity, $\langle V_c \rangle$, is a function of pH as shown in Figure 3b. The movement is maximum at pH between 7.5 and 6.5. The granulocytes can never move at large pH ($>$8) and at low pH. At this side there is a discrepancy between Fukushima et al.[34] (pH$<$4) and Nahas et al.[6] (pH$<$5.7).

The mean track velocity, $\langle V_c \rangle$, increases with increasing temperature (Figure 3c). The curve exhibits two interesting kinks, one at 22°C and and another one at 37°C. Such kinks in thermodynamic parameters are an indication that the physical state of the cell below and above the kink is different. This statement is in accordance with the abnormal change in fluidity of the plasma membrane between 25 and 27°C which we observed in our laboratory. We don't know what is happening in the cell at the second kink. However, this change at 37°C is physiologically very important since the combat against microorganisms at elevated temperatures is about twice as fast as at normal temperature (24 μm/min at 37°C to 50 μm/min at 40°C).

D. Isotropic Random Walk

The locomotion of a cell is isotropic if a cell is exposed to an environment with an isotropic symmetry, as for example, to a constant concentration of chemokinetic molecules. At every position and at every orientation the cell obtains the same information. The result is that

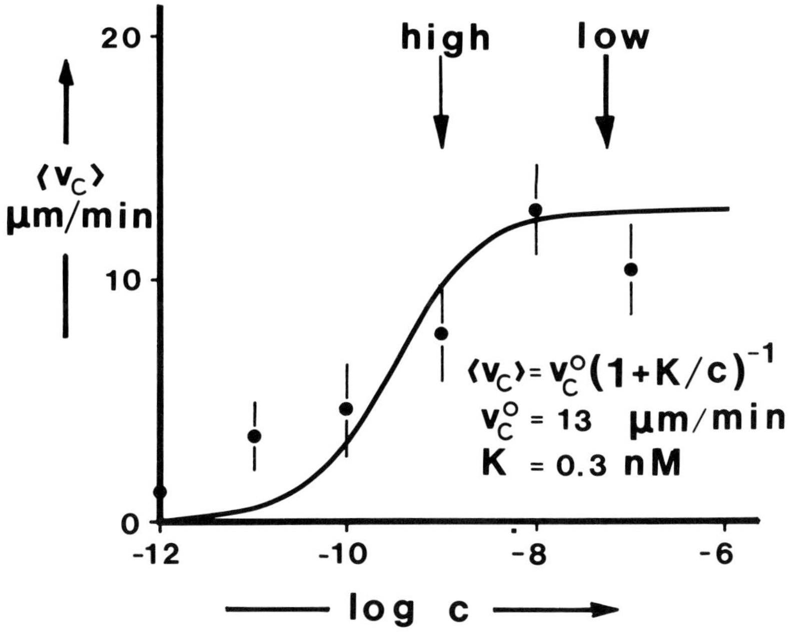

A

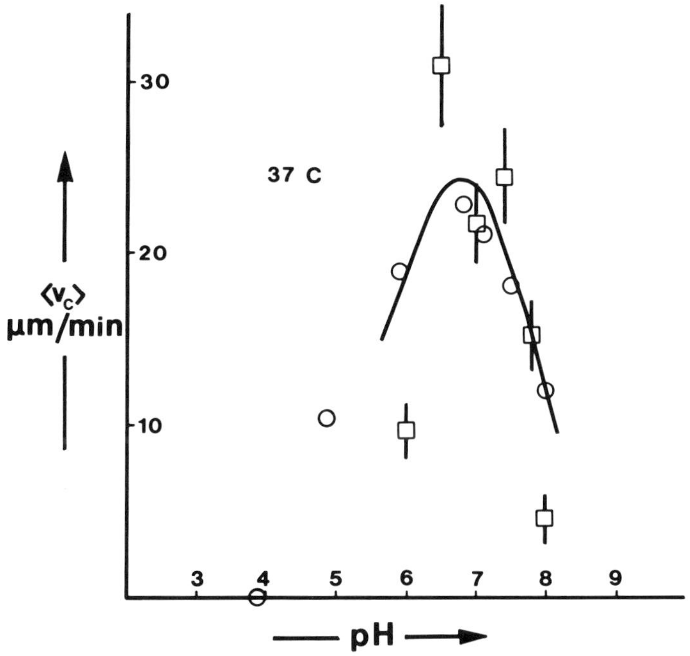

B

FIGURE 3. (A) Mean track velocity as a function of concentration of the chemokinetic molecule f-Met- Leu-Phe.[9] (B) Mean track velocity as a function of pH (References 6 and 34.) (C) Mean track velocity as a function of temperature.[6]

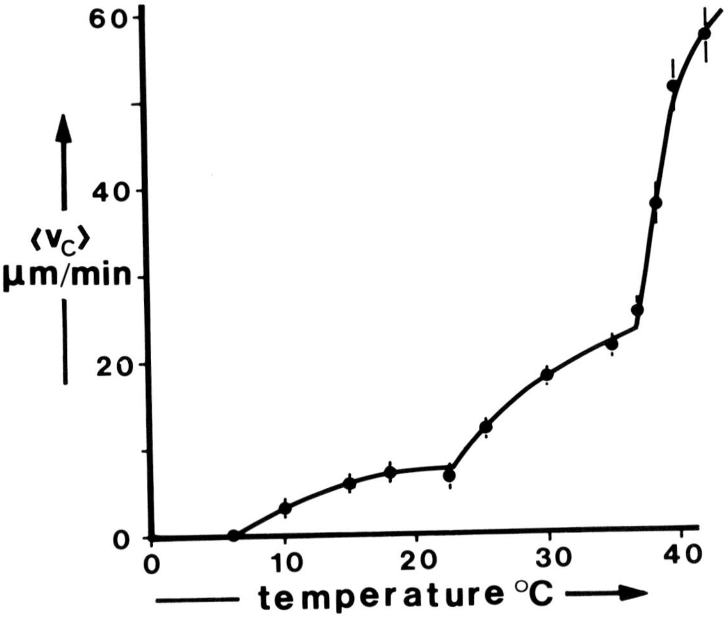

FIGURE 3C

the cell exhibits a random-walk movement in any direction. This means that the migrating cell moves in a random fashion around an arbitrarily chosen starting position. The average of the displacements of the cells, $\langle x \rangle$ and $\langle y \rangle$, in a given time, t, is zero, due to the symmetry of the cellular environment. This means that the cells migrate all the time but that the center of gravity of all the migrating cells does not change with time. In a good assay with isotropic symmetry, the mean displacements, $\langle x \rangle$, and $\langle y \rangle$, should only hover about zero, and the degree of fluctuation should be very small as compared with the size of the viewing field. The quality of the experimental assay can be tested with the mean displacements but not the random walk activity.

The mean square displacement is a nonzero value which can be used to quantify nondirected locomotion.[3,7,9,14-19] In Figure 4 the mean square displacement as a function of time is shown. Indeed, this quantity is not zero and the average of $(\langle x^2 \rangle + \langle y^2 \rangle)/2$ can be described by the Langevin equation.[9,20-22]

$$\frac{\langle x^2 \rangle + \langle y^2 \rangle}{2} = 2\,D\,\{t - \tau(1 - e^{-t/\tau})\}$$

D is the diffusion coefficient which characterizes the intensity of the random-walk process; τ is the characteristic time of the locomotion and the internal clock which governs the cellular locomotion process. More details about the random-walk process will be explained in Section III. Typical values for the diffusion coefficient and the characteristic time are 220 ± 20 µm²/min and 40 ± 5 sec, respectively.

E. Anisotropic Random Walk

The cells show an isotropic random-walk process if they are exposed to an isotropic chemical as well as physical environment. However, the random walk of the cells becomes

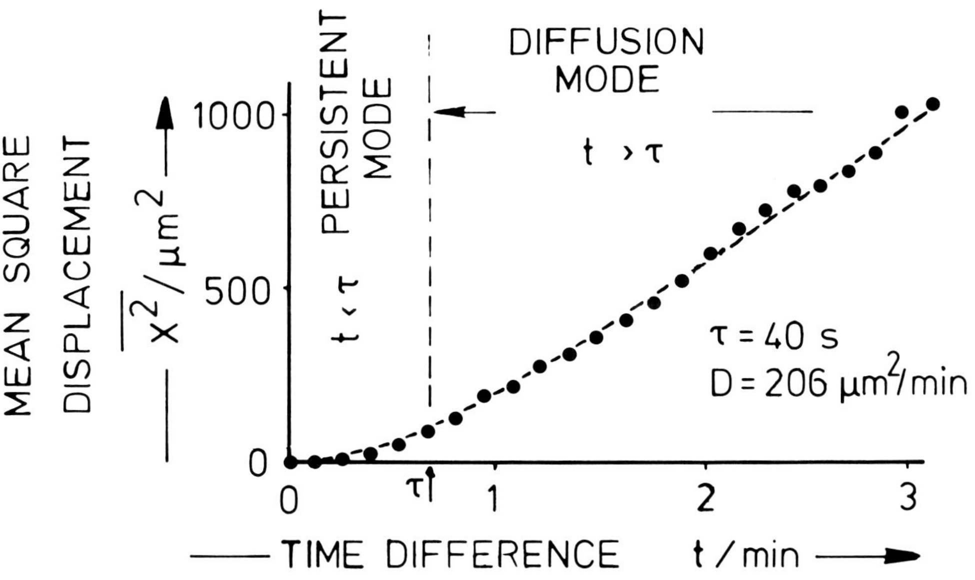

FIGURE 4. Mean square displacement of the locomoting cells as a function of time. The dots are the actual measured values. The dashed line is a theoretical prediction (Langevin equation) where the diffusion coefficient D and the characteristic time, τ, are fitting parameters. The linear region describes the diffusion mode. The extrapolation of the linear region defines the characteristic time τ.

anisotropic if the cells are exposed to an anisotropic environment.[22,23] Typical examples of such environments are, for example, grooves in a plane glass plate, parallel oriented collagen fibers, etc. The cell obtains at every position and at every orientation the same *chemical* information, since the concentration of the chemokinetic molecules is constant. However, the cells obtain at every position and at every orientation different *physical* information, since the substrate is anisotropic. The result is that the movement of the cell is anisotropic. Weiss called this type of movement ''contact guidance'', since the cell is guided by its contact to a physical structure.[24]

The locomotion of a cell is still a random-walk process, even when the cellular environment is anisotropic. The average of the displacements, <x>, and <y>, in a given time, t, is zero due to the symmetry of the cellular environment. This means that the cells migrate all the time, but the center of gravity of all the cells does not change with time. In a good assay, the mean displacements, <x>, and <y>, should only hover about zero, and the degree of fluctuation should be very small as compared with the size of the viewing field. The quality of the experimental assay can be tested using the mean displacements but not the random-walk activity.[25,26]

The contact guidance of the moving cells can be split into two parts: (1) The nematic type of contact guidance when the cell size is large in relation to the grid length of the undulated surface and (2) the smectic type of contact guidance when the cell size is small in relation to the grid distance of the substrate.

1. Nematic Contact Guidance*[26]

The cell obtains the same information at every position if the size of the cell is large compared with the characteristic length of the substrate. However, the information which the cell obtains from the substrate depends only on the orientation of the cell with respect

* The descriptions of nematic liquid crystal and the nematic contact guidance are similar.

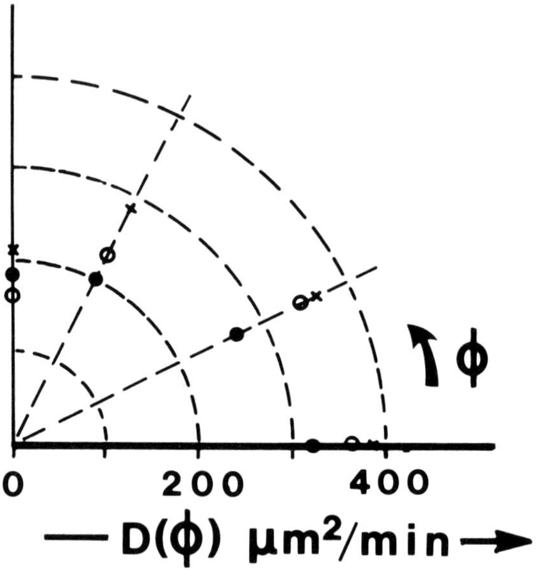

FIGURE 5. The diffusion coefficients of locomoting cells exposed to an anisotropic surrounding as an aluminum plate with parallel lines in it are angular dependent. The angular dependence of the diffusion coefficients can be described by an ellipse (x = optical grid, ○ = scratched aluminum plate, ● = stretched polyethylene foil[25,26]).

to the physical structure. Therefore, the nematic contact guidance is characterized by an angular-dependent random-walk process. In the investigated substrates, the diffusion process parallel to the lines is faster than the diffusion process perpendicular to them. The angular-dependent diffusion coefficient, as shown in Figure 5, can be described by an ellipse.

$$D(\Phi) = D_{o,g} + \langle P_2 \rangle P_2 \tag{3}$$

$$D_{o,g} = \frac{1}{2}(D_x + D_y) \tag{4}$$

$$\langle P_2 \rangle = \frac{(D_x - D_y)}{(D_x + D_y)} \tag{5}$$

$$P_2(\Phi) = 2 \cos^2\Phi - 1 \tag{6}$$

The apolar order parameter, $<P_2>$, was introduced for the quantification of the nematic contact guidance. This apolar order parameter is zero when the diffusion coefficient, D_x, measured parallel to the lines equals the diffusion constant, D_y, measured perpendicular to the lines. In this case the random-walk process is isotropic. The circle in Figure 5 characterizes an isotropic random-walk process. The apolar order parameter is one if the diffusion coefficient parallel to the lines is nonzero and the diffusion coefficient perpendicular to the lines is zero. In this case the random walk is a one-dimensional process parallel to the lines. The

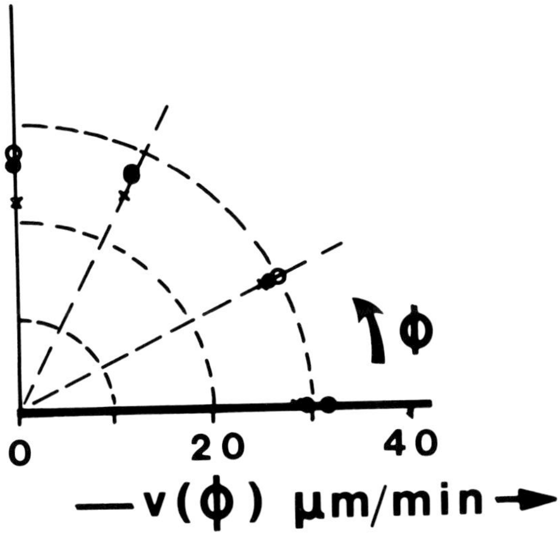

FIGURE 6. The locomotive activity of cells exposed to an anisotropic surrounding is angular dependent. The angular dependence of the locomotive activity can be described by an ellipse (x = optical grid, ○ = scratched aluminum plate, ● = stretched polyethylene foil[25,26]).

movement can be compared with a "two-way" traffic along the lines. Correspondingly, the directed movement would be a "one-way" traffic. The apolar order parameter is minus one if the diffusion coefficient parallel to the lines is zero and nonzero perpendicular to the lines. In this case, the diffusion process is a "two-way" traffic perpendicular to the lines. Typical values of the apolar order parameter are +0.3.

The nematic contact guidance is also characterized by the angular-dependent locomotive activity of the cells. In the investigated substrates, the locomotive activity parallel to the lines is faster than the locomotive activity perpendicular to them. The locomotive activity of cells exposed to an aluminum plate with parallel lines is shown in Figure 6. Again, the angular dependence of the locomotive activity can be described by an ellipse. The locomotive activity of cells exposed to a constant concentration of chemokinetic molecules and to a flat surface is also shown in Figure 5. The result is a circle.

2. Smectic Contact Guidance*[26]

The cell obtains at every position and at every orientation the same chemical information but different physical information if the size of the cell is small compared with the characteristic length of the substrate, and if the concentration of the chemokinetic molecules is constant.

The smectic contact guidance with its spatially dependent random-walk process is characterized by a spatially dependent diffusion coefficient, D(y). In a steady state, cell area density, N(y), and the diffusion coefficient, D(y), are connected by the following formula[27]

$$D(y)\frac{\partial N(y)}{\partial y} = C \tag{7}$$

where C is a constant determined by the boundary conditions. In the investigated substrates,

* The descriptions of smectic liquid crystals and the smectic contact guidance are similar.

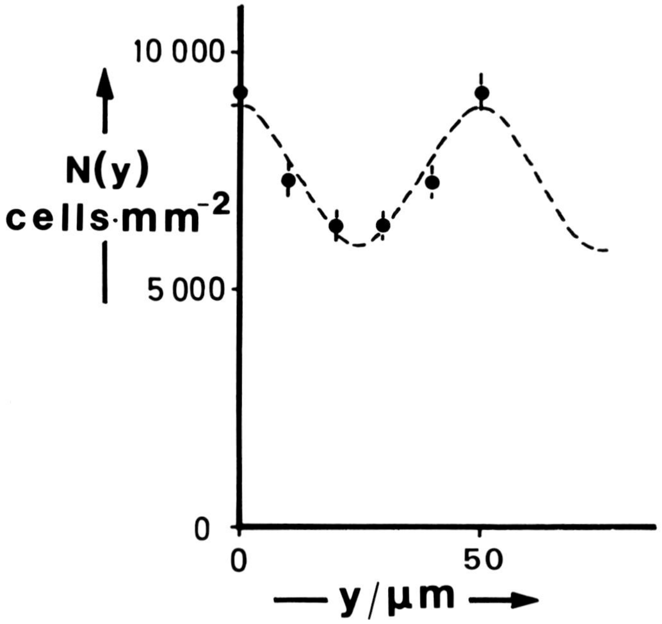

FIGURE 7. The spatially dependent cell-area concentration, N(y), is shown for cells migrating on a ground plate of a Neubauer counting chamber.[25,26] The dots are the actual measured values. The dashed line is a theoretical prediction (Equation 8).

the cells are "attracted" by the lines. A consequence is that the cell-area density is higher at the lines and lower in the flat parts of the surface. The spatial dependence of the cell-area density is shown in Figure 7. The cell area density, N(y), can be very well described by

$$N(y) = A_o(1 + \langle P_{1d} \rangle \cos \frac{2\pi y}{L} + \cdots) \tag{8}$$

$$\langle P_{1d} \rangle = \frac{A_1}{A_0} \tag{9}$$

The coefficient A_0 is the average cell area density and the coefficient A_1 describes the amplitude of the sinusoidal (cosineoidal) variation in the spatially dependent cell-area density. The characteristic length or grid length of the undulated surface is L. The spatially dependent random walk is characterized with the density-order parameter, $<P_{1d}>$. The cells do not recognize the undulated surface when A_1 is zero. In this case $<P_{1d}>$ is zero. If, on the contrary, the cells were "attracted" by the lines and concentrated by them, then A_0 would be equal to A_1 and the density-order parameter, $<P_{1d}>$, would become one. If the cells were "repulsed" from the lines and concentrated in the middle between two lines, then A_1 would equal minus A_0 and the density order parameter would become minus one. Thus, the density-order parameter quantifies how strong cells are "attracted" or "repulsed" by the lines. For example, the density-order parameter of locomoting cells on the ground plate of a Neubauer counting chamber is +0.2.[25,26] The anisotropic random-walk process will be discussed later in more detail.

F. Directed Locomotion

The directed movement of cells is obtained when the cellular environment has a polar symmetry. For example, a cell exposed to an electric field shows a directed locomotion which is referred to as galvanotaxis. On the other hand, a cell exposed to a concentration gradient of chemotactic molecules also shows a tactic locomotion which is referred to as chemotaxis. In general, the cells collect information from their environment and guide their movement in such a way that the cells drift parallel to the direction of the polar field.

The mean displacement quantifies the directed locomotion. When the direction of the polar field is parallel to the x-axis one obtains the following

$$\langle \Delta x(t) \rangle = \langle v_{II} \rangle * t \tag{10}$$

where $<v_{II}>$ is the drift velocity (= mean velocity parallel to the direction of the polar field). The center of gravity of many cells changes in time as described by Equation 10. This behavior is one biological function of the locomoting granulocytes. The cells have to find their target.

The real path of a migrating cell in a polar field is a zig-zag curve which is, on average, parallel to the direction of the polar field. Therefore, the directed locomotion is a superposition of two types of movements: (1) the drift mode characterized by Equation 10 and (2) the diffusion mode described by a random-walk process. The locomotion perpendicular to the direction of the polar field is a random-walk process, and this type of movement is described by a diffusion coefficient. In the case of necrotaxis where a target cell is lysed by a laser flash, the random walk activity is decreased from 220 ± 30 μm^2/min (diffusion coefficient without a polar field) to about 30 μm^2/min (diffusion coefficient measured perpendicular to the polar field).

The mean displacement, $<\Delta y>$, perpendicular to the preferential direction, can be used to check the experimental assay. In a good assay with a polar symmetry, the mean displacement, $<\Delta y>$, should only hover about zero and the degree of fluctuation should be very small as compared with the size of the viewing field.[28]

The mean displacement, $<\Delta x>$, parallel to the preferential direction is nonzero for cells exposed to a polar field. $<\Delta x>$ can be predicted if the mean drift velocity is known. Therefore, $<v_{II}>$ is an important parameter and needs a closer inspection. For this reason the track velocity, $v_c(t)$, is projected onto the direction of the polar field.

$$v_{II}(t) = v_c(t) * \cos \phi(t) \tag{11}$$

The angle of migration, $\phi(t)$, as well as the track velocity, $v_c(t)$, change their values with time. The average of this equation leads to the mean drift velocity. If the track velocity is independent of time one obtains

$$\langle v_{II} \rangle = \langle v_c \rangle * \langle \cos \phi \rangle \tag{12}$$

This equation is also correct if the time dependence of the track velocity is *not correlated* with the time dependence of the angle of migration. In other words, if Equation 12 holds true, then the chemokinetic activity quantified by the mean track velocity and the directionary sensing activity of the cells quantified by the average of cos ϕ are independent processes. The mean values of v_c, v_{II} and cos ϕ can be independently measured so that Equation 12 can be proved. The results show that at least for granulocytes, [7] monocytes,[29] and fibroblasts,[30] the chemokinetic activity is independent of the cellular decision activity. For this special case the mean of cos ϕ as the polar order parameter, $<P_1>$, is identical to the chemotropism index[31] $<v_{II}>/<v_c>$ and the McCutcheon index[32] (ratio between travel parallel to the direction of the polar field and the actual distance traveled along the trajectory).

The polar-order parameter is the quantity which characterizes the directed locomotion

In the macroscopic level, the migration is described in a phenomenological way, without concerning itself with the source of that motion. The aim is to find a small number of unifying laws. These laws can be applied to different cell types, even for inert particles. A typical example is the random-walk process of actively migrating cells and of inert particles. These random-walk processes are characterized by one unifying law — the diffusion law. The knowledge of these laws has the advantage of allowing one to predict how a system will behave under certain boundary conditions.

At the microscopic level, the molecular and physicochemical mechanisms underlying cellular locomotion were discussed. At the microscopic level, there is a big difference between a living cell and an inert particle or between one type of cell and another type of cell. The aim is to find the mechanisms underlying locomotion and to reconstruct the long chains of causes from these foundations up to the full range of natural events. This discussion will focus on the biophysical aspects and less on the biochemical ones, which will be dealt with elsewhere.

A. Phenomenology of Cell Locomotion

1. Cellular Environment and Type of Locomotion

The cell collects information from its environment and guides its locomotory machinery. Thus, the cellular environment can be used to characterize the type of locomotion. The response of cells to their environment can be averaged, resulting in a value of a parameter which quantifies the cellular locomotion. In addition, a detailed analysis of the cellular response provides us with information about intrinsic properties of the locomotory machinery.[2]

Chemokinesis — The chemical environment stimulates locomotion. No information is given about whether it is directed or nondirected locomotion. The strength of the locomotory activity of a migrating cell is described by the mean track velocity.

Random Walk — (1) The symmetry of the cellular environment is isotropic resulting in an *isotropic random walk*. The nondirected locomotion can be described by the diffusion law and the strength of the random-walk activity is described by the diffusion coefficient. (2) The symmetry of the cellular environment is anisotropic resulting in an *anisotropic random walk*. This type of movement occurs in contact guidance but also in chemotaxis and galvanotaxis, and the movement can be described by the diffusion law.

Contact guidance — The chemical environment is uniform but the physical environment is nonuniform and thus stimulates the cell to make an anisotropic random walk. The nematic and smectic contact guidance are quantified by the apolar-order parameter and by the density order parameter, respectively.

Chemotaxis — The chemical environment is spatially dependent (concentration gradient of chemotactic molecules) and stimulates the cellular locomotion in such a way that it is directed. The cells drift parallel to the axis of symmetry so that this type of movement is quantified by the mean drift velocity.

Galvanotaxis — The electric potential is spatially dependent resulting in an electric field. This environment stimulates the cell to a directed locomotion which is also quantified by the drift velocity.

Necrotaxis — The chemical environment and the electric potential around a dying cell is spatially dependent resulting in a polar field with a radial symmetry. This environment stimulates the cell to a directed locomotion which is also quantified by the drift velocity.

Other parameters, such as the McCutcheon index and the polar-order parameter, are also used to quantify the directed locomotion. These parameters can be related to the mean drift velocity if the cell has special intrinsic properties, as the locomotive activity and the direction sensing mechanism are independent of each other.

2. Gain of Information

One has the following general phenomenon: (1) in isotropic environments the locomotion of cells is disordered and (2) in anisotropic environments it is to some degree ordered (for example, contact guidance, chemotaxis, galvanotaxis, etc.). This means that information is involved in the cellular response. One could ask how much information would be necessary for a ''disordered'' locomotion to be transformed into an ''ordered'' one. This question can be answered if the cellular response is analyzed. One has to keep in mind that a large amount of information is hidden in the response curve. For example, when the polar-order parameter is determined from the response curve of the cells a large amount of information is lost by the averaging procedure. The response curve of the directed locomotion is the angular distribution function, $f_p(\phi)\Delta\phi$. It describes the number of cells which migrate between the angle ϕ and $\phi + \Delta\phi$. This angular distribution function has in the direction of the concentration gradient a maximum. In the case of isotropic random walk, the response curve is a horizontal line decribed by the constant f_{iso} since no direction is preferred. The question is now how much information is necessary to create from an isotropic angular distribution function, $f_{iso}\Delta\Phi$, a polar angular distribution, $f_p(\phi)\Delta\phi$. Rényi solved this problem and obtained[37-40]

$$K(f_p(\Phi); f_{iso}) = \frac{1}{\ln 2} \Sigma\, fp(\Phi)\, \Delta\Phi\, \ln \frac{f_p(\Phi)}{f_{iso}} \tag{13}$$

The formula is written in such a way that the result is given in bit. The unit ''bit'' means to *one* question *one* answer is given with either yes or no. K is the so-called gain of information.

Looking at the gain of information for directed cell locomotion, if this equation is used to determine the amount of information the cells acquire from their environment, it is found to be very low: 2.1 bits (chemotaxis: $c^{-1} * dc/dx = 2\ mm^{-1}$ and c = 5 n*M* f-Met-Leu-Phe), 2.2 bits (necrotaxis), and 1.4 bits (galvanotaxis: 0.4 V/mm). The directed movement will be better, the more information is gathered by the cells from their environment; the gathered information is a function of the strength of the polar field. The tactic movement will be better, the stronger the polar field is.

Moreover, it has been found that locomoting fibroblasts and growing fungal hyphae in an electric field also require surprisingly little information.[41] This implies that the involved sensory mechanism in different cell types is very simple. The proposed mechanism for the decision device in the sensory machinery is an instability.

Gain of information for contact guidance: if the cells were to ignore the anisotropy in the physical structure of the environment, then the locomotion would be an isotropic random walk. However, if the cells recognize the anisotropy in the physical structure the cell-area density distribution or the angular distribution function would not be a constant. By applying Equation 13 one can obtain the amount of information which the cells would collect from the anisotropic physical environment.

The nematic contact guidance was investigated on a stretched polyethylene foil (L = 0.75 μm), an optical grid (L = 5 μm), and a scratched aluminum surface (L = 11 μm). The anisotropic locomotion was quantified by the determination of angular distribution functions. The amount of information was 0.25, 0.12, and 0.12 bits for the scratched aluminum surface, the optical grid, and the stretched polyethylene foil, respectively.[25,26]

The smectic contact guidance was investigated on a ground plate of a Neubauer counting chamber (L = 50 μm). The anisotropic locomotion was quantified by the determination of the distribution function of the cell-area density. The amount of information which the cells obtained from the lines was 0.04 bits.[26]

The amount of information gathered by the cells is large when the cell can recognize many lines and the gain of information is low when the cell can recognize only a few lines. The anisotropic random walk will be better, the more lines are on the substrate. This rule is in analogy to that found in directed locomotion — the stronger the polar field the better the directed locomotion.

The important physical property which is responsible for the contact guidance is the spatial and angular-dependent surface tension of the substrate, as will be shown later. One possible way to make the surface tension anisotropic is to introduce lines.

3. Directional Memory

It has been shown that cells collect information from their environment. If the locomotion of the cells is a steady state then the cells also have to lose information. The gain of information is the amount of information which the cells gather during their memory time.

It is known that the polar-order parameter is zero for nondirected locomotion and nonzero for directed locomotion. However, there can be order of short range in the nondirected locomotion. These effects can be understood by examining the movement of single cells. At time zero, the cell moves in a certain direction, and the answer to the question as to how long it will take for this cell to lose its orientation is the memory time, τ_m, of the migrating cell. τ_m is for locomoting granulocytes on a flat glass plate between 30 and 150 sec[3,7,9] The directional memory time for migrating granulocytes on an undulated surface is in the same range, but the polar-order parameter does not reach zero as expected. This means that the moving cells on an undulated surface have a short-term and a long-term directional memory. The latter is not a real directional memory of the cell. It only reflects the fact that the cells obtained information all the time from the undulated surface.

The directional-memory time is influenced by microtubuli since colchicin-treated cells have a shorter directional memory than control cells.[42] It is likely that the memory is located in the molecular architecture of the plasma membrane and in the microtubuli. Building up a new state (= new direction of locomotion) depends on the previous state of the plasma membrane.

4. Chemotaxis, Polar-Order Parameter, and Cooperativity

The chemotactic response is one basic biological function of granulocytes. The polar-order parameter is the basic difference between the directed and nondirected locomotion and is therefore of fundamental interest. New insights into cellular locomotion can be obtained when the angular distribution function of the migrating cells is investigated instead of the average of cos ϕ. It is clear that a large amount of information is lost during this averaging procedure.

The angular distribution function of the directed locomotion is a broad curve with its maximum in the direction of the applied polar field. The angular distribution function can be approximated quite well by a gaussian curve.[3,30] However, the gaussian distribution has not the same symmetry as the experimental assay and, therefore, it is not a good description. For example, the physical state of the migrating cells is unchanged when the coordinate system is rotated by an angle of 360°, but the gaussian distribution is altered if the coordinate system is rotated by an angle of 360°. The measured angular distribution function can be described by the following function which is in accordance with the symmetry of the system.

$$f_p(\phi)\Delta\phi = e^{(a + b \cos \phi + \dots)}\Delta\phi \tag{14}$$

Figure 10 shows that the measured angular distribution function is very well approximated by the two terms of the Fourier series in Equation 14. Only one of the two unknown coefficients a and b has to be determined since the angular distribution function is calibrated. The unknown coefficient b is obtained by fitting Equation 14 to the measured angular

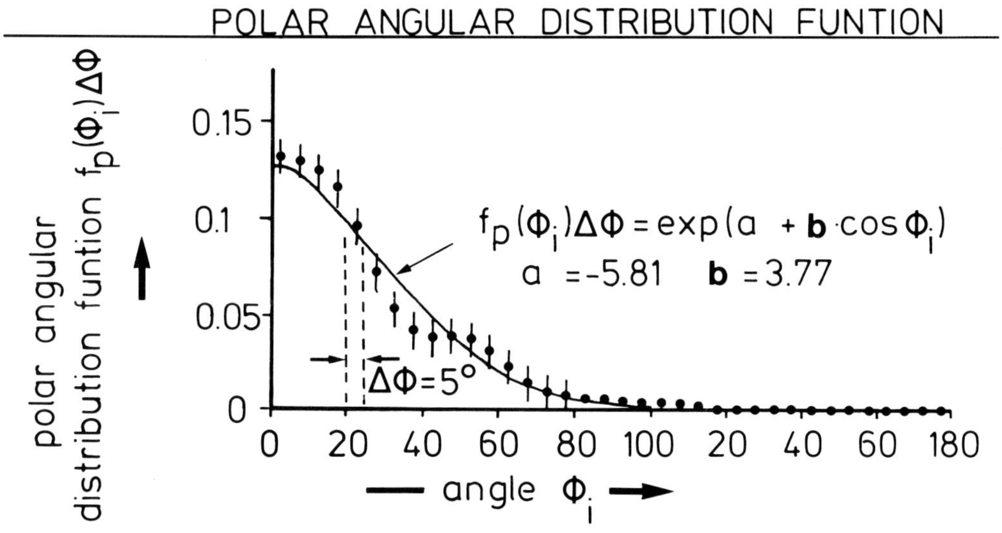

FIGURE 10. The angular distribution function obtained from locomoting cells exposed to a polar field.

distribution function. If one repeats this procedure for other angular distribution functions obtained for different values of the polar field strength, then the final result is the coefficient b as a function of the polar field strength (Figure 11). The result can be described by the following power law

$$b = (K_{CT} * \frac{1}{c}\frac{dc}{dx})^h \tag{15}$$

The slope of the straight line seen in Figure 11 determines the cooperativity coefficient, h. The value K_{CT} is obtained for b = 1. The meaning of K_{CT} and h will be discussed later.

The polar-order parameter, as the average of cos ϕ, can be calculated from Equation 14. The result is

$$\langle P_1 \rangle = \langle \cos \phi \rangle = I_1(b)/I_0(b) \tag{16}$$

where I_0 and I_1 are hyperbolic Bessel functions. The final result is described by Equation 15 and Equation 16 and shown in Figure 8.

The calculation of the polar-order parameter, $<P_1>$, as a function of the strength of the polar field is based on the measured angular distribution function. Therefore, *any* directed movement of *any* cell (whose movement can be quantified by an angular distribution function) can be analyzed. The method has already been applied to granulocytes migrating in different polar fields: (1) chemotaxis [43] (K_{CT} = -2.2 mm and h = 1.0 for 5 μM f-Met-Met-Met), and (2) galvanotaxis[33,33a,43] (K_G^{-1} = -0.20 V/mm and h = 1.0). The polar-order parameter, $<P_1>$, is negative since the electric field goes from + to − and the granulocytes migrate to the anode (+). The method has also been applied to other cell types, such as fibroblasts moving in an electric field[30] (K_G^{-1} = 0.28 V/mm and h = 2). Fibroblasts drift towards the cathode (−) so that the polar-order parameter is positive. The method can also be applied for cellular growth. Cellular growth can also be investigated provided the growth process can be quantitated by an angular distribution function. The growth of fungal hyphae in an electric field was investigated. The galvanotactic constant for this growing process was 0.3 V/mm.[41]

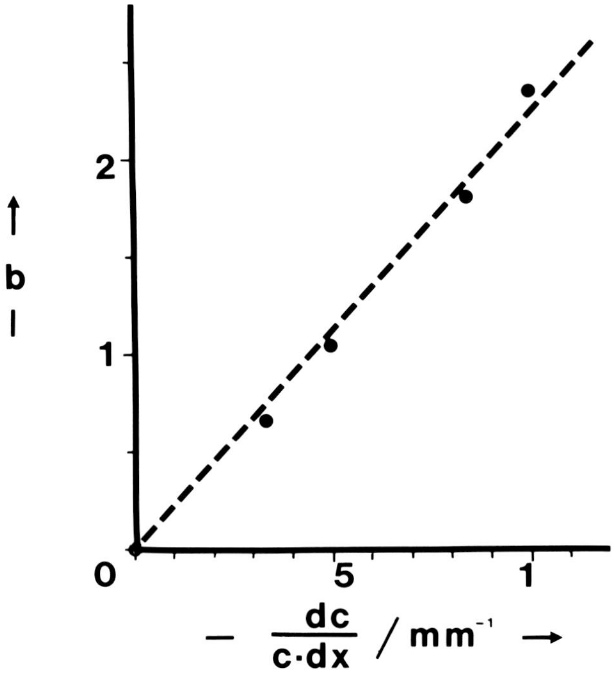

FIGURE 11. The relation between the Fourier coefficient b and the strength of the concentration gradient of chemotactic molecules is shown. The published angular distribution functions for cells exposed to f-Met-Met-Met were evaluated.[10]

The results of studies on directed locomotion will now be explained in detail. As previously pointed out, the directed locomotion chemotaxis, galvanotaxis, etc. can be treated in analogy to enzyme kinetics, with which every one is familiar. The concentration gradient of chemotactic molecules, the electric field strength, etc. which are applied to the cell, have the same meaning as the substrate. The substrate concentration divided by the Michaelis-Menten constant, K_m, is a dimensionless number which reflects the natural range of the enzyme sensitivity to a substrate. In the same way, the range of cellular sensitivity in a dimensionless scale can be introduced if the concentration gradient, grad ln c, is multiplied by the chemotactic constant, K_{CT}, or if the electric field strength is multiplied by the galvanotactic constant, K_G.

The output of the enzyme reaction is the newly formed product. The velocity of this product formation is described by the Michaelis-Menten equation. The actual velocity of the enzyme reaction divided by its maximum velocity can be treated as an order parameter of the reaction, P_{enzyme}. If the substrate concentration equals the value of the Michaelis-Menten constant, then the velocity of the product formation is half maximum, and P_{enzyme} is 0.5. The output of the cellular system is the directed locomotion quantified by the polar-order parameter, $<P_1>$. If the concentration gradient, c^{-1}* dc/dx, equals the value of K_{CT}^{-1} or if the applied electric field strength equals K_G^{-1}, then the polar-order parameter is given by $I_1(1)/I_0(1)$, which is 0.446 . . . .

The enzyme reaction can be cooperative. This means there are deviations from the normal (= hyperbolic) Michaelis-Menten curve. These deviations can be described by the cooperativity coefficient, h. The normal Michaelis-Menten curve is obtained for h = 1. For h > 1, P_{enzyme} as a function of the substrate concentration is a sigmoid curve. The cellular response of granulocytes to concentration gradients or to electric fields is noncooperative.

The polar-order parameter, $<P_1>$, as a function of the normalized field (K_{CT} * (c^{-1}* dc/dx), or K_G * E) is a hyperbolic-like curve (see Figure 8). The mechanism for directed locomotion is obviously different between granulocytes and fibroblasts since fibroblasts have a cooperative galvanotactic response.[30]

In the case of enzyme kinetics this means that the noncooperative enzyme reaction is a nonregulatory to enzyme and that the cooperative enzyme is a regulary enzyme. At positive cooperativity, the binding of one substrate molecule at one site enhances the binding of subsequent molecules at other sites. A typical example is the binding of oxygen molecules to hemoglobin. The binding of one oxygen molecule to hemoglobin enhances the binding of subsequent oxygen molecules. Sigmoid curves are very steep in the middle part, so that rather small increases in substrate can cause a very large acceleration of the rate of catalysis much more than is afforded by a single nonregulatory enzyme obeying the hyperbolic Michaelis-Menten relationship. In a nonregulatory enzyme, a 32-fold increase in substrate is required to bring the order parameter, $P_{enzyme,}$ from 0.15 to 0.85. In a regulatory enzyme with a cooperativity coefficient of two, a 5.6-fold increase in substrate is required to bring P_{enzyme} from 0.15 to 0.85.

Positive cooperativity can be important for directed cell movement. At positive cooperativity, the increase in the polar-order parameter, $<P_1>$, induced by one increment of the concentration gradient is low if the total number of increments is small (= low value of the polar field). However, the increase in the polar-order parameter induced by one increment at high level of increments (= high value of the polar field) is large. Sigmoid curves are very steep in the middle part, so that a rather small increase in the field can cause a large increase in the polar-order parameter much more than is afforded by a simple noncooperative response curve. In a noncooperative response curve, a 12.3-fold increase in the concentration gradient is required to bring the polar order parameter, $<P_1>$, from 0.15 to 0.85. In a cooperative response curve with a cooperativity coefficient of two a 3.4-fold increase of the concentration gradient is required to bring the polar-order parameter, $<P_1>$, from 0.15 to 0.85.

The cooperative response curve of cell movement may have physiological importance. One can think of many reasons why a cooperative curve would be good for invading cells in inflammation. For example, the reaction of the cells towards the site of inflammation is very weak far away from the inflammation (K_{CT} * c^{-1} * dc/dx < 1); however, near the inflammation the reaction of the cells to the inflammation is very strong (K_{CT} * c^{-1} * dc/dx > 1). Therefore, the cooperative response curve enables the system to have a strong reaction, but this reaction is only localized around the inflammation. The reaction of cells having a noncooperative response curve is not so strong near the inflammatory site as for cells having cooperativity. This means the fight against invading microorganisms is not so fast and not so effective for cells having no cooperativity as with those having cooperativity. On the other hand, the reaction of the migrating cells far away from the inflammation is stronger for cells having noncooperativity. This means the inflammatory reaction of noncooperatively working cells is spread over a large region.

The chemotactic response curve of granulocytes is not cooperative. However, the chemotactic constant does not have a fixed value.[43] It depends on the absolute concentration of the chemokinetic molecules. The cellular response of granulocytes with a concentration-dependent chemotactic constant has similarities to the response of cells with a cooperative response. Therefore, inflammation needs a closer inspection: mammalian tissue damaged or invaded by a microorganism releases molecules and ions and the released substances guide the granulocytes to the site of inflammation. The cellular decision is a function of dc/dx. However, obviously the cell cannot measure the absolute value of dc/dx, but rather the normalized value c^{-1} * dc/dx. The disadvantage of this effect is obvious: the relative sensitivity decreases with increasing c. Near the inflammation site the absolute concentration of chemotactic molecules is high, and, consequently, the relative sensitivity to dc/dx is low. Granulocytes overcome this disadvantage at low concentration, since the chemotactic constant increases with increasing concentration.[43] Receptors with different sensitivities to the chemotactic molecule could explain the concentration dependence of the chemotactic constant.

5. Contact Guidance

It is well known that directed locomotion in a concentration gradient of chemotactic molecules is based on the specific binding of molecules to the membrane-bound receptors. The question is how can a cell recognize different structural details of its surrounding. The exact nature of this measuring system is still unknown, though certain aspects of this mechanism may be discussed.[26]

1. It could be hypothesized that the measuring system comprises a membrane-bound receptor. This concept is very unlikely since a single receptor is much too small compared to the extension of the undulated surface. Furthermore, the receptors are highly specialized and can thus hardly account for the cellular response to different surfaces.
2. The basis of the measuring system of structural details could be the pattern of forces which are involved in galvanotaxis. Details of the galvanotactic response are unknown.
3. Another basis of the measuring system of structural details could be a system which acts at the level of cell adhesion. This model has been earlier discussed by Weiss.[24]
4. Another possible basis of the measuring device of structural details could be elastic strain fields in the plasma membrane which are induced by bending the membrane.

The leading front, as an essential element of the recognition process, can morphologically be distinguished from the rest of the plasma membrane, suggesting that a molecular decomposition does take place. The origin of this decomposition is not clear. It could be an elastic strain field (point 4), or a change in the surface tension (point 3), or an electric field (point 2). The experimental evidence supports Weiss' model (point 2).[26]

The nematic contact guidance is characterized by the angular-dependent track velocity and the angular-dependent random walk. The question is what is the basis for this angular dependence? A detailed analysis of the diffusion process yields the answer.

It can be shown that the diffusion coefficient is proportional to the characteristic time of locomotion and the mean square of the track velocity.[21] If the diffusion coefficient divided by the square of the velocity were a constant, then the anisotropic random walk process could be simply explained by the anisotropy in the speed of the cell. In Figure 12, $2D(\phi)/\langle v(\phi)^2 \rangle$ is plotted vs. $(2\cos^2\phi - 1)$. The values are distributed along straight lines. An almost horizontal line is evident for cells moving on an optical grid or a scratched aluminum surface, indicating that the nematic contact guidance is mainly due to the anisotropy in the track velocity. The track velocity itself is dependent on the surface tension between a cell and the surface of a substrate and this surface tension is related to the contact angle between these two. A drop of medium on an undulated surface show that the contact angle is angular dependent. Thus, these results support the observations of Weiss, who showed that the smectic contact guidance is determined by the anisotropy of the surface tension.

6. Virtual Force and Torque

Cells exposed to a polar field result in a directed movement: the cells are oriented on average parallel to the direction of the polar field. The cellular response resulting in an orientation is described by the polar angular distribution function. We will show how the virtual torque can be derived from the distribution function.

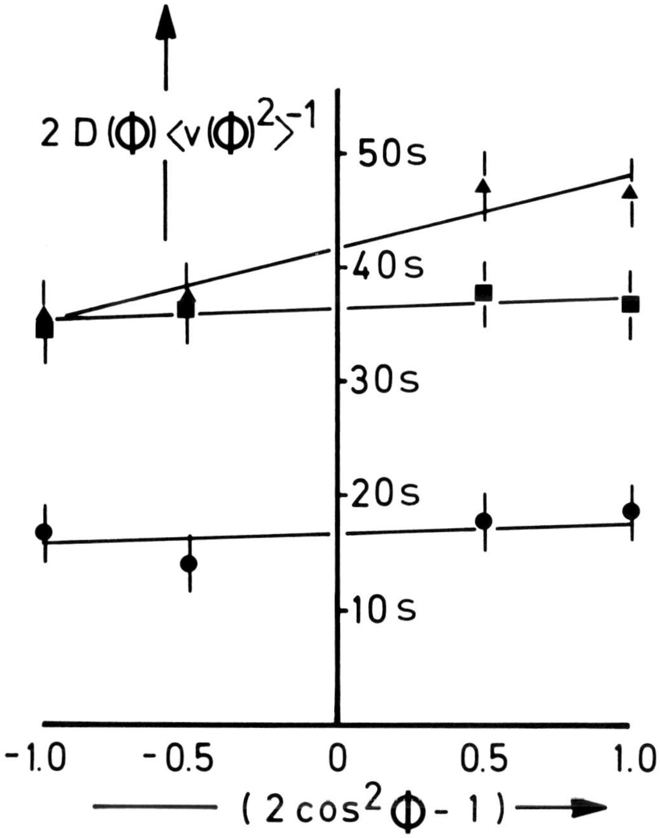

FIGURE 12. The diffusion coefficient divided by the square of the velocity is shown as a function of the angle ϕ. (▲ = stretched polyethylene foil; ● = scratched aluminum plate; ■ = optical grid).

Cells exposed to an anisotropic physical structure adopt in an anisotropic random walk. In smectic contact guidance, the cells are attracted by the lines and the cellular response is described by the cell-area distribution function. In nematic contact guidance, the cells are oriented by the lines and the cellular response is described by the apolar angular distribution function.

These distribution functions can be expressed by new functions, $V(y)$ and $V(\phi)$, without losing information

$$N(y) = e^{V(y)} \tag{17}$$

$$f_p(\phi)\Delta\phi = e^{V(\phi)}\Delta\phi \tag{18}$$

The new functions, $V(y)$ and $V(\phi)$, are referred to as the generating functions, which in physical systems have the meaning of a normalized energy.[39]

In the case of small values for the order parameter ($\langle P_{1d}\rangle < 1$, $\langle P_2\rangle < 1$) the generating functions for the smectic and nematic contact guidance are

$$V(y) = \ln A_0 + \langle P_{1d}\rangle \cos(2\pi y/L) + \ldots \tag{19}$$

$$V(\phi) = C_0 + \langle P_2\rangle (2\cos^2\phi - 1) + \ldots \tag{20}$$

where C_0 is a constant, its value being determined by the calibration procedure of the angular distribution function. The generating function for the chemotactic as well as galvanotactic response is defined by Equation 14.

The virtual force, F(y), and the virtual torque, M(ϕ), can be obtained from the generating functions

$$F(y) = -dV(y)/dy \tag{21}$$

$$M(\phi) = -dM(\phi)/d\phi \tag{22}$$

The virtual force has the dimension of the inverse of a length as a thermodynamic force, while the virtual torque has no dimension.

For example, one obtains for the chemotaxis and galvanotaxis these simple expressions.

$$M(\phi) = \left(K_{CT} * \frac{1}{c}\frac{dc}{dx}\right)^h * \sin\phi \tag{23}$$

$$M(\phi) = (K_{CT} * E)^h * \sin\phi \tag{24}$$

In the case of noncooperative cells ($h = 1$), the virtual torque is proportional to the natural sensitivity of the cell to the polar field and to sin ϕ. For cooperative cells ($h > 1$), the virtual torque is stronger in high polar fields and weaker in small polar fields as compared with those of noncooperative cells.

B. Intrinsic Properties of the Moving Cell: Internal Programs of the Cell

1. Change in Direction of Locomotion

A moving granulocyte is a correlated walker. For short times ($t < \tau$), locomotion persists. This correlated locomotion can be seen directly because the granulocyte normally moves in approximately straight paths which are separated by discrete changes in direction. A much better method is to construct a time-dependent angular distribution function, and this reveals something about the inherent properties of the cell.

It is possible to determine the short-range effects if the paths of the cells are known. To do this, an origin is chosen and the direction of locomotion at this time is considered as "preferential". Then the direction of locomotion at time t_k is determined. The angle φ (t_k) between the direction of locomotion and the "preferential" direction is determined. The procedure is repeated many times for different origins and different cells resulting in the time-dependent angular distribution function $f(\varphi\ t_k)\Delta\varphi$ (Figure 13). For $t_k = 0$ sec, $f(\varphi; t_k)$ $\Delta\ \varphi$ shows a peak at $\varphi = 0°$. For $t_k = 16$ sec, the central peak is still evident, but another peak is visible at about ±35°. This trend continues with increasing time. At $t_k = 32$ sec, the central peak has disappeared, and only the peak at about ±35° is visible.[3] (See Figure 13).

The correlated locomotion defines approximately straight paths which are separated by discrete changes in direction. The transit time (= characteristic time of the internal clock) is about 40 sec and the rotation angle fluctuates around 35°. If this turn angle (= new direction of migration created by the previous direction) were to be determined by a stochastic process, then every angle between 0° and 360° would have the same probability and one would expect a horizontal line as the angular distribution function f ($\varphi\ t_k$)$\Delta\varphi$ ($\varphi = / \neq 0$). Such a behavior was not found. Thus, the turn angle is not determined by a stochastic process. The consequence is that the turn angle is created by an internal program of the cell.

In colchicine-treated cells, the satellite peaks which characterize the turn angle, are very broad and their maximum is between 40° and 60°. This means that the internal program

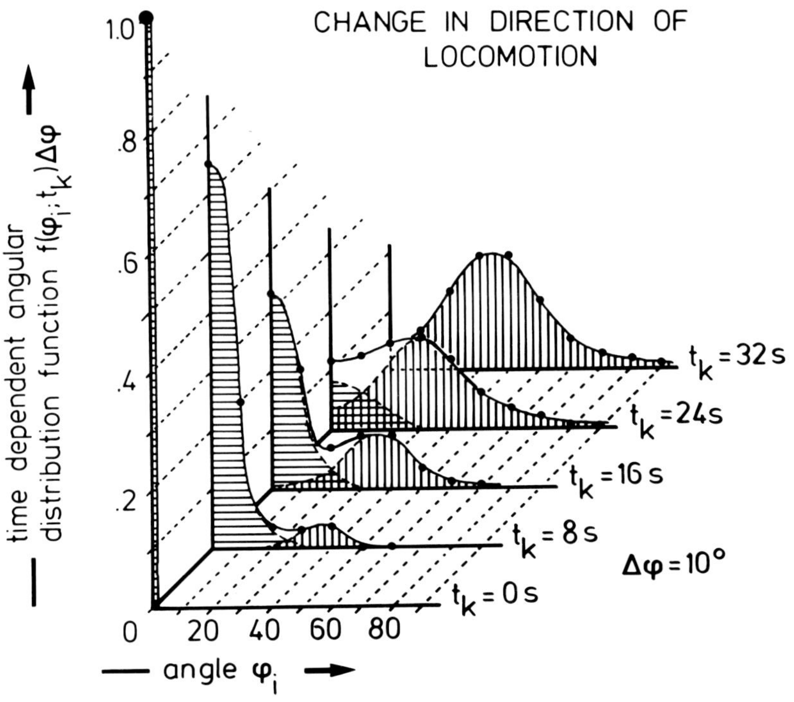

FIGURE 13. The change of direction of 54 moving granulocytes is demonstrated by the time-dependent angular distribution function $f(\phi;t_k)\ \phi$. The symmetry of the assay is isotropic. The probability that a cell has not changed its direction in the time t_k is given by the area of the peak at $\phi = 0$ (horizontal hatching). The probability that a cell has changed its direction is given by the peak at $\phi = \pm 35^\circ$ (vertical hatching).

responsible for the turn angle is more random in the case of cells with disrupted microtubuli than in the case of control cells. Cells treated with cytochalasin B, have the same time-dependent angular distribution function as the control cells. The density of the microfilaments is not important for the internal program that is responsible for the change in direction.

2. Locomotive activity

The locomotive activity is quantified by the mean track velocity. However, we have to keep in mind that much of the information is lost during the averaging procedure. A detailed analysis of the locomotive response provides us with intrinsic properties of the locomotive machinery.

The cellular locomotion with its discrete changes in direction can be described by the following differential equation

$$dv/dt = f \tag{25}$$

In the case of an inert particle, this unknown function, f, is a force divided by the mass of this particle and this force has its origin in the environment of the inert particle. However, in the case of a cell, this unknown function, f, depends not only on the internal programs of the cell but also on the cellular environment. Since the cell is not accelerated all the time, the unknown function, f, can be expressed as[42,44]

$$f = -\beta v + \Gamma(t) \tag{26}$$

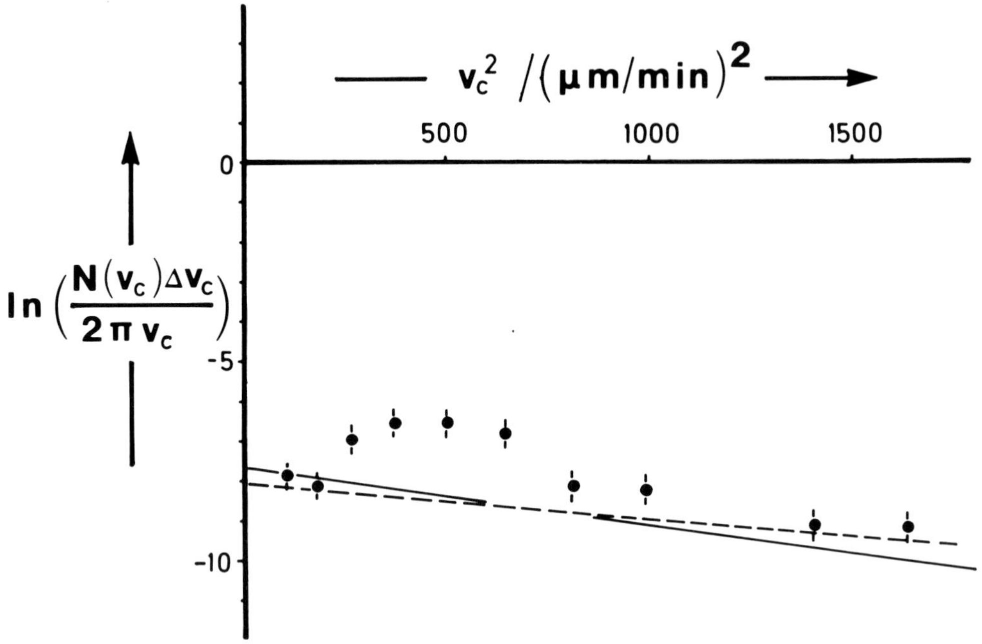

FIGURE 14. Demonstration of the internal program for the chemokinetic activity. The predicted straight line behavior is calculated by using either the diffusion coefficient, D, and the characteristic time, τ, (dashed line), or the mean square velocity, $<v_c^2>$, (solid line).

β^{-1} is a constant with the dimension of time. The time-dependent fluctuation in the cell movement is described by the Langevin force $\Gamma(t)$. The average of $\Gamma(t)$ is zero. If we assume that $\Gamma(t)$ is a white noise force, then the random walk activity as well as the locomotive activity can be calculated when the spectral density of the white noise is described by 2q:

1. One obtains for the mean square displacement, $<x^2>$ or $<y^2>$, as a function of time t in the Langevin equation. The diffusion coefficient D is now expressed by the spectral density of the white noise, q, and the characteristic time, τ

$$D = q\,\tau^2/2 \tag{27}$$

$$\tau = \beta^{-1} \tag{28}$$

The fit of the Langevin equation to the experimental points is shown in Figure 4. The good agreement demonstrates that the derived differential equation has the right mathematical structure.

2. One obtains for the velocity distribution function, $N(v_c)\Delta v_c$

$$N(v_c)\Delta v_c = \frac{2\beta q v_c}{q} * \exp(-\frac{\beta v_c^2}{q})\,\Delta v_c \tag{29}$$

If the logarithm of the velocity distribution function divided by the track velocity is plotted as a function of the square of the track velocity then one expects accordingly Equation 29 to be a straight line. Such a plot is shown in Figure 14. The experimental data points do not follow the predicted straight-line behavior. This indicates that the white noise assumption for the Langevin force, $\Gamma(t)$, is not correct. This means that the track velocity of the migrating cell is not made by a random process. Hence, an internal program of the cell is responsible for the chemokinetic activity.

The very complex locomotory machinery of the cell can be described phenomenologically by a potential function, $P(v_c)$. The form of this function depends on the concentration of the chemokinetic stimuli (Figure 15): for very low concentrations, the cell do not migrate. The absolute minimum of $P(v_c)$ is at $v_c = 0$. The stationary locomotion is described by the curvature of this minimum. For very high concentrations of the chemokinetic stimuli, the cell migrates with its mean track velocity. At this situation, the absolute minimum is at $<v_c>$. The velocity distribution function $N(v_c)$ v_c is determined by the curvature of this minimum. The two minima of the potential function also explain the cellular behavior at medium concentrations of the chemokinetic stimuli. Thus, some of the cells migrate with track velocities comparable with that obtained at high concentration. Another group of cells does not migrate. The chemokinetic activity of a moving cell is closely related to a first-order phase transition. In this case the potential function is the free energy.

The locomotory machinery of a cell is guided by a cellular program. The existence of this program is demonstrated by inspecting the velocity distribution function. Microfilaments are important elements of the locomotory machinery, since cytochalasin B, which disrupts microfilaments, alters the locomotory activity. Microtubuli are not important elements of the locomotory machinery since colchicine, which disrupts microtubuli, does not alter the locomotory activity.[42]

3. Internal Clock

The locomotive machinery of the cell is guided by an internal clock. The existence of this internal clock is best demonstrated by inspecting the mean-square displacement or the time-dependent angular distribution function. The characteristic time, τ, of this internal clock is approximately 40 sec.

The characteristic time, which is the interval between two successive changes in the direction of locomotion is found to be the same for the control cells as for cytochalasin B and colchicine-treated cells. Since colchicine disturbs microtubules, and since cytochalasin B disrupts microfilaments, these suggest that microtubules and microfilaments are not important for the determination of the characteristic time of the locomoting cell.

There exists a retardation time (or refractory period) during which the cell is insensitive to external signals. It is the period after the cellular decision where the cellular decision cannot be altered. The existence of the retardation time is best demonstrated by applying time-dependent polar fields.[45,46] The retardation time is approximately 20 sec.

4. Directional Memory

The directional memory is an intrinsic property of the migrating cell.[3] The directional memory time is between 40 and 150 sec depending on cell treatment. Cells with no microtubuli or disrupt microtubuli have a diminished memory time.

5. Direction-Sensing Mechanism

An intrinsic property of the migrating cell is that the direction-sensing mechanism is independent of the locomotive activity of the cell. The direction-sensing mechanism still works even when the cell has no or disrupted microtubuli.

The mechanism that we propose for the decision device in the sensory machinery is an instability having two different states. In every cycle of migration, the cell decides once for one state out of two. A granulocyte exposed to a polar field as a concentration gradient decides between two directions $\alpha+$ and $\alpha-$ (state one: $\alpha+ = +35°$, and state two: $\alpha- = -35°$). If the cell always makes the right decision, then the amount of information per

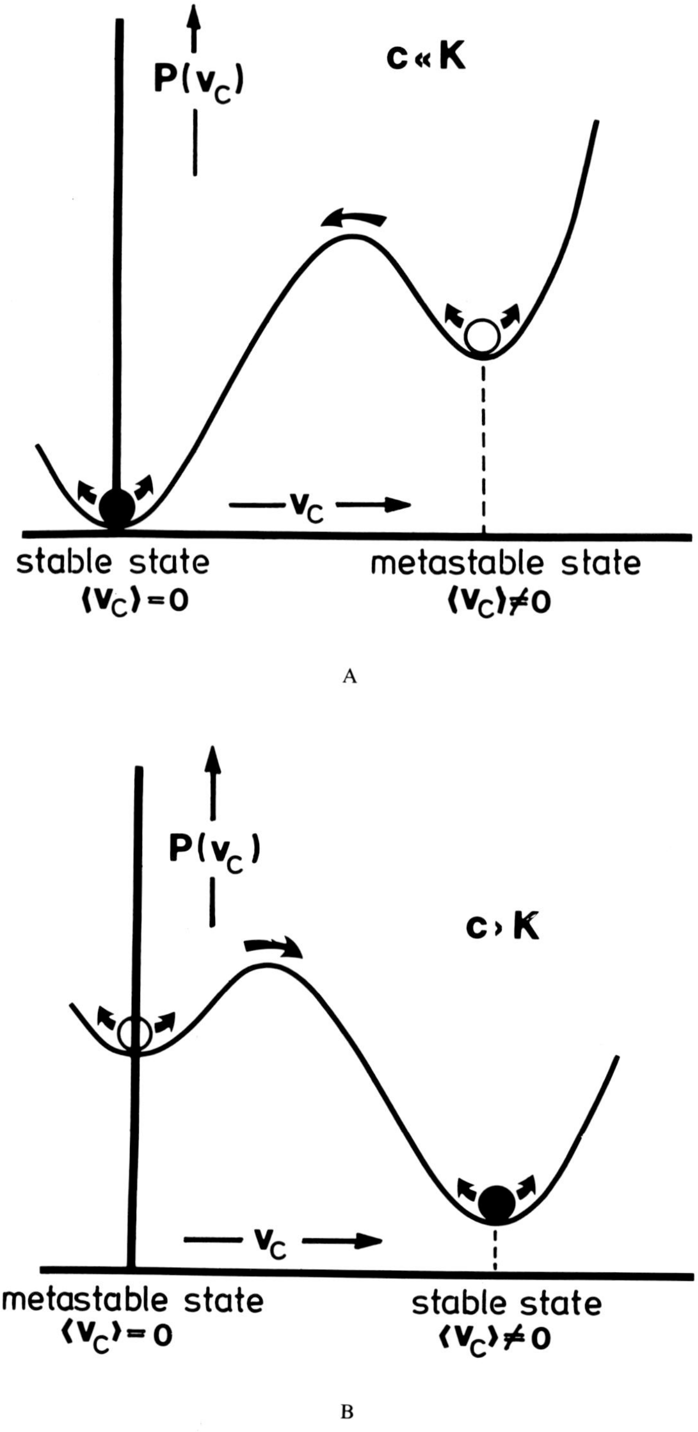

FIGURE 15. Schematic representation of potential curves which are responsible for the chemokinetic activity of the cells: (A) low and (B) high concentration of the chemokinetic molecule.

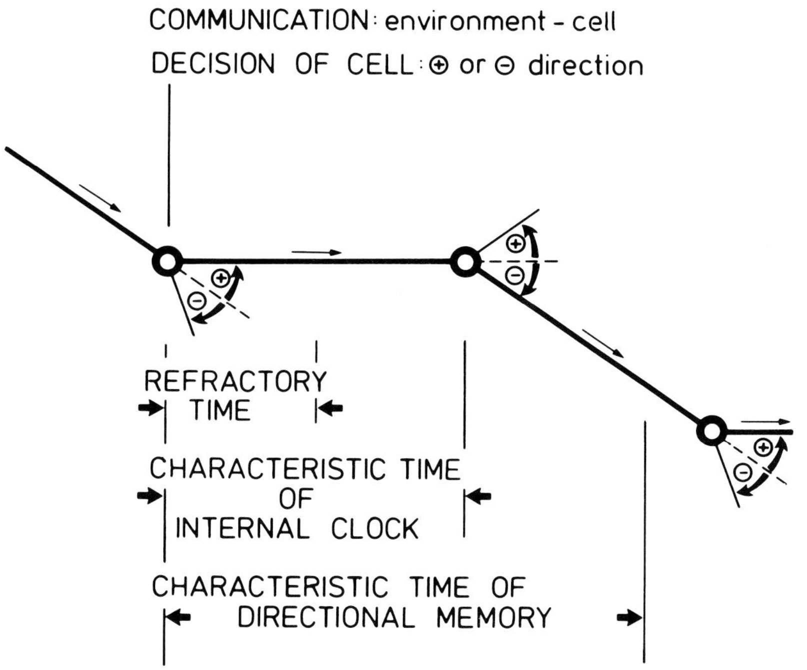

FIGURE 16. Schematic representation of the locomotion: the cell communicates with its environment (○) and digitates the analog signal from the environment into a + or − direction. (In the isotropic symmetry, the + and − are equally probable; in the anisotropic symmetry, the + and − direction have different probabilities.) The cell can not communicate with its environment during the refractory time. The time length of one moving state is determined by the characteristic time of the cellular clock. The migrating granulocyte has a directional memory. Its characteristic time is nearly twice the characteristic time of the internal clock.

decision must be one. Therefore, we can conclude that granulocytes migrating in a strong chemotactic or necrotactic gradient make no mistake.

The moving granulocyte acts like an analog-digital converter: the chemical environment of the cell is the analog signal since the concentration gradient can have different values. The cellular response is digitated.

6. *Concluding Remarks*

A schematic representation of the locomotion is shown in Figure 16: the cell communicates with its environment (○) for less then 8 sec and digitates the analog signal from the environment into a + or − direction. In the isotropic symmetry, the + and − are equally probable. In the anisotropic symmetry, the + and − direction have different probabilities. The cell cannot communicate with its environment during the refractory time. The time length of one moving state is determined by the characteristic time of the cellular internal clock. The locomoting granulocyte has a directional memory. Its characteristic time is nearly twice the characteristic time of the internal clock.

The mechanisms involved in a cellular response that requires so little information may resemble an unstable system. Examples of instability are oscillating chemical reactions and autoelectrophoresis of molecules in the plane of the membrane. Any mechanism that involves a strong feedback tends to be an unstable state and may produce strong responses to little input. This is a phenomenological description, and a complete understanding of the reactions involved requires considerable effort.

The detailed analysis of cell locomotion leads to new insights into cellular movement.

The practical importance of the biophysical investigation lies in the detailed analysis of the experimental conditions, and, moreover, it provides a more objective description of cell locomotion, which may be useful in the definition of granulocyte dysfunctions (see next section). Furthermore, the beauty of the physical technique is that it is not restricted to the locomotion of granulocytes. Other cell types or clusters of cell or even whole animals can be investigated.

IV. BIOPHYSICS OF DISTURBED TACTIC CELLULAR ACTIVITY

A. Virus and Chemotaxis

Invasion of a tissue by pathogenic bacteria results in the production of an inflammatory exudate which in the early phase consists of an abundant number of granulocytes. By contrast, as known from human pathology, most virus-induced tissue lesions are characterized histologically by a mononuclear inflammatory cell exudate with only few or no granulocytes.[47]

As an example, the Echo virus, type 9, strain A. Barty, used in the experiments of Bültmann et al.,[48] induced nonpurulent murine polymyositis. This enterovirus, originally isolated from a patient suffering from aseptic meningitis, was investigated in vitro experiments with granulocytes. The cell movement is disturbed by this virus, whereas other cell functions, like phagocytosis and intracellular killing of live staphylococci, remain intact. The disturbed cell movement was investigated in a Zigmond chamber where the tripeptide f-Met-Leu-Phe (10 n*M*/mm) was used as a chemotactic molecule.[9,28,49]

The objective here is to investigate the locomotion of the virus-treated cells in order to learn about the microscopic and macroscopic properties of the disturbed locomotive mechanism of the cell.

The directed locomotion of granulocytes is quantified phenomenologically by using the value of the polar-order parameter, $<P_1>$, which changes as a function of the virus treatment. The polar-order parameter of the virus-treated cells approached to zero with increasing time.[28] However, the track velocity of the virus-treated cells was unaffected by the virus. One has to conclude that only the direction-sensing mechanism and not the locomotive activity is disturbed by the virus.

One biological function of the migrating granulocytes is to drift towards an increasing concentration of chemotactic molecules. This drift process depends, as already mentioned, on the locomotive activity of the cells and on the direction-sensing mechanism of the cells. If the latter is disturbed, the migrating granulocytes have lost their ability to direct their movement towards the microorganism-invaded tissue. Occasionally, a few granulocytes will come to the target cells during their nondirected locomotion.

The membrane-bound receptors which are specific for chemotactic molecules are important elements of the direction-sensing mechanism of the locomoting cells. If the virus would bind to these receptors, the directed locomotion could be disturbed. But this mechanism is not valid since Pietrzyk[50] found that the Echo 9 virus strain A. Barty does not bind to the f-Met-Leu-Phe receptor.

The important cellular elements for the directed locomotion are the plasma membrane and the cytoplasma.[42] Both could be disturbed by the virus. However, it is unlikely that the cytoplasma is altered by the virus, since the directed locomotion can also be disturbed if noninfective Echo 9 virus is used. One has to draw the conclusion that the virus somehow alters the plasma membrane, resulting in a disturbed sensing mechanism.

It is known that the virus can change the molecular architecture and the fluidity of the membrane.[49] One has to keep in mind that the fluidity can also be changed with temperature, although the migrating cells do not lose their sensing mechanism at the altered temperature. Therefore, the fluidity is not a very significant parameter in this problem. Changes in the molecular architecture of the plasma membrane are hard to quantify, and it is even harder to explain the disturbed molecular architecture in terms of disturbed sensing mechanism. For this reason we concentrated our efforts on the phenomenology of the virus-cell interaction.

B. Virus-Induced Order-Disorder Transition

The directed locomotion of granulocytes is quantified by using the value of the polar-order parameter, $<P_1>$. However, we have to keep in mind that much information is lost by using an averaging procedure. Therefore, it is necessary to investigate the polar angular distribution function to gain new insights in the virus-cell interaction.[3,9,49]

Following virus treatment, the originally homogeneous population of moving granulocytes splits into two subpopulations. The cells of one subpopulation still exhibit the normal chemotactic response (= subpopulation of the ordered state, $<P_1>_{ORDER} = 0.8$ to 0.85) and the cells of the other subpopulation have totally lost their chemotactic response (= subpopulation of the disordered state, $<P_1>_{DISORDER} = 0$). The cells in the disordered state are in the stable state, whereas the cells of the ordered state are in a metastable state. Cells from the latter subpopulation can lose their chemotactic ability at any time. The number of cells in the subpopulation ordered state decreases with increasing time. The lifetime, τ_{live}, of the cells in the ordered state is a constant independent of time since the polar order parameter, $<P_1>$, decays exponentially with time at a constant viral dose, a_{virus}.

The cells of the subpopulation ordered state can be described in analogy to the behavior of radioactive elements. The exponential decay of the cells in the ordered state is only valid for a population of cells, not for a single cell. Therefore, it is not possible to predict the moment at which the single cell will change its state from the ordered one to the disordered one. We can only say how probable it is that a cell may change its state within the time t from the ordered to the disordered one. This is in analogy to a single radioactive element where the transition cannot be predicted either. It is only possible to say how probable such a transition is.*

The existence of the two subpopulations indicates that a cell from the ordered state and a cell from the disordered state are physically different (even if there is no difference in the chemical composition). The molecular architecture of the plasma membrane is probably different for cells belonging to different subgroups.

In order to determine the cellular mechanisms involved in the disturbed direction-sensing mechanism, it is useful to describe the observed movement mathematically because mathematics may make it possible to obtain information that is not immediately obvious from the data.

The lifetime, τ_{live}, of the exponential law describing the order-disorder transition depends on the viral dose, a_{virus}, and the relation between the lifetime and the viral dose obeys a law similar to the Weber-Fechner law in physiology.[3,9,49]

$$(\tau_{live}^0/\tau_{live})^2 = \ln(a_{virus}/a_{virus}^0) \qquad (30)$$

with $\tau_{live}^0 = 2.55$ hr, and $a_{virus}^0 = 0.8$ plaque-forming units (pfu) per cell.

There is no disturbance in the chemotactic response for sufficiently small viral doses ($a_{virus} < a_{virus}^0$). Above the threshold viral dose, a_{virus}^0, the directed locomotion is never that of the stable cellular state.

The lifetime of the ordered state as a function of the viral dose contains in an implicit way the information on the disturbed direction-sensing mechanism. This information is not

* The analogy between radioactive elements and virus-treated cells allows us to make a prediction for the following experiment: we take virus-treated cells which migrate in a polar field. Two subpopulations are determined by simultaneous observation of all migrating cells. Then all the cells in the disordered state are rapidly removed. At this moment, all the remaining cells are in the ordered state, but if we wait long enough the population of cells will split again into two subpopulations, ordered and disordered.

available on a molecular, but on a phenomenological, level. To obtain an insight into the virus-induced disturbance it is necessary to make a phenomenological model. A condition for its validity is that it conforms to Equation 30.

Equation 30 can be derived on the basis of the following simple ideas:

1. The disturbed direction-sensing mechanism can be described phenomenologically by an unknown function, g_{virus}, which can be approximated in the following way: a change in the virus concentration, a_{virus}, will induce a change in the phenomenological function, g_{virus}. A linear relation between dg_{virus} and da_{virus} is not expected since new viruses at low virus concentration have many possible ways of disturbing the direction-sensing mechanism. Therefore, a large increase in g_{virus} is expected at low viral dose. At high virus concentration new viruses have only few ways in which they can disturb the direction-sensing mechanism, because most of the disturbances have already taken place. Such behavior could be described by

$$dg_{virus} = A * \frac{da_{virus}}{a_{virus}} \qquad (31)$$

where A is a constant.

2. The disturbed direction-sensing mechanism can be characterized by the lifetime, τ_{live}, of the ordered state. The probability, w, of changing from the ordered state into the disordered one, is the inverse of the lifetime, τ_{live}. This probability, w, depends on the state of the disturbed direction-sensing system which is described by the unknown function, g_{virus}. A change in g_{virus} will induce a change in w. Again a linear relation between dg_{virus} and dw is not expected. A change in g_{virus} will induce a small change in the probability w if w is already large and will induce a large change in the probability w if w is small. Such behavior could be described as

$$dg_{virus} = B * w * dw \qquad (32)$$

where B is a constant.

Equation 30 as the final result is obtained when dg_{virus} is eliminated from Equations 31 and 32 and then integrated from a_{virus}^0 to a_{virus} and from 0 to w [$(\tau_{live}^0)^2 = B/(2A)$].

The basic assumption in this model for the disturbed direction-sensing system is an unspecific disturbance which can be interpreted as virus-induced noise in the data-processing system of the migrating cell. If there is a low level of noise in the data-processing system, the cell can recognize its environment and has therefore the capacity for tactic movements. However, if there is a high level of noise in the data processing system, the cell never recognizes its environment properly and therefore loses this capacity for tactic movements.

As mentioned above, the migrating granulocyte can also be regarded as an analog-digital converter. If the level of the virus-induced noise is small compared with the environmental signal, the output signal of the analog-digital converter is correct with respect to the environmental signal. However, if the level of the virus-induced noise is large, then the output signal of the analog-digital converter is stochastic with respect to the virus-induced noise.

The direction-sensing mechanism is only disturbed for large enough viral doses ($a_{virus} > a_{virus}^0$). The threshold viral dose, a_{virus}^0, determined from the directed locomotion of granulocytes is 0.8 pfu/cell. The infectivity of Echo 9 virus was titrated in a confluent monolayer of African-green monkey kidney cells where the virus-infected cells as plague forming units were counted. The threshold virus dose for the African-green monkey kidney cells is 1 pfu/cell. It is surprising that there are similarities between these different experiments and different types of cells. This is a strong hint that the virus interaction with the human

* This equation does not hold for $a_{virus} = 0$.

granulocyte as well as with the monkey kidney cell takes place at the same or at a similar level.

The order-disorder transition of granulocytes was investigated in vitro with the Echo 9 virus. Other viruses, e.g., influenza,[51-53] measles,[54] Herpes simplex,[55] and hepatitis virus[56] also induce disturbances of the chemotactic response of human granulocytes and/or monocytes* under in vitro and in vivo conditions. Obviously, different types of viruses can induce noise in the data-processing system of human leukocytes. The next question is, whether the order-disorder transition is restricted to leukocytes. It is known that the rhythm of in vitro beating heart cells can be disturbed by viruses. Coxsackie B virus infection can be accompanied by dysfunction of the heart — disturbance of the rhythm and a dilatation of the heart.[47] This dysfunction can be explained by virus-induced noise in the data-processing system of the heart.

ACKNOWLEDGMENTS

I am grateful for fruitful discussions with M. Bessis and A. de Boisfleury-Chevance. This work was supported by grants for the Deutsche Forschungsgemeinschaft, NATO, and Fond der chemischen Industrie.

* The locomotion of monocytes and of granulocytes is very similar. But the locomotive activity of monocytes is low and the directional memory time is large compared with those of granulocytes[32].

REFERENCES

1. **Wilkinson, P. C.,** *Chemotaxis and Inflammation,* J. & A. Churchill London, 1974, chap. 1.
2. **Gruler, H.,** Physical basis for defining terms used to describe cell movement phenomena, *Blood Cells,* 10, 123, 1984.
3. **Gruler, H. and Bültmann, B. C.,** Analysis of cell movement, *Blood Cells,* 10, 61, 1984.
4. **Benedek, G. B. and Villars, F. M. H.,** *Physics with Illustrative Examples from Medicine and Biology,* Vol. 1, Addison-Wesley Publishing, Reading, Mass., 1974, chap. 1.
5. **Verschueren, H. and Vanlarebeke, N.,** A new model for the quantitative analysis of cell movements in vitro: definition of a shape change factor, *Cytometry,* 5, 557, 1984.
6. **Nahas, G. G., Tanniers, M. L., and Lennon, J. F.,** Direct measurement of leukocyte motility: effects of pH and temperature, *Proc. Soc. Exp. Biol. Med.,* 138, 350, 1971.
7. **Gruler, H.,** Cell movement analysis in necrotactic assay, *Blood Cells,* 10, 107, 1984.
8. **Ramsey, W. S.,** Analysis of individual leucocyte behavior during chemotaxis, *Exp. Cell Res.,* 70, 129, 1972.
9. **Gruler, H. and Bültmann, B. D.,** Virus-induced order-disorder transition of moving human leukocytes, *Il Nuovo Cimento,* Suppl. 3D, 152, 1984.
10. **Zigmond, S. H.,** Ability of polymorphonuclear leukocytes to orient in gradients of chemotactic factors, *J. Cell Biol.,* 75, 606, 1977.
11. **Becker, E. L., Showell, H. J., Naccache, P. H., and Sha'afi, R.,** Enzymes in granulocyte movement: preliminary evidence for the involvement of Na^+, K^+ ATPase, in *Leukocyte Chemotaxis Methods, Physiology, and Clinical Implications,* Gallin, J. I. and Quie, P. G., Eds., Raven Press, New York, 1978, 113.
12. **Maher, J., Martell, J. V., Brantley, B. A., Cox, E. B., Niedel, J. E., and Rosse, W. F.,** The response of human neutrophils to a chemotactic tripeptide (*N*-formyl-methionyl-leucyl-phenylalanine) studied by microcinematography, *Blood,* 64, 221, 1984.
13. **Snyderman, R.,** Pharmacologic manipulation of leukocyte chemotaxis. Present knowledge and future trends, *Am. J. Med.,* 31, 10, 1983.
14. **Einstein, A.,** Über die von der molekularkinetischen Theorie der Wärme geforderete Bewegung von in ruhenden Flüssigkeiten suspendierten Teilchen, *Ann. Physik,* 17, 549, 1905.
15. **von Smoluchowski, M.,** Zur kinetischen Theorie der Brownschen Molekularbewegung und der Suspension, *Ann. Physik,* 21, 756, 1906.

16. **Peterson, S. C. and Noble, P. B.,** A two-dimensional random walk analysis of human granulocyte movement, *Biophys. J.*, 12, 1048, 1972.
17. **Alt, W.,** Biased random walk models for chemotaxis and related diffusion approximations, *Math. Biol.*, 9, 147, 1980.
18. **Hall, R. L. and Peterson, S. C.,** Trajectories of human granuloclytes, *Biophys. J.*, 25, 365, 1979.
19. **Wilkinson, P. C. and Allan, R. B.,** Assay systems for measuring leukocyte locomotion: an overview, in *Leukocyte Chemotaxis,* Gallin, J. I. and Quie, P. G., Eds., Raven Press, New York, 1978, 1.
20. **Fürth, R.,** Die Brownsche Bewegung bei Berücksichtigung einer Persistenz der Bewegungsrichtung. Mit Anwendung auf die Bewegung lebender Infusorien, *Z. Physik,* 2, 244, 1920.
21. **Reif, F.,** *Fundamentals of Statistical Physics and Thermal Physics,* McGraw-Hill, New York, 1965, chap. 15.
22. **Wilkinson, P. C. and Lackie,** Adhesion and locomotion of neutrophil leucocytes on 2-D substrate and in 3-D matrices, in *White Blood Cells Mechanics: Basic Science and Clinical Aspects,* Alan, R. Liss, New York, 1984.
23. **Haston, W. S., Shields, J. M., and Wilkinson, P. C.,** Lymphocyte locomotion and attachment on two-dimensional surfaces and in three-dimensional matrices, *J. Cell Biol.*, 92, 747, 1982.
24. **Weiss, P.,** Guiding principles in cell locomotion and cell aggregation, *Exp. Cell Res.*, Suppl. 8, 260, 1961.
25. **Matthes, T.,** Die Bewegung neutrophiler Granulozyten in einer physikalisch anisotropen Umgebung, dissertation, University of Ulm. Ulm, F. R. G., 1984.
26. **Matthes, T. and Gruler, H.,** Analysis of cell locomotion: contact guidance of human polymorphonuclear leukocytes, *Eur. Biophys. J.*, 15, 343, 1988.
27. **Jost, W. and Hauffe, K.,** *Diffusion: Methoden der Messung und Auswertung,* Steinkopf Verlag, Darmstadt, 1972, chap. 1.
28. **Bültmann, B. D. and Gruler, H.,** Analysis of the directed and nondirected movement of human granulocytes: influence of temperature and Echo 9 virus on *N*-formyl methionyl leucyl phenylalanine-induced chemokinesis and chemotaxis, *J. Cell Biol.*, 96, 1708, 1983.
29. **de Boisfleury-Chevance, A., Rapp, B., and Gruler, H.,** Analysis of cell locomotion: neutrophile, monocyte, and lymphocyte, *Blood Cells,* in press.
30. **Gruler, H. and Nuccitelli, R.,** New insights into galvanotaxis and other directed cell movements: an analysis of the translocation distribution function, in *Ionic Currents in Development,* Nuccitelli, R., Ed., Alan R. Liss, New York, 1986, 337.
31. **Nossal, R. and Zigmond, S. H.,** Chemotropism indices for polymorphonuclear leukocytes, *Biophys. J.*, 16, 1171, 1976.
32. **McCutcheon, M.,** Chemotaxis in Leukocytes, *Physiol. Rev.*, 26, 319, 1946.
33. **Rapp, B.,** Untersuchungen zur Lokomotion von Leukozyten und zur Galvanotaxis von Granulozyten, Diplomarbeit, University of Ulm, Ulm, F. R. G., 1986.

33a. **Rapp, B., de Boisfleury-Chevance, A., and Gruler, H.,** Galvanotaxis of human granulocytes. Dose-response curve, *Eur. Biophys. J.*, in press.

34. **Fukushima, K., Senda, N., Inui, H., Tamai, Y., and Murakami, Y.,** *Med. J. Osaka Univ.*, 4, 195, 1953.
35. **Dineur, E.,** Note sur la sensibilité des leucocytes a l'electricité. *Bull. Seances Soc. Belge Microscopie (Bruxelles),* 18, 113, 1891.
36. **Monguio, J.,** Über die polare Wirkung des galvanischen Stromes auf Leukozyten, *Z. Biol.*, 93, 553, 1933.
37. **Rényi, A.,** *Wahrscheinlichkeitsrechnung,* VEB Verlag, Berlin, 1966.
38. **Schlögel, F.,** Produced entropy in quantum statistics, *Z. Physik,* 249, 1, 1971.
39. **Haken, H.,** *Synergetics,* Springer Verlag, Berlin, 1983.
40. **Gruler, H. and Bültmann, B. C.,** Analysis of cell movement, *Blood Cells,* 10, 61, 1984.
41. **Gow, N. A. R. and Gruler, H.,** Directed growth of fungal hyphe in an electric field. A biophysical analysis, will be published.
42. **de Boisfluery-Chevance, A. and Gruler, H.,** Chemokinese and necrotaxis of human granulocytes: the important cellular organelles, *Z. Naturforsch.*, 42C, 1126, 1987.
43. **Gruler, H.,** Cell movement and symmetry of the cellular environment, *Z. Naturforsch.*, in press.
44. **Risken, H.,** *Fokker-Planck Equation,* Springer Verlag, Berlin, 1985, chap. 1.
45. **Gerish, G. and Keller, H.-U.,** Chemotactic reorientation of granulocytes stimulated with micropipettes containing f-Met-Leu-Phe, *J. Cell Sci.*, 52, 1, 1981.
46. **Fukushima, K., Senda, N., Ishigami, S., Murakami, Y., and Nishian, K.,** Dynamic pattern in the movement of leukocytes. II. The behaviour of neutrophil immediately after commencement and removal of stimulation, *Med. J. Osaka Univ.*, 5, 47, 1954.
47. **Budding, G. J.,** in *Viral and Rickettsial Infections of Man,* Horsefall, F. L. and Tamm, I., Eds., Lippincott, Philadelphia, 1969.

48. **Bültmann, B. D., Galle, J., Eggers, H. J., and Haferkamp, O.,** Ultrastructural degenerative changes of age-dependent Echo 9 virus-induced polymyositis in infant mice, *Virchows Arch. (Cell Pathol.)*, 35, 177, 1981.
49. **Bültmann, B. D., Haferkamp, O., Egggers, H. J., and Gruler, H.,** Echo 9 virus-induced order-disorder transition of chemotactic response of human polymorphonuclear leukocytes: phenomenology and molecular biology, *Blood Cells*, 10, 1984.
50. **Pietrzyk, C.,** Department of Pathology, University of Ulm, Ulm, F. R. G., private communication, 1984.
51. **Busse, W. W., Cooper, W., Warshauer, D. M., Dick, E. C., Wallow, I. H. L., and Albrecht, R.,** *Annu. Rev. Respir. Dis.*, 119, 561, 1979.
52. **Larson, H. E. and Blades, R.,** *Lancet*, 1, 650, 1976.
53. **Schlesinger, J. J. and Blades, R.,** *Lancet*, 1, 283, 1976.
54. **Anderson, R., Sher, R., Rabson, A. R., and Koormhof, H. F.,** *J. Pediatr.*, 89, 27, 1976.
55. **Rabson, A. R., Whiting, D. A., Anderson, R., Glover, A., and White, J.,** *J. Infect. Dis.*, 135, 133, 1977.
56. **Maliulo, B. and BenziCapelli R.,** *N. Engl. J. Med.*, 293, 303, 1975.

Chapter 4

THE MOLECULAR MECHANISM OF OXYGEN REDUCTION BY THE NEUTROPHIL OXIDASE

Andrew Richard Cross and Owen T. G. Jones

TABLE OF CONTENTS

I. Introduction 98

II. The Composition of the Superoxide Generating System of PMN 98
A. Cytochrome b 99
B. Flavoprotein 100
C. Ubiquinone 101
D. Other Requirements 101

III. Reactions of the Neutrophil Oxidase 101
A. Dye Reductase Activities 101
B. Electron Transfer Reactions within the Oxidase 102
C. Inhibition of the Oxidase 104
D. Purification of the Oxidase 104

References 106

I. INTRODUCTION

The normal function of the professional phagocytes, polymorphonuclear leukocytes (neutrophils, PMN), monocytes, macrophages, and eosinophils is to protect the individual from infection by seeking out and destroying invading microbes. These phagocytic cells possess a membrane-bound enzyme system which can be activated to produce toxic oxygen radicals in response to a wide variety of stimuli. The function of this enzyme system is to participate in the destruction of microbes, parasites, or tumors by directing the radicals against the target cells. While the beneficial effects of these oxygen radicals is well known, it has become clear that inappropriate production of oxygen radicals can be severe deleterious effects in several inflammatory diseases, including rheumatoid arthritis, Crohn's disease, and adult respiratory distress syndrome (ARDS).[1-3] In this article we shall describe the composition and properties of this oxygen radical generating system.

II. THE COMPOSITION OF THE SUPEROXIDE GENERATING SYSTEM OF PMN

Identification and characterization of the enzymic system responsible for the production of superoxide has been the subject of intense investigation in recent years. The "increased respiration of phagocytosis" (the respiratory burst) was reported by Baldridge and Gerard in 1933[4] and was thought to result from the stimulation of mitochondrial respiration necessary to provide energy for phagocytosis. The system responsible for the respiratory burst was, however, shown to be nonmitochondrial when it was found that the increased oxygen uptake was resistant to the inhibitors of mitochondrial respiration, cyanide, and antimycin A.[5] From subsequent attempts to characterize this phagocytosis-triggered oxygen consuming system a variety of oxidases were described. They were variously and sometimes conflictingly described as NADH-dependent, soluble[6,7] or membrane-bound[8] oxidases and NADPH-dependent, particulate[9,10] or NAD(P)H-dependent, particulate enzymes.[11,12] In 1968 Paul and Sbarra[13] showed clearly that hydrogen peroxide was produced by stimulated phagocytes and in 1973 Babior and co-workers,[14] established that superoxide was a major product of the oxidase. It is now generally accepted that the enzyme is membrane bound, exhibits a preference for NADPH (k_m =45 μM) over NADH (K_m = 450 μM), and converts oxygen to its one electron-reduced product, superoxide.

$$NADPH + H^+ + 2O_2 \longrightarrow NADP^+ + 2H^+ + 2O_2^-$$

The hydrogen peroxide arises from subsequent dismutation of the superoxide.

$$2O_2^- + 2H^+ \longrightarrow H_2O_2 + O_2$$

The enzyme activity is almost undetectable in resting (unstimulated) phagocytes, but increases dramatically upon stimulation. In patients with the rare genetic disorder, chronic granulomatous disease (CGD), there is a severe predisposition to chronic recurrent infection. The neutrophils from these patients phagocytose normally but the respiratory burst is absent and NADPH oxidase activity (and radical production) is undetectable, indicating that the oxidase has an important microbicidal function.

Much evidence suggests that the oxidase system is located on the plasma membrane of the PMN and becomes incorporated into the wall of the developing phagosome during phagocytosis. Briggs et al.[15] used a cytochemical method to show that superoxide production was localized on the plasma membrane following activation by a nonphagocytosable stimulus; superoxide was found inside the phagolysozome when a particulate stimulus was used. On subcellular fractionation NADPH oxidase activity was associated with the plasma membrane fraction.[16,17] Granule-depleted neutrophil cytoplasts, which are virtually devoid of subcellular organelles, are capable of mounting a respiratory burst equivalent to that of the whole cell (on the basis of membrane area),[18,19] strongly suggesting that intracellular membrane components are not involved. The plasma membrane location of the oxidase ensures that superoxide is produced in the immediate vicinity of the phagocytosed microbe.

Segal et al.[20] were the first to suggest that the oxidase was a multicomponent electron transport chain and subsequently three redox components have been proposed: a flavin adenine dinucleotide (FAD) containing flavoprotein, a b-type cytochrome, and ubiquinone.

A. Cytochrome b

A novel cytrochrome b in (equine) neutrophils was first reported by Hattori in 1961,[21] and in 1966 Shinigawa et al.[22] suggested it might be involved in the respiratory burst as it was autooxidizable and bound carbon monoxide, which are properties generally characteristic of cytochrome oxidases. This work was largely forgotten until it was discovered by Segal and co-workers that the cytochrome was absent from some[20,23] but not all[24] patients with chronic granulomatous disease. Density gradient centrifugation of neutrophil homogenates showed that the cytochrome had a dual localization in both the plasma membrane and the specific granules. The cytochrome became incorporated into the phagolysozomal membrane during phagocytosis of latex beads (which can be isolated by flotation from neutrophil homogenates).[25,26] Under anaerobic conditions the cytochrome became reduced on stimulation of the cells and was reoxidized when air was readmitted;[27] this would be expected for a component participating in the transfer of electrons from NADPH to O_2. It was not reduced in the cells of those CGD patients who did have the cytochrome,[28] suggesting that an electron transport protein functioning between NADPH and the cytochrome was absent or defective. Subsequent work has shown that the cytochrome b-negative form of the disease follows an X-linked form of inheritance and the cytochrome b-positive disease an autosomal form of inheritance.[29] In female carriers of the X-linked disease the concentration of cytochrome is variable (conforming with the Lyon hypothesis) and is directly proportional to the capacity of the cells to produce O_2^-. In vitro complementation studies where monocytes from X-linked CGD patients and autosomal recessive CGD patients were fused, with restoration of oxidase function, elegantly supported these findings.[30]

The suggested involvement of the cytochrome b in the superoxide-generating system led to further investigation of its properties. The cytochrome has an extremely low oxidation-reduction potential (-245 mV),[31] lower than any other mammalian cytochrome b and sufficiently low to reduce oxygen to superoxide ($E_{m7} = -160$ mV).[32] This unique property has proved valuable in quantifying its concentration in cells containing other cytochromes and we have designated it cytochrome b_{-245} for this reason (others have used the terminology cytochrome b_{558} or b_{559}). The cytochrome is present only in professional phagocytes: neutrophils, monocytes, macrophages and eosinophils.[33] Its synthesis is induced during the DMSO-elicited maturation of a promyelocytic cell-line (HL-60), concomitant with the ability to produce superoxide.[34] We have recently shown that cytochrome b_{-245} is also expressed in virus transformed lymphocytes which are capable of superoxide production.[35] It is absent from all other nonphagocytic cell types which have been examined and is present in a wide variety of animal phagocytes. Cytochrome b_{-245} forms a light-dissociable complex with carbon monoxide and, although its affinity for this ligand is relatively low ($K_m = 1.18$ m*M*);[36] this indicates that the haem prosthetic group is sufficiently exposed to bind oxygen. The cytochrome also binds pyridine, isocyanides, and imidazole which act as inhibitors of the oxidase.[37,38] Stopped-flow kinetic experiments show that chemically reduced cytochrome b_{-245} is reoxidized by oxygen very rapidly ($t_{1/2} = 4.7$ msec), [36,39] which suggests that the cytochrome is the terminal component in the oxidase system.

Table 1
PROPERTIES OF THE SOLUBILIZED SUPEROXIDE-GENERATING SYSTEM OF PORCINE NEUTROPHILS

Property	Assay Value
Protein content	1 to 3 $mg^{-1} \cdot m\ell^{-1}$
NADPH-dependent O_2^- formation	365 $nmol \cdot min^{-1} \cdot mg^{-1}$
K_m NADPH	45 μM
K_m NADH	450 μM
Cytochrome b_{-245} content	331 $pmol \cdot min^{-1} \cdot mg^{-1}$
FAD content	421 $pmol \cdot min^{-1} \cdot mg^{-1}$
Em_7 (cytochrome)	−238 mv
Stability	
Half-life at 4°C	32 hr
Half-life at 25°C	2.5 min[a]

[a] Loss of haem spectrum above 30°C.

and dye reduction were different; there was no inhibition of oxygen uptake in the presence of artificial electron acceptors; and the partially purified oxidase did not reduce dye acceptors even in anaerobiosis. Further evidence has come from the use of the specific inhibitor of NADPH oxidase, (diphenylene iodonium, DPI) which is thought to react at the level of the flavoprotein.[81] There is no inhibition of dye reductase activities by DPI under conditions where superoxide production is 99% abolished, which strongly suggests that dye reductase activities are not associated with the superoxide generating system itself.[82] It seems likely, therefore, that the neutrophil superoxide generating system does not possess intrinsic dye reductase activity under normal circumstances, in agreement with Bellavite et al.[80] and that diaphorase activity is a contaminant of oxidase preparations. One explanation for the contradictory observations may be the finding of Gabig and Lefker[55] that little dye reductase activity was associated with the solubilized enzyme, but that after separating the flavoprotein moiety from the cytochrome b_{-245}, significant dye reductase activity appeared, associated with the flavoprotein.

B. Electron Transfer Reactions within the Oxidase

Most of the work in identification of the electron transfer reaction and kinetics within the oxidase has been made possible by the use of a very active, optically clear solubilized enzyme from neutrophils. Plasma membranes isolated from phorbol myristate acetate-stimulated pig neutrophils extracted in 0.25% (v/v) Lubrol PX and 0.25% (v/v) sodium deoxycholate gave a preparation of oxidase with a specific activity of about 400 nmol O_2^- $min^{-1} \cdot mg$ $protein^{-1}$ containing almost equimolar amounts of FAD and cytochrome b_{-245}.[37] The properties of the solubilized enzyme are summarized in Table 1.

Addition of NADPH to an anaerobic solution of the oxidase results in the slow, incomplete reduction of both FAD and cytochrome b_{-245} giving 50% reduction after 30 min incubation.[36,37,47,54] Such slow rates of electron transfer are orders of magnitude too slow to account for the known rate of electron transfer of electrons to O_2 under aerobic conditions, and for some time this kinetic problem was the principal obstacle in the acceptance of cytochrome b_{-245} as a component of the O_2^- -generating system. However, this kinetic anomaly has been resolved by investigation of the kinetics of electron transfer within the complex in the *presence* of oxygen. This has shown that *the presence of oxygen itself is necessary for rapid transfer of electrons from NADPH into the complex.* This phenomenon is similar to the effect of oxygen on mitochondrial cytochrome oxidase (cytochrome a/a_3) where intramo-

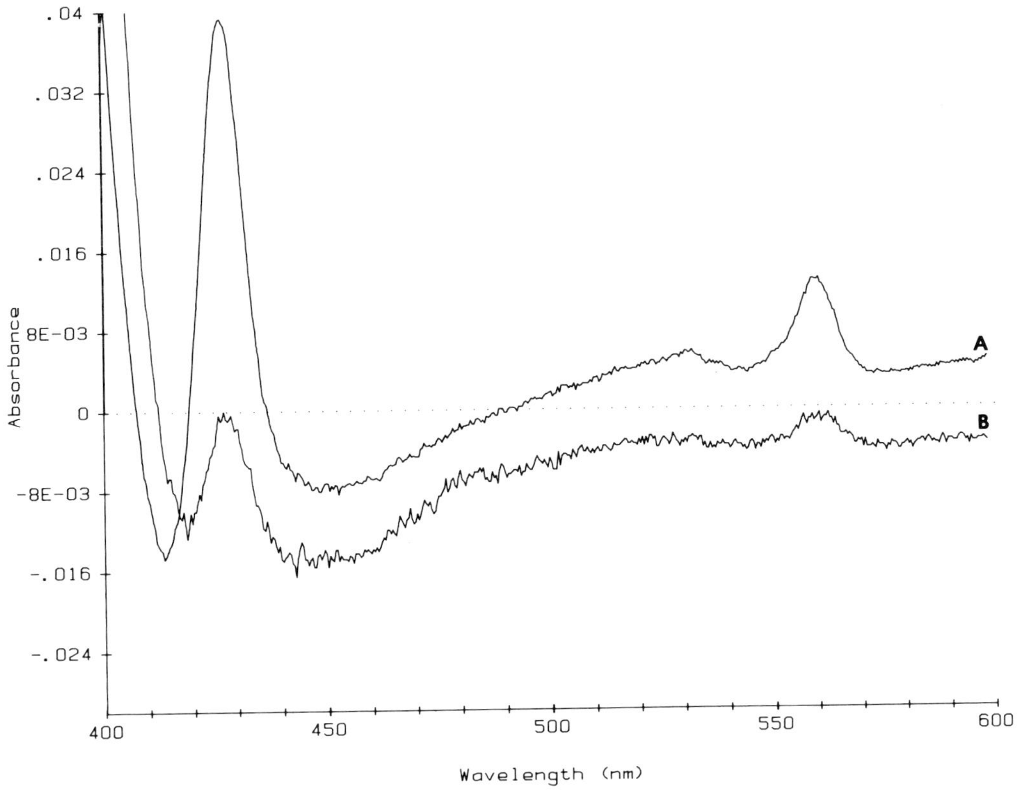

FIGURE 1. Difference absorption spectra showing the aerobic steady state reduction of cytochrome b_{-245} by NADPH in a detergent solubilized preparation of the superoxide generating oxidase of stimulated pig neutrophils. (B) After spectra of the oxidized enzyme had been recorded, 20 μℓ of 20 m*M* NADPH was added (final concentration 1 m*M*), spectra recorded within 1 min and the oxidized spectrum subtracted. The absorbance has been expanded fourfold for clarity. (A) Complete reduction of cytochrome b_{-245} by addition of a few crystals of sodium dithionite to (B). Cytochrome b_{-245} content was determined by $\Delta A_{559\text{-}540} = 21.6\ mM^{-1} \cdot cm^{-1}$.[36]

lecular electron transfer is 3 orders of magnitude slower in the absence of oxygen than in its presence.[83] Kinetic experiments determined the parameters v_1, v_2, and k_2 shown below. v_1 is the rate of steady-state electron flow from NADPH to cytochrome b_{-245}; v_2 is the steady-state rate of reduction of oxygen, and k_2 is the pseudo first-order rate constant for the reduction of oxygen by cytochrome b_{-245} (the rate constant for reduced cytochrome oxidation).

$$\text{NADPH} \xrightarrow[\text{FAD}]{v_1} \text{Cytochrome b}_{-245} \xrightarrow[k_2]{} O_2 \xrightarrow{v_2} O_2^-$$

1. The first-order rate constant for the reduction of oxygen by cytochrome b_{-245} (k_z) was determined by partly reducing the cytochrome b_{-245} in an anaerobic solution of the oxidase with sodium dithionite. On mixing with O_2 saturated buffer in a stopped-flow spectrophotometer, the reduced cytochrome b_{-245} was oxidized with a pseudo first-order rate constant (k_2) of 147 sec^{-1}.[39]

2. The rate of electron flow from NADPH into cytochrome b_{-245} (v_1) can be calculated from the steady state level of reduction of cytochrome b_{-245} in the presence of oxygen and NADPH. The extent of reduction of the cytochrome is typically around 9% (Figure 1). By

using the value for k_2 (above) we can calculate the rate of electron flow (v_1) into and out of cytochrome b by using the expression:

$$v_1 = k_2 \times \frac{[\text{reduced cytochrome}]}{[\text{total cytochrome}]} \cdot \text{mol}^{-1} \text{ cyt. b}$$

$$= 147 \text{ sec}^{-1} \times \frac{9}{100} \cdot \text{mol}^{-1} \text{ cyt. b}$$

$$= 13.23 \text{ sec}^{-1} \cdot \text{mol}^{-1} \text{ cyt. b}$$

Clearly this should equal the rate of superoxide generation (v_2) if the scheme is correct.

3. Determination of the overall rate of superoxide production (v_2) for a particular preparation of the oxidase by the superoxide dismutase inhibitable rate of cytochrome c reduction gave a measured rate of 13.03 $\text{sec}^{-1}\cdot\text{mol}^{-1}$ cyt. b, almost exactly the calculated rate of electron transfer. These findings indicate very strongly that every electron that flows from NADPH to oxygen passes through cytochrome b_{-245} and that the electron transport chain is unbranched. It has been postulated that the FAD semiquinone (formed by the one electron transfer from $FADH_2$ to cytochrome b_{-245}) might react directly with O_2 to give O_2^-. This now appears not to be the case.

In agreement with this scheme of linear electron flow, where FAD is between NADPH and cytochrome b_{-245}, FAD is relatively more reduced in the steady state than is cytochrome b_{-245}; 40% as opposed to 9% (Figure 1). The extents of reduction of both components vary with the activity of the preparation, but there is a linear relationship between the extent of cytochrome b_{-245} reduction and the rate of O_2^- formation. When NADPH is added to detergent extracts of plasma membranes from unstimulated neutrophils there is no production of O_2^- and no reduction of either cytochrome b_{-245} or FAD. This strongly suggests the site of regulation of the oxidase is at the level of (or prior to) the flavoprotein.

C. Inhibition of the Oxidase

One interesting feature of the oxidase is its susceptibility to a wide range of inhibitors. (Table 2) These can broadly be divided into two groups, those which inhibit the activation of the oxidase in whole cells (but do not inhibit the isolated enzyme) and those which exert a direct inhibitory effect. The latter group has provided insights into the function of the oxidase and can in turn be divided into classes.

1. Chelators such as EDTA which remove essential metal ions have demonstrated a Mg^{2+} requirement (common in NADPH requiring enzymes for nucleotide binding).
2. Sulfhydryl reagents such as pCMB which at very low concentrations inhibit superoxide production and produce a characteristic shift in the midpoint redox potential of the cytochrome b_{-245}[49] without affecting flavoprotein reduction.
3. Flavin analogues such as 5-deaza-FAD.
4. NADPH analogues such as ADP which are competitive inhibitors.
5. Haem liganding agents such as pyridine.
6. Iodonium compounds which probably react covalently with the flavoprotein component of the oxidase.
7. Other inhibitors whose mechanism of action is unknown.

D. Purification of the Oxidase

The NADPH oxidase has been purified by several groups of workers but with apparently different polypeptide compositions: FAD and cytochrome b_{-245} contents, ranging from high FAD, low cytochrome content;[57] low FAD, high cytochrome b_{-245};[44,56] to no detectable

Table 2
INHIBITORS OF THE SUPEROXIDE GENERATING OXIDASE

Compound	Direct(D) or indirect(I) (activation)	Possible site of action	Ref.
Chelators	I and D	(i) Removal of Mg^{2+}	71,84
NEM	D[a] and I	(ii) Sulfhydryl reagent	85-89
pCMB	D(low conc) and I	(ii) Sulfhydryl reagent (affects cyt. b_{-245})	39, 49, 85
5-deaza-FAD	D	(iii) FAD analogue	47, 73
Trifluorperazine	D (+ I?)	(iii) FAD analogue?	73, 90, 91
Quinacrine	D	(iii) FAD analogue	39, 49, 92
Phenothiazines (incl. chlorpromazine)	D (+I)	(iii) (+Indirect on membrane fluidity)	90,93
Quercetin and other flavonoids[b]	D and I	(vii)? Flavin analogues?	66,94-96
Ab to P-450 reductase	D	Cross-reaction with flavoprotein?	97
ADP	D	(iv) NADPH analogue	98
Phenylbutazone	Hexose monophosphate pathway	Reduces NADPH availability	99
Pyridine	D	(v)Reversible haem ligand	38
Imidazole	D	(v)Reversible haem ligand	38
Butyl isocyanide	D	(v) Haem ligand	39
Iodonium compounds	D	(vi) Reaction with flavoprotein	81, 82
SHAM	D	(vii) ?	94
Tropolone	D	(vii) ?	66
Lapachol	D	(vii) ?	66
B Thujaplin	D	(vii) ?	66
Juglone	D	(vii) ?	66
CTAB	D	(vii) ?	39
NDGA[b]	I ?	(vii) ?	66,100
Trithienylbutane dione	?	(vii) ?	101
Cross-linkers	I	Receptor signaling	102
Colchicine	I	?	103
Cytochalasin B	I	?	104, 105
Organotin compounds	I	?	106
Anaesthetics (Halothane)	?	?	93, 107
Adenosine	I	?	108
Opiate antagonists (Naloxone)	?	?	109
Polymyxin B	I	(pKC inhibition?)	110
cAMP	I	Initiates inactivation ?	111
pKC inhibitors	I	Prevent activation	112, 113
NSAIDS (piroxicam, aspirin, indomethacin, ibuprofen)	I	?	114-116
Lysozomotropic weak bases	I?	?	117
Doxycycline and other tetracyclines	?	?	118
Chlorhexidine	?	?	119
Antiheumatics	?	?	120
Auranofin	?	?	115
Phenidone	?	?	121
p-Bromophenacyl bromide	?	?	121
Compounds which are noninhibitory to the isolated oxidase			
Cyanide and azide		Cytochrome oxidase inhibitors	5
Rotenone		NADH dehydrogenase inhibitor	

Table 2 (continued)
INHIBITORS OF THE SUPEROXIDE GENERATING OXIDASE

Compound	Direct(D) or indirect(I) (activation)	Possible site of action	Ref.
Antimycin		bc_1 Complex	5
Myxothiazole		bc_1 Complex	98
Arsenite		Multiple sites of inhibition	
Sulphite		Inhibits flavoprotein oxidases	98
Flurbiprofen[c]		Cyclooxygenase inhibitor	98
Verapamil[c]			98
Indomethacin		Cyclooxygenase inhibitor	98, 121
Acetylsalicylate		Cyclooxygenase inhibitor	121
Pargyline		Monoamine oxidase inhibitor	98

[a] Some species differences.
[b] These compounds cause severe interference in the cytochrome c assay; they do not inhibit the isolated pig enzyme.
[c] Tested on isolated enzyme only.

cytochrome or FAD.[122] The differences in these reports is probably the result of the difficulty in accurately determining the amounts of these cofactors in very small quantities of the purified material.

REFERENCES

1. **Lunec, J., Blake, D. R., McCleary, S. J., Brailsford, S., and Bacon, P. A.,** Self-perpetuating mechanisms of immunoglobulin G aggregation in rheumatoid inflammation, *J. Clin. Invest.*, 76, 2084, 1985.
2. **Bulkley, G. B.,** The role of oxygen free radicals in human disease process, *Surgery,* 94, 407, 1983.
3. **Tate, R. M. and Repine, J. E.,** Neutrophils and adult respiratory distress syndrome (ARDS), *Am. Rev. Respir. Dis.*, 128, 552, 1983.
4. **Baldridge, C. W. and Gerard, R. W.,** The extra respiration of phagocytosis, *Am. J. Physiol.*, 103, 235, 1933.
5. **Sbarra, A. J. and Karnovsky, M. L.,** The biochemical basis of phagocytosis. I. Metabolic changes during the ingestion of particles by polymorphonuclear leukocytes, *J. Biol. Chem.*, 234, 1355, 1959.
6. **Baehner, R. I. and Karnovsky, M. L.,** Deficiency of reduced NADH oxidase in CGD, *Science,* 162, 1277, 1968.
7. **Badwey, J. A. and Karnovsky, M. L.,** Production of superoxide and hydrogen peroxide by an NADH oxidase in guinea-pig polymorphonuclear leukocytes: modulation by nucleotides and divalent cations, *J. Biol. Chem.*, 254, 11, 530, 1979.
8. **Segal, A. W. and Peters, T. J.,** Characterization of the enzyme defect in CGD, *Lancet,* 1, 1363, 1976.
9. **Patriaca, P., Cramer, R., Moncalvo, S., Rossi, F., and Romeo, D.,** Enzymic basis of metabolic stimulation of leukocytes during phagocytosis: the role of activated NADPH oxidase, *Arch. Biochem. Biophys.*, 145, 255, 1971.
10. **Hohn, D. C. and Lehrer, R. I.,** NADPH oxidase deficiency in X-linked chronic granulomatous disease, *J. Clin. Invest.*, 55, 707, 1975.
11. **Iyer, G. Y. N. and Quastel, J. H.,** NADPH and NADH oxidation by guinea-pig PMN leukocytes, *Can. J. Biochem. Physiol.*, 41, 427, 1963.
12. **Curnutte, J. T., Kipnes, R. S., and Babior, B. M.,** Defect in pyridine nucleotide dependent superoxide production by a particulate fraction from the granulocytes of patients with chronic granulomatous disease, *N. Engl. J. Med.*, 293, 628, 1975.
13. **Paul, B. and Sbarra, A. J.,** The role of the phagocyte in host-parasite interaction. XIII. The direct quantitative estimation of H_2O_2 in phagocytosing cells, *Biochim. Biophys. Acta,* 156, 168, 1968.
14. **Babior, B. M., Kipnes, R. S., and Curnutte, J. T.,** Biological defense mechanisms: the production of superoxide, a potential bactericidal agent, *J. Clin. Invest.*, 52, 741, 1973.

15. **Briggs, R. J., Drath, D. B., Karnovsky, M. L., and Karnovsky, M. J.**, Localization of NADH oxidase on the surface of human polymorphonuclear leukocytes by a new cytochemical method, *J. Cell Biol.*, 67, 566, 1975.
16. **Dewald, B., Baggiolini, M., Curnutte, J. T., and Babior, B. M.**, Subcellular localization of the superoxide forming enzyme in human neutrophils, *J. Clin. Invest.*, 61, 21, 1979.
17. **Yamaguchi, T., Sato, K., Shimada, K., and Kakinuma, K.**, Subcellular localization of O_2^- generating enzyme in guinea-pig polymorphonuclear leukocytes; fractionation of subcellular particles by using a percoll density gradient, *J. Biochem.*, 91, 31, 1982.
18. **Roos, D., Voetman, A. A., and Meerhof, L. J.**, Functional activity of enucleated human polymorphonuclear leukocytes, *J. Cell Biol.*, 97, 368, 1983.
19. **Lutter, R., van Zweiten, R., Weening, R. S., Hamers, M. N., and Roos, D.**, Cytochrome b, flavins and ubiquinone-50 in enucleated human neutrophils (polymorphonuclear leukocyte cytoplasts), *J. Biol. Chem.*, 259, 9603, 1984.
20. **Segal, A. W., Jones, O. T. G., Webster, D., and Allison, A. C.**, Absence of a newly described cytochrome b from neutrophils of patients with CGD, *Lancet*, 2, 446, 1978.
21. **Hattori, H.**, Studies on the labile, stable NADI oxidase and peroxidase staining reactions in the isolated particles of the horse granulocyte., *Nagoya J. Med. Sci.*, 23, 362, 1961.
22. **Shinigawa, Y., Tanaka, C., Teraoka, A., and Shinigawa, Y.**, A new cytochrome in neutrophilic granules of rabbit leukocytes, *J. Biochem.*, 59, 622, 1966.
23. **Segal, A. W. and Jones, O. T. G.**, Novel cytochrome b system in phagocytic vacuoles of human phagocytes, *Nature (London)*, 276, 515, 1978.
24. **Borregaard, N., Johansen, K. S., Taudorff, E., and Wandall, J. H.**, Cytochrome b is present in neutrophils from patients with CGD, *Lancet*, 1, 949, 1979.
25. **Segal, A. W. and Jones, O. T. G.**, The subcellular distribution and some properties of the cytochrome b component of the microbicidal oxidase system of human neutrophils. *Biochem. J.*, 182, 181, 1979.
26. **Segal A. W. and Jones, O. T. G.**, Rapid incorporation of the human neutrophil plasma membrane cytochrome b into phagocytic vacuoles, *Biochem. Biophys. Res. Commun.*, 92, 710, 1980.
27. **Segal A. W. and Jones, O. T. G.**, Reduction and subsequent oxidation of a cytochrome b of human neutrophils after stimulation with phorbol myristate acetate, *Biochem. Biophys. Res. Commun.*, 88, 130, 1979.
28. **Segal A. W. and Jones, O. T. G.**, Absence of cytochrome b reduction in stimulated neutrophils from both female and male patients with CGD, *FEBS Lett.*, 110, 111, 1980.
29. **Segal, A. W., Cross, A. R., Garcia, R., Borregaard, N., Valerius, N. H., Soothill, J. F., and Jones, O. T. G.**, Absence of cytochrome b $_{-245}$ in CGD: A multicentre European evaluation of its incidence and relevance, *N. Engl. J. Med.*, 308, 245, 1983.
30. **Hamers, M. N., de Boer, M., Meerhof, L. J., Weening, R. S., and Roos, D.**, Complementation in monocyte hybrids reveals genetic heterogeneity in CGD, *Nature (London)*, 307, 553, 1984.
31. **Cross, A. R., Jones, O. T. G., Harper, A. M., and Segal, A. W.**, Oxidation reduction properties of the cytochrome b found in the plasma membrane fraction of human neutrophils: a possible oxidase in the respiratory burst, *Biochem. J.*, 194, 599, 1981.
32. **Wood, P. M.**, The redox potential of the system oxygen-superoxide, *FEBS Lett.*, 44, 22, 1974.
33. **Segal, A. W., Garcia, R., Goldstone, A. H., Cross, A. R., and Jones, O. T. G.**, Cytochrome b $_{-245}$ of neutrophils is also present in human monocytes, macrophages and eosinophils, *Biochem. J.*, 196, 363, 1981.
34. **Roberts, P. J., Cross, A. R., Jones, O. T. G., and Segal, A. W.**, Development of cytochrome b $_{-245}$ and an active oxidase system in association with maturation of a human promyelocytic (HL-60) cell line, *J. Cell Biol.*, 95, 720, 1982.
35. **Maly, F.-E., Cross, A. R., Jones, O. T. G., Wolf-Vorbeck, G., Walker, C., Dahinden, C. A., and De Weck, A. L.**, Biochemical and functional properties of the superoxide generating system of B cell lines: structural homology with the phagocytic oxidase and triggering via surface immunoglobulin, *J. Immunol.*, 140, 2334, 1988.
36. **Cross, A. R., Higson, F. K., Jones, O. T. G., Harper, A. M., and Segal, A. W.**, The enzymic reduction and kinetics of oxidation of the cytochrome b $_{-245}$ of neutrophils, *Biochem. J.*, 204, 479, 1982.
37. **Cross, A. R., Parkinson, J. F., and Jones, O. T. G.**, The superoxide generating oxidase of leukocytes: NADPH-dependent reduction of flavin and cytochrome b in solubilized preparations, *Biochem. J.*, 223, 337, 1984.
38. **Iizuka, T., Kanegasaki, S., Makino, R., Tanaka, T., and Ishimura, Y.**, Pyridine and imidazole reversibly inhibit the respiratory burst in porcine and human neutrophils: evidence for the involvement of cytochrome b_{558} in the reaction, *Biochem, Biophys. Res. Commun.*, 130, 621, 1985.
39. **Cross, A. R., Parkinson, J. F., and Jones, O. T. G.**, Mechanism of the superoxide-producing oxidase of neutrophils: O_2 is necessary for the fast reduction of cytochrome b $_{-245}$ by NADPH, *Biochem. J.*, 226, 881, 1985.

40. **Harper, A. M., Dunne, M. J., and Segal, A. W.,** Purification of cytochrome b $_{-245}$ from human neutrophils, *Biochem. J.*, 219, 519, 1984.
41. **Lutter, R., van Schaik, M., van Zweiten, R., Wever, R., Roos, D., and Hamers, M. N.,** Purification and partial characterization of the b-type cytochrome from human polymorphonuclear leukocytes, *J. Biol. Chem.*, 260, 2237, 1985.
42. **Pember, S. O., Heyl, B. L., Kincade, J. M., and Lambeth, J. D.,** Cytochrome b_{558} from (bovine) granulocytes: partial purification from Triton X-114 extracts and properties of the isolated enzyme, *J. Biol. Chem.*, 259, 10, 590, 1984.
43. **Serra, M. C., Bellavite, P., Davoli, A., Bannister, J. V., and Rossi, F.,** Isolation from neutrophil membranes of a complex containing active NADPH oxidase and cytochrome b $_{-245}$, *Biochim. Biophys. Acta*, 788, 138, 1984.
44. **Bellavite, P., Papini, E., Zeni, L., Della Bianca, V., and Rossi, F.,** Studies on the nature and activation of the O_2^- -forming NADPH-oxidase of leukocytes. Identification of a phosphorylated component of the active enzyme, *Free Radical Res. Commun.*, 1, 11, 1985.
45. **Babior, B. M. and Kipnes, R. S.,** Superoxide-forming enzyme from human neutrophils: Evidence for a flavin requirement, *Blood*, 50, 517, 1977.
46. **Babior, B. M. and Peters, W.,** The superoxide producing enzyme of human neutrophils: Further properties, *J. Biol. Chem.*, 256, 2321, 1981.
47. **Light, D. R., Walsh, C., O'Callaghan, A. M., Goetzl, E. J., and Tauber, A. I.,** Characteristics of the cofactor requirements of the superoxide generating NADPH oxidase of human polymorphonuclear leukocytes, *Biochemistry*, 20, 1468, 1981.
48. **Wakeyama, H., Takeshige, K., and Minikami, S.,** NADPH-dependent reduction of 2,6,dichlorophenol indophenol by the phagocytic vesicles of pig polymorphonuclear leucocytes, *Biochem. J.*, 210, 577, 1983.
49. **Bellavite, P., Cross, A. R., Serra, M. C., Davoli, A., Jones, O. T. G., and Rossi, F.,** The cytochrome b and flavin content and properties of the superoxide forming NADPH oxidase solubilized from activated neutrophils, *Biochim. Biophys. Acta*, 746, 40, 1983.
50. **Cross, A, R., Jones, O. T. G., Garcia, R., and Segal, A. W.,** The association of FAD with the cytochrome b_{-245} of human neutrophils, *Biochem. J.*, 208, 759, 1982.
51. **Gabig, T. G.,** The NADPH-dependent O_2^--generating oxidase from human neutrophils: identification of a flavoprotein component that is deficient in a patient with CGD, *J. Biol. Chem.*, 258, 6352, 1983.
52. **Gabig, T. G. and Lefker, B. A.,** Deficient flavoprotein component of the NADPH-dependent O_2^--generating oxidase in the neutrophils from three male patients with chronic granulomatous disease, *J. Clin. Invest.*, 73, 701, 1984.
53. **Ohno, Y., Buescher, E. S., Roberts, R., Metcalf, J. A., and Gallin, J. I.,** Re- evaluation of cytochrome b and flavin adenine dinucleotide in neutrophils from patients with chronic granulomatous disease with probable autosomal recessive inheritance of cytochrome b deficiency, *Blood*, 67, 1132, 1986.
54. **Morel, F., Doussiere, J., Stasia, M-J., and Vignais, P. V.,** The respiratory burst of bovine neutrophils: role of a b-type cytochrome and coenzyme specificity, *Eur. J. Biochem.*, 152, 699, 1985.
55. **Gabig, T. G. and Lefker, B. A.,** Catalytic properties of the resolved flavoprotein and cytochrome b components of the NADPH-dependent O_2^--generating oxidase from human neutrophils, *Biochem. Biophys. Res. Commun.*, 118, 430, 1984.
56. **Bellavite, P., Jones, O. T. G., Cross, A. R., Papini, E., and Rossi, F.,** Composition of partially purified NADPH oxidase from pig neutrophils, *Biochem. J.*, 223, 639, 1984.
57. **Glass, G. A., DeLisle, D. M., DeTogni, P., Gabig, T. G., Magee, B. H., Markert, M., and Babior, B. M.,** The respiratory burst oxidase of human neutrophils: further studies of the purified enzyme, *J. Biol. Chem.*, 261, 13247, 1986.
58. **Wakeyama, H., Takeshige, K., Takanagi, R., and Minikami, S.,** Superoxide-forming NADPH-oxidase preparation of porcine polymorphonuclear leukocytes, *Biochem. J.*, 205, 593, 1982.
59. **Kakinuma, K., Kaneda, M., Chiba, T., and Ohnishi, T.,** Electron spin resonance studies on a flavoprotein in neutrophil plasma membranes: redox potentials of the flavin and its participation in NADPH oxidase, *J. Biol. Chem.*, 261, 9426, 1986.
60. **Massey, V. and Hemmerich, P.,** Active site probes of flavoproteins, *Biochem. Soc. Trans.*, 8, 246, 1980.
61. **Rossi, F., Bellavite, P., and Papini, E.,** Respiratory response of phagocytes: terminal NADPH oxidase and the mechanisms of its action, in *Biochemistry of Macrophages*, Pitman, London, 1986, 172.
62. **Peterson, J. A., Ebel, R. E., O'Keeffe, P. H., Matsubara, T., and Esterbrook, R. W.,** Temperature dependence of cytochrome P-450 reduction. A model for NADPH-cytochrome P-450 reductase: cytochrome P-450 interaction, *J. Biol. Chem.*, 251, 4010, 1976.
63. **Millard, J. A., Gerard, K. W., and Schneider, D. L.,** The isolation from rat peritoneal leukocytes of plasma membrane enriched in alkaline phosphatase and a b-type cytochrome, *Biochem. Biophys. Res. Commun.*, 90, 312, 1979.
64. **Crawford, D. R. and Schneider, D. L.,** Evidence that a quinone may be required for the production of superoxide and hydrogen peroxide in neutrophils, *Biochem. Biophys. Res. Commun.*, 99, 1277, 1981.

65. **Sloan, E. P., Crawford, D. R., and Schneider, D. L.**, Isolation of plasma-membrane from human neutrophils and determination of cytochrome b and quinone content, *J. Exp. Med.*, 153, 1316, 1981.
66. **Crawford, D. R. and Schneider, D. L.**, Identification of UQ_{50} in human neutrophils and its role in microbicidal events, *J. Biol. Chem.*, 257, 6662, 1982.
67. **Cunningham, C. C., DeChatelet, L. R., Spach, P. I., Parce, J. W., Thomas, M. J., Lees, C. J., and Shirley, P. S.**, Identification and quantification of electron transport components in human polymorphonuclear leukocytes, *Biochim. Biophys. Acta*, 682, 430, 1982.
68. **Cross, A. R., Jones, O. T. G., Garcia, R., and Segal, A. W.**, The subcellular localization of ubiquinone in human neutrophils, *Biochem. J.*, 216, 765, 1983.
69. **Szarkowska, C. and Klingenberg, M.**, On the role of ubiquinone in mitochondria: spectrophotometric and chemical measurements of its redox reactions, *Biochem. Z.*, 338, 674, 1963.
70. **Gabig, T. G. and Babior, B. M.**, The O_2^--forming oxidase responsible for the respiratory burst in human neutrophils: properties of the solubilized enzyme, *J. Biol. Chem.*, 254, 9070, 1979.
71. **Green, T. R., Wu, D. E., and Wirtz, M. K.**, The O_2^--generating oxidoreductase of human neutrophils: evidence for an obligatory requirement for Ca^{2+} and Mg^{2+} for catalytic activity, *Biochem. Biophys. Res. Commun.*, 110, 973, 1983.
72. **Suzuki, H., Pabst, M. J., and Johnston, R. B. Jr.**, Enhancement by Ca^{2+} or Mg^{2+} of catalytic activity of the superoxide-producing NADPH oxidase in membrane fractions of human neutrophils and monocytes, *J. Biol. Chem.*, 260, 3635, 1985.
73. **Parkinson, J. F.**, NADPH-oxidase. Characterisation of the Solubilised Enzyme from Pig Neutrophils, Ph.D. thesis, University of Bristol, Bristol, U.K., 1985.
74. **Green, T. R. and Schaefer, R. E.**, Intrinsic dichlorophenol indophenol reductase activity associated with superoxide generating oxidoreductase of human granulocytes, *Biochemistry*, 20, 7483, 1981.
75. **Green, T. R. and Wu, D. E.**, Detection of NADPH diaphorase activity associated with human neutrophil NADPH-O_2 oxidoreductase activity, *FEBS Lett.*, 179, 82, 1985.
76. **Sakane, F., Takahashi, K., and Koyama, J.**, Separation of three menadione-dependent O_2^--generating pyridine nucleotide oxidizing enzymes in guinea-pig polymorphonuclear phagocytes, *J. Biochem.*, 94, 931, 1983.
77. **Nisimoto, Y., Wilson, E., Heyl, B. L., and Lambeth, J. D.**, NADH dehydrogenase from bovine neutrophil membranes: purification and properties, *J. Biol. Chem.*, 261, 285, 1986.
78. **Tauber, A. I., Wright, J., Higson, F. K., Edelman, S. A., and Waxman, D. J.**, Purification and characterization of human neutrophil NADH-cytochrome B_5 reductase, *Blood*, 66, 673, 1985.
79. **Mackler, B., Person, R., Davis, K. A., and Ochs, H.**, Studies of pyridine nucleotide oxidizing enzymes from human neutrophils, *Biochem. Int.*, 11, 319, 1985.
80. **Bellavite, P., Della Bianca, V., Serra, M. C., Papini, E., and Rossi, F.**, NADPH oxidase of neutrophils forms superoxide anion but does not reduce cytochrome c and dichlorophenol indophenol, *FEBS Lett.*, 170, 157, 1984.
81. **Cross, A. R. and Jones, O. T. G.**, The effect of the inhibitor diphenylene iodonium on the superoxide generating system of neutrophils. Specific labelling of a component polypeptide of the oxidase, *Biochem. J.*, 237, 111, 1986.
82. **Cross, A. R.**, The inhibitory effects of some iodonium compounds on the superoxide generating system of neutrophils and their failure to inhibit diaphorase activity, *Biochem. Pharmacol.*, 36, 489, 1987.
83. **Antonini, E., Brunori, M., Colosimo, A., Greenwood, C., and Wilson, M. T.**, Oxygen 'pulsed' cytochrome c oxidase: Functional properties and catalytic relevance, *Proc. Natl. Acad. Sci. U.S.A.*, 74, 3128, 1977.
84. **Parkinson, J. F. and Cross, A. R.**, unpublished.
85. **Yamashita, T.**, Effect of maleimide derivatives, sulfhydryl raegents, on stimulation of neutrophil superoxide anion generation with Con A, *FEBS Lett.*, 164, 267, 1983.
86. **Yamashita, T., Someya, A., and Tsuuzawa-Kido, Y.**, Response of superoxide anion production by guinea-pig eosinophils to various soluble stimuli: comparison to neutrophils, *Eur. J. Biochem.*, 145, 71, 1985.
87. **Higson, F. K.**, The Superoxide-Generating System of Animal Blood Neutrophils, Ph.D. thesis, University of Bristol, Bristol, U.K., 1983.
88. **Newberger, P. E., Chovaniec, M. E., and Cohen, H. J.**, Activity and activation of the granulocyte superoxide-generating system, *Blood*, 55, 85, 1980.
89. **Zimmerli, W., Lew, P. D., Cohen, H. J., and Waldvogel, F. A.**, Comparative superoxide-generating system of granulocytes from blood and peritoneal exudates, *Infect. Immun.*, 46, 625, 1984.
90. **Ochs, P. L. and Reed, P. W.**, Inhibition of the neutrophil oxidative burst and degranulation by phenothiazines, *Biochem. Biophys. Res. Commun.*, 102, 958, 1981.
91. **Bellavite, P., Serra, M. C., Davoli, A., Bannister, J. V., and Rossi, F.**, The NADPH oxidase of guinea-pig polymorphonuclear leucocytes: properties of the deoxycholate extracted enzyme, *Mol. Cell. Biochem.*, 52, 17, 1983.

92. **Tauber, A. I. and Simons, E. R.**, Dissociation of human neutrophil membrane depolarization, respiratory burst stimulation and phospholipid metabolism by qinacrine, *FEBS Lett.*, 156, 161, 1985.
93. **Cohen, H. J., Chovaniec, M. E., and Ellis, S. E.**, Chlorpromazine inhibition of granulocyte superoxide production, *Blood,* 56, 23, 1980.
94. **Crawford, D. R. and Schneider, D. L.**, Ubiquinone content and respiratory burst activity of latex filled phagolysozomes isolated from human neutrophils and evidence for the probable involvement of a third granule, *J. Biol. Chem.*, 258, 5363, 1983.
95. **Long, G. D., De Chatelet, L., O'Flaherty, J. T., McCall, C. E., Bass, D. A., Shirley, P. S., and Parce, J. W.**, Effects of quercetin on magnesium-dependent adenosine triphosphate and the metabolism of human polymorphonuclear leukocytes, *Blood.*, 57, 561, 1981.
96. **Pagonis, C., Tauber, A. I., Pavlotsky, N., and Simons, E. R.**, Flavonoid impairment of neutrophil response, *Biochem. Pharmacol.*, 35, 237, 1986.
97. **Takayama, H., Iwaki, S., Tamoto, K., and Koyama, J.**, Inhibition of the O_2^--generating NADPH oxidase of guinea-pig polymorphonuclear leukocytes by rabbit antibody to homologous liver NADPH-cytochrome c (P-450) reductase, *Biochim. Biophys. Acta,* 799, 151, 1984.
98. **Cross, A. R. and Jones, O. T. G.**, unpublished, 1983.
99. **Strauss, R. R., Paul, B., and Sbarra, A. J.**, Effect of phenylbutazone on phagocytosis and intracellular killing by guinea-pig polymorphonuclear phagocytes, *J. Bact.*, 96, 1982, 1968.
100. **Rossi, F., Della Bianca, V., and Bellavite, P.**, Inhibition of the respiratory burst and of phagocytosis by nordihydroguariaretic acid in neutrophils, *FEBS Lett.*, 127, 183, 1981.
101. **Beswick, P. H. and Slater, T. F.**, The 'metabolic burst' in bovine blood polymorphonuclear leucocytes when stimulated by phagocytosis, *Biochem. Soc. Trans.*, 5, 1299, 1977.
102. **Aviram, I., Simons, E. R., and Babior, B. M.**, Reversible blockade of the respiratory burst in human neutrophils by a cleavable cross-linking agent, *J. Biol. Chem.*, 259, 306, 1984.
103. **Malawista, S. E. and Bodel, P. T.**, Dissociation by colchicine of phagocytosis from increased oxygen consumption in human leukocytes, *J. Clin. Invest.*, 46, 786, 1967.
104. **Malawista, S. E., Gee, J. B. L., and Bensch, K. G.**, Cytochalasin B reversibly inhibits phagocytosis: functional, metabolic and ultrastructural effects in human blood leukocytes and rabbit alveolar macrophages, *Yale J. Biol. Med.*, 44, 286, 1971.
105. **Goldstein, I. M., Roos, D., Kaplan, H. B., and Weissmann, G.**, Complement and immunoglobulins stimulate superoxide production by human leukocytes independently of phagocytosis, *J. Clin. Invest.*, 56, 1155, 1975.
106. **Matsui, H., Wada, O., Ushijima, Y., and Akuzawa, T.**, Triphenyltin chloride inhibits superoxide production by human neutrophils stimulated with a surface active agent, *FEBS Lett.*, 164, 251, 1983.
107. **Welch, W. D.**, Halothane reversibly inhibits human neutrophil bacterial killing, *Anaethesiology,* 55, 650, 1981.
108. **Cronstein, B. N., Kramer, S. B., Weissmann, G., and Hirschorn, R.**, Adenosine: A physiological modulator of O_2^- generation by human neutrophils, *J. Exp. Med.*, 158, 1160, 1983.
109. **Simkins, C. O., Ives, N., Tate, E., and Johnson, M.**, Naloxone inhibits O_2^- release from human neutrophils., *Life Sci.*, 37, 1381, 1985.
110. **Naccache, P. H., Molski, M. M., and Sha'afi, R. I.**, Polymyxin B inhibits phorbol 12-myristate 13-acetate, but not chemotactic factor, induced effects in rabbit neutrophils, *FEBS Lett.*, 193, 227, 1985.
111. **Nakagawara, A. and Minikami, S.**, Generation of superoxide anions by leukocytes treated with cytochalasin E, *Biochem. Biophys. Res. Commun.*, 64, 760, 1975.
112. **Gerard, C., McPhail, L. C., Marfat, A., Stimler-Gerard, N. D., Bass, D. A., and McCall, C. E.**, Role of protein kinases in stimulation of human polymorphonuclear leukocyte oxidative metabolism by various agonists: differential effects of a novel protein kinase inhibitor, *J. Clin. Invest.*, 77, 61, 1986.
113. **Cooke, E. and Hallett, M. B.**, The role of C-kinase in the physiological activation of the neutrophil oxidase: evidence from using pharmacological manipulation of C-kinase activity in whole cells, *Biochem. J.*, 232, 323, 1985.
114. **Biemond, P., Swaak, A. J. G., Penders, J. M. A., Beindorf, C. M., and Koster, J. F.**, Superoxide production by polymorphonuclear leukocytes in rheumatoid arthritis and osteoarthritis: in vivo inhibition by the antirheumatic drug piroxicam due to interference with the activation of the NADPH-oxidase, *Ann. Rheum. Dis.*, 45, 249, 1986.
115. **Hurst, N. P., French, J. K., Bell, A. L., Nuki, G., O'Donnell, M. L., Betts, W. H., and Cleland, L. G.**, Differential effects of mepacrine, chloroquine and hydroxychloroquine on superoxide anion generation, phospholipid methylation and arachidonic acid release by human blood monocytes, *Biochem. Pharmacol.*, 35, 3083, 1986.
116. **Kaplan, H. B., Edelson, H. S., Korchak, H. M., Given, W. P., Abramson, S., and Weissmann, G.**, Effects of non-steroidal anti-inflammatory agents on human neutrophil functions *in vitro* and *in vivo, Biochem. Pharmacol.*, 33, 371, 1984.

117. **Styrt, B. and Klempner, M. S.,** Inhibition of neutrophil oxidative metabolism by lysosomotropic weak bases, *Blood,* 67, 334, 1986.
118. **Forsgren, A., Schmeling, D., and Quie, P. G.,** Effect of tetracycline on the phagocytic function of human leukocytes, *J. Infect. Dis.,* 130, 412, 1974.
119. **Goultschin, J. and Levy, H.,** Inhibition of superoxide generation by human polymorphonuclear leukocytes with chlorhexidine: its possible relationship to periodontal disease, *J. Periodontol.,* 57, 422, 1986.
120. **Minta, J. O. and Williams, M. D.,** Interactions of anti-rheumatic drugs with the superoxide generating system of activated human polymorphonuclear leukocytes, *J. Rheumatol.,* 13, 498, 1986.
121. **Tsunawaki, S. and Nathan, C. F.,** Release of arachidonate and reduction of oxygen: independent metabolic bursts of the mouse peritoneal macrophage, *J. Biol. Chem.,* 261, 11563, 1986.
122. **Doussiere, J. and Vignais, P. V.,** Purification and properties of an O_2^- generating oxidase from bovine polymorphonuclear neutrophils, *Biochemistry,* 24, 7231, 1985.

Chapter 5

MOBILIZATION, METABOLISM, AND BIOLOGICAL EFFECTS OF EICOSANOIDS IN POLYMORPHONUCLEAR LEUKOCYTES

Paul H. Naccache, Ramadan I. Sha'afi, and Pierre Borgeat*

TABLE OF CONTENTS

I. Introduction 114

II. Stimulated Arachidonic Acid Release in Neutrophils 114

III. Arachidonic Acid Metabolism in Neutrophils 116
- A. The 5-Lipoxygenase Pathway 116
- B. The 15-Lipoxygenase Pathway 119
- C. The 12-Lipoxygenase Pathway 119
- D. Distinctive Lipoxygenase Product Profiles in Human Granulocyte Subsets and Cell Lines 119
- E. The double Dioxygenation of Arachidonic Acid in the Neutrophil 120
- F. Modulation of Leukotriene Synthesis 120
- G. The Metabolism of Lipoxygenase Products in the Neutrophil 121
- H. The Enzymes of the Lipoxygenase Pathways 123

IV. Biological Effects and Transduction Mechanisms 124
- A. Functional Responses to Arachidonic Acid 124
- B. Functional Responses to Leukotriene B_4 and other Lipoxygenase Metabolites 125
  - 1. Locomotion and Chemotaxis 125
  - 2. Adherence to Artificial Surfaces and to Endothelium 125
  - 3. Aggregation 125
  - 4. Degranulation 126
  - 5. Oxidative Metabolism 126
  - 6. Expression of Surface Antigens and Receptors 126
- C. Effect of Dietary Ω3-Fatty Acids on Neutrophil Function 126
- D. Signal Transduction 126
  - 1. Receptors 127
    - a. Characteristics 127
    - b. Localization 127
    - c. Association with the Cytoskeleton 127
  - 2. Guanine Nucleotide-Binding Proteins 128
  - 3. Does Leukotriene Activate Phospholipase C? 128
  - 4. Calcium Mobilization 128
  - 5. Cytoskeletal Response 129
  - 6. The Na+/H+ Antiport 129
  - 7. The Role of Protein Kinase C 130
  - 8. Cyclic AMP 130

References 130

* Supported in part by MRC (Canada) and NCI (Canada) grants (PHN, PB) and by NIH grant #GM-37694 (RIS).

concentration of free calcium. This rise in calcium can come from a release of calcium by IP_3 and/or a net uptake from the outside medium. Furthermore, this increase is necessary, and provided it is high enough, is sufficient for PLA_2 activation. This calcium-activation of PLA_2 can be modulated by protein kinase C activation and/or the perturbation of the environment in which the PLA_2 is located.

III. ARACHIDONIC ACID METABOLISM IN NEUTROPHILS

The metabolism of arachidonic acid in neutrophils has been the subject of extensive investigations in the last 10 years. The structure elucidation of the major metabolite of arachidonic acid in glycogen-elicited rabbit peritoneal neutrophils as the 5*S*-hydroxy-6,8,11,14(*E*,*Z*,*Z*,*Z*)-eicosatetraenoic acid first revealed the presence of the 5-lipoxygenase pathway in these cells.[22]

It is also in the course of these studies on rabbit peritoneal neutrophils that the leukotrienes, a group of arachidonic acid metabolites derived from the 5-lipoxygenase pathway, were discovered.[1,23,24] The trivial name "leukotriene" was introduced because the substances were discovered in leukocytes and because these compounds are characterized by the presence of a conjugated triene unit (see Figure 2) which account for the typical UV spectrum of these substances.[25] Some leukotrienes were then identified as the bioactive components of the "slow reacting substance of Anaphylaxis" (SRS-A).[26] This finding and the observations of the potent inflammatory and myotropic activities of the leukotrienes have stimulated a great deal of interest for these substances in many research fields related to allergy and inflammation.[27]

Currently, it is well accepted that lipoxygenases are the dominant metabolic pathways of arachidonic acid in neutrophils. The formation of cyclooxygenase products has also been reported in neutrophils; for instance, neutrophils were shown to produce PGE_2 and TXB_2.[28] However, the most recent observations indicate that the amount of lipoxygenase products formed by human and rabbit neutrophils largely exceed the amount of cyclooxygenase products.[211] For these reasons, the present discussion will focus on the lipoxygenase pathways of the neutrophils.

A. The 5-Lipoxygenase Pathway

As mentioned above, arachidonic acid is present mainly in ester forms in the cells. Its metabolism through the various dioxygenases first involves the action of various lipases which release the fatty acid in the cellular milieu.

The 5-lipoxygenase catalyses the specific dioxygenation of arachidonic acid at position C-5 and initiates the synthesis of leukotrienes (Figure 2). Arachidonic acid is first transformed into (5*S*)-5-hydroperoxy-(*E*,*Z*,*Z*,*Z*)-6,8,11,14-eicosatetraenoic acid (5-HPETE). Intact neutrophils rapidly metabolize 5-HPETE into (5*S*)-5-hydroxy-(*E*,*Z*,*Z*,*Z*)-6,8,11,14-eicosatetraenoic acid (5-HETE) or (5*S*,6S)- 5(6)-oxido-(*E*,*E*,*Z*,*Z*)-7,9,11,14-eicosatetraenoic acid (leukotriene A_4).[1,22-24] 5-HETE is likely the product of the action of a peroxidase on 5-HPETE. The enzyme responsible for this particular reaction has not been characterized.

The formation of leukotriene A_4 from 5-HPETE is enzymatic and involves the stereospecific removal of the 10D(R)hydrogen atom (of 5-HPETE) and loss of water.[29] The biosynthesis of leukotriene A_4 in rabbit peritoneal neutrophils incubated with arachidonic acid is maximal within 90 sec and is followed by rapid enzymatic and nonenzymatic conversions into various more polar compounds.[24] As expected for an allylic epoxide, leukotriene A_4 is highly unstable and undergoes facile nucleophilic substitution (hydrolysis products are described below); in aqueous buffer at pH 7.4 and 25°C, the time for 50% decomposition is less than 10 sec, and leukotriene A_4 instantaneously hydrolyzes at acidic pH; leukotriene A_4 is stabilized under alkaline conditions and upon binding to serum albumin.[30] The formation

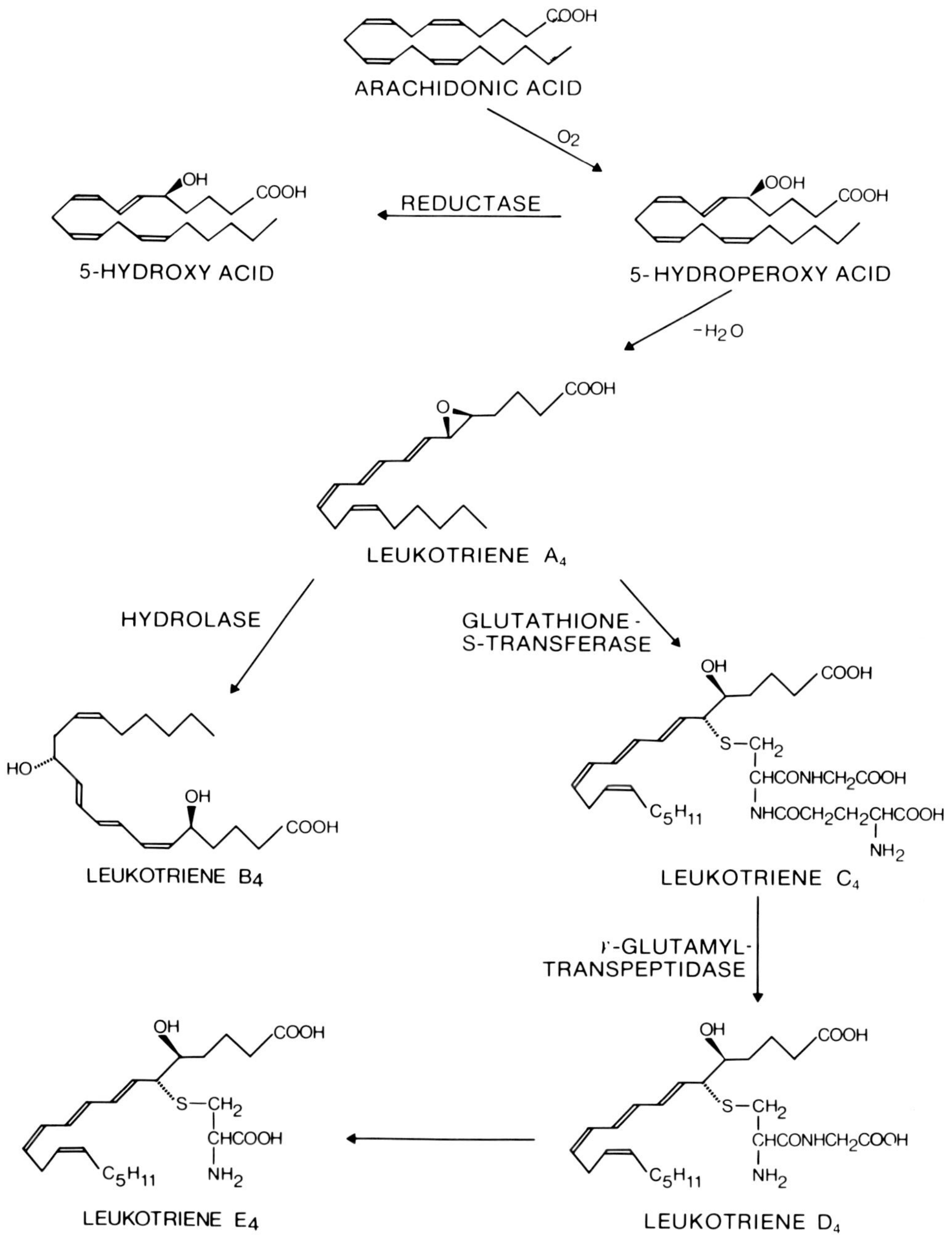

FIGURE 2. Scheme of the transformation of arachidonic acid through the 5-lipoxygenase pathway. Products of the nonenzymatic hydrolysis of leukotriene A_4, as well as metabolites of leukotrienes are not shown (see Reference 104).

of leukotriene A_4 in biological systems can be demonstrated by trapping with methanol and measurement of the stable methanolysis products epimers at C-12,(5*S*,12*SR*)-5-hydroxy-12-*O*-methyl-(*E*,*E*,*E*,*Z*)-6,8,10,14-eicosatetraenoic acids.[24]

The (5*S*,12*R*)-5,12-dihydroxy-(*E*,*E*,*E*,*Z*)-6,8,10,14-eicosatetraenoic acid (6-*trans*-leukotriene B_4) and (5*S*,12*S*)-5,12-dihydroxy- (*E*,*E*,*E*,*Z*)-6,8,10,14-eicosatetraenoic acid (6-*trans*-12-epi-leukotriene B_4) are the products of the nonenzymatic hydrolysis of leukotriene A_4, which occurs spontaneously in aqueous medium. These compounds are considered by-products in leukotriene A_4 biosynthesis.[23,24]

Hydrolysis (nonenzymatic) of leukotriene A_4 also results in the formation of two isomeric 5,6-dihydroxy-7,9,11,14-eicosatetraenoic acids (5,6-diHETEs). The mechanism of formation of these dihydroxy acids, is in part similar to that of the 5,12-dihydroxyacids with reaction at position C-6 of the carbonium ion intermediate;[23,24] thus, these compounds also appear to be by-products in leukotriene A_4 biosynthesis.

The products of the nonenzymatic hydrolysis of leukotriene A_4 show little biological activities;[31] their presence in biological material reflects 5-lipoxygenase activity and leukotriene A_4 synthesis.

It is noteworthy that the structure identification of these four isomeric dihydroxy acids in initial studies on the leukotriene pathway was a key step to the discovery of leukotriene A_4.

Leukotriene A_4 also undergoes enzymatic conversions leading to various compounds with potent biological properties (Figure 2); (5*S*,12*R*)-5,12-dihydroxy-(*Z*,*E*,*E*,*Z*)-6,8,10,14-eicosatetraenoic acid (leukotriene B_4) is the product of the enzymatic hydrolysis of leukotriene A_4. It is formed in human neutrophils incubated with the ionophore A23187[32] and other neutrophil stimuli such as the formylated chemotactic peptides.[4] The direct addition of synthetic leukotriene A_4 to a suspension of human leukocytes clearly demonstrated that leukotriene A_4 is an intermediate in leukotriene B_4 synthesis.[33] The second pathway of leukotriene A_4 metabolism involves a glutathione transferase; (5*S*,6*R*)-5-hydroxy-6-*S*-glutathionyl-(*E*,*E*,*Z*,*Z*)-7,9,11,14-eicosatetraenoic acid (leukotriene C_4) is the product of the conjugation of glutathione and leukotriene A_4.[34]

Leukotriene C_4 is formed in human neutrophils stimulated with A23187 (or other agents) or incubated directly with precursor leukotriene A_4;[35] inhibition of glutathione synthesis or trapping of the tripeptide led to decreased leukotriene C_4 synthesis in RBL-1 cells and macrophages.[36,37]

The profiles of 5-lipoxygenase products are similar in neutrophils from various species under given conditions of stimulation. For example, human, rabbit, rat, and guinea pig neutrophils will release 5-HETE, leukotriene B_4, and variable amounts of the diHETEs derived from nonenzymatic hydrolysis of leukotriene A_4.[23,38] However, the profiles of products vary drastically according to incubation conditions. This is well illustrated by the profiles of products obtained on stimulation of human blood neutrophils with A23187 or the chemotactic peptide f-Met-Leu-Phe.[4,38] A23187 induces massive arachidonic acid release and leukotriene synthesis. f-Met-Leu-Phe is a physiological stimulus acting through activation of membrane receptors; its effect on leukotriene synthesis is of much lower magnitude than that of the A23187, and this affects not only the quantity but also the profile of 5-lipoxygenase products obtained. Indeed in human neutrophils the leukotriene A_4 hydrolase is the limiting step in the synthesis of leukotriene B_4.[38] In ionophore-stimulated cells the 5-lipoxygenase will produce amounts of leukotriene A_4 that exceed the catalytic capacity of leukotriene A_4 hydrolase; consequently, beside substantial amounts of leukotriene B_4, nonenzymatic conversion of leukotriene A_4 will occur (formation 6-*trans*-leukotriene B_4, 6-*trans*-12-epi-leukotriene B_4, and two 5,6-diHETEs epimers). Under f-Met-Leu-Phe stimulation, the neutrophils produce smaller amounts of leukotriene A_4, which is almost quantitatively transformed into leukotriene B_4 without significant nonenzymatic hydrolysis of the epoxy acid. In addition, in human neutrophils, leukotriene B_4 is rapidly converted into the Ω-oxydation products 20-hydroxy- and 20-carboxy-leukotriene B_4 (see II.G).[39]

Given the limited capacity of the leukotriene B_4 Ω-oxydation system, the proportion of the leukotriene B_4 metabolized will be much higher in f-Met-Leu-Phe stimulated cells than in the A23187-stimulated cells.[4] Other factors, such as the addition of exogenous arachidonic acid to incubation media or the presence of albumin or plasma, will cause major changes in the profile of 5-lipoxygenase products generated by neutrophils.[38,40]

The question of 5-lipoxygenase product release in the incubation media by neutrophils has been addressed by some investigators. Williams et al.[41] reported that in zymosan-stimulated human neutrophils, most of the leukotriene B_4 and its metabolites formed were retained in the cells, as opposed to A23187-stimulated cells, which release a much greater proportion of the 5-lipoxygenase products formed, presumably because of alteration of the plasma membrane induced by the ionophore.

B. The 15-Lipoxygenase Pathway

In addition to the 5-lipoxygenase activity described above, neutrophils from certain species have been shown to contain a 15-lipoxygenase. The (15*S*)-15-hydroperoxy-(*Z,Z,Z,E*)-5,8,11,13-eicosatetraenoic acid (15-HPETE) is the product of the action of 15-lipoxygenase on arachidonic acid. It is formed in human and rabbit neutrophils incubated with arachidonic acid and it is rapidly reduced (enzymatically) into 15-HETE.[32,42,43]

Recently, a group of compounds derived from 15-lipoxygenase and carrying a conjugated triene unit has been reported. The biosynthesis of these compounds involves the formation of the unstable (14*S*,15*S*)-14(15)-oxido-(*Z,Z,E,E,*)-5,8,10,12- eicosatetraenoic acid from 15-HPETE.[44] The 14,15-epoxy acid could be generated in neutrophils either from arachidonic acid or directly from 15-HPETE;[45,46] its mechanism of formation is believed to be similar to that of leukotriene A_4. The enzymatic system involved has not been characterized, and it is not excluded that the same enzyme catalyzes the cyclization of both 5-HPETE and 15-HPETE into allylic epoxides.

Similarly to leukotriene A_4, the allylic 14,15-epoxy acid is highly unstable and undergoes facile hydrolysis into several dihydroxy acids, among which are (8*S*,15*S*)-8,15-dihydroxy-(*Z,E,E,E*)-5,9,11,13-eicosatetraenoic acid and (8*R*,15*S*)-8,15-dihydroxy-(*Z,E,E,E*)-5,9,11,13-eicosatetraenoic acid. The hydrolysis (nonenzymatic) occurs spontaneously in aqueous media. The two compounds are generated in incubation mixtures of neutrophils with arachidonic acid or 15-HPETE. The mechanism of formation of the dihydroxy acids appears to be similar to that of 6-*trans*-leukotriene B_4 and 6-*trans*-12-epi-leukotriene B_4 from leukotriene A_4. The nonenzymatic hydrolysis of the 14,15-epoxy acid also leads to the isomeric (14*RS*,15*S*)-dihydroxy-(*Z,Z,E,E*)-5,8,10,12-eicosatetraenoic acids; their mechanism of formation appears to be similar of that of the 5,6-dihydroxy acids. Other isomeric 8,15-dihydroxy acids and 14,15-dihydroxy acids are produced by the 15-lipoxygenase.[43,44]

It is noteworthy that the products of the nonenzymatic hydrolysis of the 14,15-epoxy acid have been isolated in incubation mixtures of human or porcine leukocytes with arachidonic acid or 15-HPETE, whereas compounds corresponding to leukotriene B_4, leukotriene C_4, or leukotriene E_4 (formed enzymatically) in the 5-lipoxygenase product series have not been detected. However, the direct addition of the synthetic 14,15-epoxy acid to RBL-1 cell suspensions led to the formation of the glutathione conjugate, i.e., 14-*S*-glutathionyl-15-hydroxy-5,8,10,12-eicosatetraenoic acid.[47]

C. The 12-Lipoxygenase Pathway

The human platelet 12-lipoxygenase was the first mammalian lipoxygenase ever described,[48] and platelets appear to be the only source of this enzyme in human blood cells. However, porcine blood neutrophils contain a very high level of 12-lipoxygenase activity.[49] The 12-lipoxygenase catalyses the transformation of arachidonic acid into (12*S*)-12-hydroperoxy-(*Z,Z,E,Z*)-5,8,10,14-eicosatetraenoic acid, which is rapidly reduced to the corresponding alcohol[48] or transformed into a 12-keto-eicosatetraenoic acid,[50] 12-oxo-dodecatrienoic acid[50,51] and 10-hydroxy-11,12-epoxy-eicosatrienoic acid.[52]

D. Distinctive Lipoxygenase Product Profiles in Human Granulocyte Subsets and Cell Lines

Suspensions of human polymorphonuclear leukocytes are frequently taken as neutrophil suspensions, given the fact that eosinophils and basophils represent very few cells in such suspensions obtained from a normal donor. In the past few years, several observations have suggested that among granulocytes, eosinophils are the major source of leukotriene C_4 and 15-lipoxygenase products. More recently, using highly purified neutrophil and eosinophil preparations from normal human blood, it was clearly demonstrated that eosinophils indeed produce large amounts of leukotriene C_4 and little leukotriene B_4, whereas the neutrophils produce mainly leukotriene B_4 and little leukotriene C_4, if any. The same studies also demonstrated that the 15-lipoxygenase was largely an eosinophil enzyme, and that contrary to neutrophils, eosinophils do not metabolize leukotriene B_4 through the Ω-oxydation pathway (see III.G.)[53-55] It is therefore clear that the metabolism of arachidonic acid in neutrophils and eosinophils shows striking differences.

Most certainly because of the difficulties in obtaining human blood basophils, very few studies have appeared on the lipoxygenase-mediated transformation of arachidonic acid in these cells. However, it seems that human basophils produce leukotriene C_4;[56,57] the studies do not permit detailed comparisons with neutrophils and eosinophils.

The human promyelocytic cell line HL-60 has recently been used for various types of investigation. These cells can be induced to differentiate into granulocytic cells by treatment with dimethylsulfoxide. Before differentiation, the cells show little lipoxygenase and cyclooxygenase activities. However, in the course of differentiation, the 5-lipoxygenase, the 15-lipoxygenase and the cyclooxygenase activities appear,[58-60] together with the responsiveness to leukotriene B_4.[61] It is noteworthy that the profile of arachidonic acid products in the differentiated (granulocytic) HL-60 is very different from the profile obtained from normal blood neutrophils. Indeed the HL-60, contrary to the blood neutrophil, has an important cyclooxygenase activity. In addition, the HL-60 cells produce leukotriene C_4 and leukotriene D_4 in larger amounts than leukotriene B_4; the metabolism of leukotriene B_4 (Ω-oxydation) is also weak in comparison to the human blood neutrophils.

Another cell line that has been used by many investigators for arachidonic acid metabolism studies is the rat basophilic leukemia cell line RBL- 1. The RBL-1 cells produce the 5-lipoxygenase products leukotriene B_4 and leukotriene C_4; they also have 12-lipoxygenase and cyclooxygenase activities.[62]

E. The Double Dioxygenation of Arachidonic Acid in the Neutrophils

Arachidonic acid has several reaction sites for lipoxygenation, and reaction at one site does not necessarily exclude reaction at other sites. Given the fact that neutrophils contain two (or three) lipoxygenases (5 and 15 for human neutrophils and eosinophils, and 5,12, and 15 for porcine neutrophils), and that in addition neutrophil suspensions can be contaminated with platelets (which contain a 12-lipoxygenase), compounds resulting from the double dioxygenation of arachidonic acid can be formed in appropriate experimental conditions. For example, (5*S*,12*S*)-5,12-dihydroxy-(*E,Z,E,Z,*)-6,8,10,14-eicosatetraenoic acid results of the successive reaction of arachidonic acid with the 5- and the 12-lipoxygenases. It can be formed by the incubation of A23187-activated neutrophils with 12-HETE or platelets, or inversely by incubation of platelets with 5-HETE.[49,63] Other compounds such as the (5*S*,15*S*)-5,15-dihydroxy-(*E,Z,Z,E*)- 6,8,11,13-eicosatetraenoic acid[64,65] or the lipoxins described more recently[66-68] are the products of the double dioxygenation of arachidonic acid by the 5-and the 15-lipoxygenases.

F. Modulation of Leukotriene Synthesis

Because of the putative roles of leukotrienes in human diseases, the regulation of leukotriene synthesis is presently the subject of extensive investigations. The data available at this time suggest that the regulation of the 5-lipoxygenase pathway is more complex than that of other arachidonic acid metabolic pathways

As mentioned previously (see Section II), the formation of the various arachidonic acid metabolites is dependent on substrate availability. The concentration of free arachidonic acid in cells under normal conditions is low (as compared to the total amount of fatty acid present in the form of esters). The formation of cyclooxygenase and lipoxygenase products is thus initially controlled by the activity of the various lipases which make arachidonic acid available to metabolizing enzymes. The effect of the A23187, which unspecifically stimulates the release of arachidonic acid (likely through stimulation of Ca^{2+}-dependent phospholipase A_2) and the synthesis of cyclooxygenase and lipoxygenase products in a variety of systems, supports this concept.[69]

The synthesis of 5-HETE and leukotrienes also depends on substrate availability. However, in many systems, such as in human blood neutrophils (neutrophils or eosinophils), monocytes, and guinea pig and human lungs, the 5-lipoxygenase will not readily transform exogenous arachidonic acid (or only to a small extent), whereas incubation of these cells or tissues in the presence of A23187 leads to synthesis of substantial amounts of 5-HETE and leukotrienes from endogenous arachidonic acid.[32,70,71] These observations were of primary importance since they clearly indicated that the ionophore not only causes the release of arachidonic acid but also activates the 5-lipoxygenase (a Ca^{2+}-dependent enzyme) involved in the further transformation of the fatty acid. Thus, most cells produce little 5-lipoxygenase products in response to exogenous arachidonic acid, and it is clear that in addition to substrate, the enzyme requires activation. This constitutes a major difference between the 5-lipoxygenase and other dioxygenases and implies the existence of additional regulatory mechanisms for the synthesis of leukotrienes. It also implies that the 5-lipoxygenase exists in two different forms, reflecting a different status of activation. The mechanisms responsible for the activation of the 5-lipoxygenase are not understood at this time. On the one hand, it is well known that agents that can activate phagocytes, basophils, and mast cells, such as the chemotactic peptide f-Met-Leu-Phe, PAF, opsonized zymosan (phagocytosis), IgE (monomeric or aggregated), or immune complexes, through interactions with surface receptors, also induce the synthesis of 5-lipoxygenase products from endogenous substrate. On the other hand, it is also generally accepted that the 5-lipoxygenase is a Ca^{2+}-dependent enzyme (see III.H). It is therefore possible that alterations of Ca^{2+} metabolism induced by cell activation (Ca^{2+} influx or release of intracellular pools) is involved in the activation of the 5-lipoxygenase. Other factors that have been shown recently to regulate the activity of the purified 5-lipoxygenase (see III.H) might also be implicated.

It is worth mentioning here that contrary to human blood neutrophils, eosinophils, and monocytes (as well as many other 5-lipoxygenase containing cells from various species), rabbit neutrophils are fully responsive to exogenous substrates (i.e., are not further activated by A23187). In fact, rabbit neutrophils (blood or peritoneal) produce twice as much 5-lipoxygenase products on arachidonic acid (30 μM, 5 min) challenge than on A23187 (2 μM, 5 min) challenge.[212] Under the two same incubation conditions, human blood neutrophils produce twenty to fifty times more of the 5-lipoxygenase products on A23187 challenge (Figure 3). This demonstrates a striking difference in the metabolism of arachidonic acid in human and rabbit neutrophils and suggests that the regulatory mechanisms of the synthesis of leukotrienes in these cells could be quite different.

Other factors that can modulate the synthesis of LTs in neutrophils include the various HETEs, mainly 15-HETE, which is an inhibitor of LT synthesis (IC50 $\approx$ 5 μM),[72] interaction with activated platelets, which can release 12-HPETE, an activator of the 5-lipoxygenase,[73,74] and diets where arachidonic acid is partially replaced by eicosapentaenoic acid (marine diet or fish oil supplement) or 5,8,11-eicosatrienoic acid (essential fatty acid deficiency) in cellular lipids.[75,76]

G. The Metabolism of Lipoxygenase Products in the Neutrophils

The first metabolic process of lipoxygenase products reported in neutrophils was the to the margination of the neutrophils that accompanies the intravenous injection of chemotactic factors.[144,145] Leukotriene B_4, and to progressively smaller extent, 20-OH leukotriene B_4 and 20-COOH leukotriene B_4, [31,124] and the 5- and 12-HETEs[146] have been shown to induce an aggregation response from the neutrophils. The neutrophil aggregatory properties of arachidonic acid are also generally ascribed to its lipoxygenase metabolites.[147-150] The aggregation response to the various fatty acids is, it should be noted, of a small magnitude compared to that induced by peptidic chemoattractants.

4. Degranulation

Several lipoxygenase products have been shown to possess secretory activity towards the neutrophils, including leukotriene B_4, [131,151] and at relatively high concentrations, lipoxin A.[152] In contrast, the chemotactically active 8,15-diHETE was found to be inactive with respect to its ability to provoke a degranulation response.[135] The magnitude of the exocytocis induced by leukotriene B_4 is significantly smaller than that observed with f-Met-Leu-Phe and, in contrast to the sigmoidal shape of the secretory response of the latter stimulus, appears to be directly related to the dose of the fatty acid used (e.g., Reference 131).

5. Oxidative Metabolism

The lipoxygenase metabolites of arachidonic acid are, at best, poor activators of the respiratory burst of the neutrophils. Superoxide production as well as increased chemiluminescence in response to leukotriene B_4 have, however, been reported in some,[8,153] but not all, [129,154] studies. The magnitude of these effects is 2 to 5 times smaller than those produced by f-Met-Leu-Phe, and the dose required elevated compared to those affecting neutrophil locomotion. On the other hand, leukotriene B_4 has been shown to potentiate the chemotactic peptide-stimulated oxidative metabolism of the neutrophils.[154]

6. Expression of Surface Antigens and Receptors

Leukotriene B_4 (and 5-HETE) enhance (as measured by rosetting techniques) the expression of C3b and, to a lesser extent, IgG-Fc receptors on human neutrophils in vitro.[155]

C. Effect of Dietary Ω-3 Fatty Acids on Neutrophil Function

Increasing by dietary means the content of the neutrophil membranes in eicosapentaenoic and docohexaenoic acids has been shown to significantly affect some but not all of the functions of these cells. Neutrophil chemotaxis and adherence to endothelial cells as stimulated by leukotriene B_4 has thus been found to be impaired in neutrophils from volunteers whose diet was supplemented with capsules (5 to 6 g/day) of Ω-3 fatty acids. The generation of leukotrienes in response to the A23187 was similarly impaired.[156] On the other hand, the results of a previous report, using a lower dose of Ω-3 fatty acids (1 to 2 g/day), indicated that the aggregation, chemotactic and superoxide production responses to f-Met-Leu-Phe were unaffected.[157] It remains to be seen whether these results reflect stimuli specificities or methodological differences.

D. Signal Transduction

Signal transduction in leukotriene-stimulated neutrophils resembles in most aspects that defined for other agonists of this cell type. This includes the presence of receptors, the mediatory role of G proteins, the activation of the phospholipase C specific for the polyphosphoinositides, increases in the level of calcium, and a reorganization of the cytoskeletal structures of the cells. The similarity between the apparent mode of action of leukotriene B_4 and that of other neutrophil stimuli such as f-Met-Leu-Phe is not complete however. Unresolved questions remain concerning, among others, the functional significance of the activation of the "polyphosphoinositide cycle" by leukotriene B_4, the details of the coordination of the various stimulated biochemical events in order to lead to the performance of the neutrophils' functions, and also the potential role of the lipoxygenase in the mediation of the effects of exogenous chemotactic factors.

1. Receptors

The presence of specific receptors for leukotriene B_4 was deduced early after the description of its biological activities in view of the cell specificity of the effects of the eicosanoid, the low concentrations necessary for the initiation of the latter, of the observation of deactivation following preincubation of the cells with leukotriene B_4 or with f-Met-Leu-Phe, and of the stereospecificity of the responses.

a. Characteristics

Rapid, saturable, and reversible binding of leukotriene B_4 in human neutrophils has been demonstrated.[126,158-163] Leukotriene B_4 receptors have also been detected in dimethylsulfoxide differentiated HL-60 cells.[166] Evidence for curvilinear Scatchard plots in neutrophils has recently been obtained.[160,162] The binding constants of the high-affinity sites are of the order of 0.5 n*M* (0.27 to 1.96 $\times$ 10^4 sites per cell) while the low-affinity sites have binding constants of about 61 to 541 n*M* (11 to 45 $\times$ 10^4 sites per cell). Guanine nucleotides modulate the binding affinities of the leukotriene B_4 receptors in much the same way that they affect that of the formylated peptides; the addition of GTP to a membrane suspension reduces the K_m by about sixfold without affecting significantly the total number of binding sites.[159] Cross-linking experiments have demonstrated that the binding of leukotriene B_4 is associated with a 60-kdalton protein. The latter can be detected both in intact cells and in plasma membrane enriched fractions.[159]

Binding exhibits structural specificity in accordance with the biological activity of the leukotriene B_4 analogues, i.e., 20-OH-leukotriene B_4 is 16 times and 5*S*, 12*S*DHETE 300 times less effective than native leukotriene B_4 in displacing labeled leukotriene B_4, while the stereoisomers of leukotriene B_4 are essentially inactive.[160] Two binding sites for 20-OH leukotriene B_4 on human neutrophils have also been demonstrated with affinities equal to 14 and 100 n*M*. Binding to the former sites can be displaced by leukotriene B_4 as well as by 20 OH-leukotriene B_4 but not by 15 HETE, 5(*S*), 12(*S*)-leukotriene B_4 or the 6-*trans* stereoisomer of leukotriene B_4.[164] Physiological evidence (inhibition of neutrophil degranulation) has been presented indicating that the leukotriene B_4 derivative, leukotriene B_4-dimethylamide, behaved as a receptor antagonist.[208] Previous exposure to leukotriene B_4 results in the loss of the high, but not of the low, affinity sites.[162] This high-affinity binding site is presumably the chemotactically relevant one. The low-affinity sites have been related to the secretory response of the cells to leukotriene B_4. Leukotriene B_4 receptors can be down-regulated (decrease in receptor number) by pretreatment of whole cells and cytoplasts, but not isolated plasma membranes, with protein kinase C activators such as phorbol esters and mezerein.[165]

b. Localization

Fractionation studies in human neutrophils have demonstrated a concurrent enrichment of plasma membrane markers and of binding activity of leukotriene B_4.[82,160] These results indicate that the bulk of the leukotriene B_4 binding sites is located on the plasma membranes of the neutrophils. Though not directly supported by the present evidence, the possibility of intracellular (possibly endoplasmic reticulum associated) leukotriene B_4 binding sites cannot be totally ruled out.

c. Association with the Cytoskeleton

Leukotriene B_4 becomes rapidly (<1 min) associated with a Triton X-100 insoluble pellet

(the cytoskeleton) in ra
B_4 itself, by 20-OH l
preincubation of the c
4-alpha-phorbol 12,13
B_4 with the cytoskele
B_4 receptors by protei
esters on the associat
underlies the effects o
that a specific associa
be involved in signal

2. Guanine Nucleotid

Neutrophil stimulat
to inhibition by pertu
and chemotaxis[168] as
actin[170] by pertussis t
and Holian[172] have als
of phosphatidylinosito
view of the mechanis
preted as evidence fo
excitation-response co

3. Does Leukotriene l

Apparently conflict
B_4 on the activity of
Volpi et al.[174] and Br
levels of labeled pol
rabbit neutrophils. Le
inositol phosphates in
in human neutrophils
phages all found evi
phosphates) by leukot
that elicited by the fo
n*M*). Measurements (
inositide breakdown.[1
though less efficientl
poor activity of leuko
responses that have be
results leave open the
concentrations (incluc
hydrolysis of the pol

4. Calcium Mobilizat

Leukotriene B_4 is
Leukotriene B_4 has th
the level of free calci
changes in the fluores
redistribution of inter
B_4 metabolites and ar
dose dependencies an
B_4 stimulated mobili

26. **Murphy, R. C., Hammarström, S., and Samuelsson, B.,** Leukotriene C: a slow-reacting substance from murine mastocytoma cells, *Proc. Natl. Acad. Sci. U.S.A.*, 76, 4275, 1979.
27. **Samuelsson, B.,** The leukotrienes, highly biologically active substances involved in allergy and inflammation, *Angew. Chem. Int. Ed. Engl.*, 21, 902, 1982.
28. **Morley, J., Bray, M. A., Jones, R. W., Nugteren, D. H., and van Dorp, D. A.,** Prostaglandin and thromboxane production by human and guinea-pig macrophages and leucocytes, *Prostaglandins,* 17, 729, 1979.
29. **Maas, R. L., Ingram, C. D., Taber, D. F., Oates, J. A., and Brash, A. R.,** Stereospecific removal of the D^r hydrogen atom at the 10-carbon of arachidonic acid in the biosynthesis of leukotriene A_4 by human leukocytes, *J. Biol. Chem.*, 257, 13,525, 1982.
30. **Fitzpatrick, F. A., Morton, D. R., and Wynalda, M. A.,** Albumin stabilizes leukotriene A_4, *J. Biol. Chem.*, 257, 4680, 1982.
31. **Naccache, P. H., Molski, T. F. P., Becker, E. L., Borgeat, P., Picard, S., Vallerand, P., and Sha'afi, R. I.,** Specificity of the effect of lipoxygenase metabolites of arachidonic acid on calcium homeostasis in neutrophils: correlation with functional activity, *J. Biol. Chem.*, 257, 8608, 1982.
32. **Borgeat, P. and Samuelsson, B.,** Metabolism of arachidonic acid in polymorphonuclear leukocytes. Effects of the ionophore A23187, *Proc. Natl. Acad. Sci. U.S.A.*, 76, 2148, 1979.
33. **Radmark, O., Malmsten, C., Samuelsson, B., Clark, D. A., Goto, G., Marfat, A., and Corey, E. J.,** Leukotriene A: stereochemistry and enzymatic conversion to leukotriene B, *Biochem. Biophys. Res. Commun.*, 92, 954, 1980.
34. **Hammarström, S., Murphy, R. C., Samuelsson, B., Clark, D. A., Mioskowski, C., and Corey, E. J.,** Structure of leukotriene C: identification of the amino acid part, *Biochem. Biophys. Res. Commun.*, 91, 1266, 1979.
35. **Hammarström, S. and Samuelsson, B.,** Detection of leukotriene A_4 as an intermediate in the biosynthesis of leukotriene C_4 and D_4, *FEBS Lett.*, 22, 83, 1980.
36. **Parker, C. W., Fischman, C. M., and Wedner, H. J.,** Relationship of biosynthesis of slow reacting substance to intracellular glutathione concentrations, *Proc. Natl. Acad. Sci. U.S.A.*, 77, 6870, 1980.
37. **Rouzer, C. A., Scott, W. A., Griffith, O. W., Hamill, A. L., and Cohn, Z. A.,** Depletion of glutathione selectively inhibits synthesis of leukotriene C by macrophages, *Proc. Natl. Acad. Sci. U.S.A.*, 78, 2532, 1981.
38. **Fruteau de Laclos, B., Braquet, P., and Borgeat, P.,** Characteristics of leukotriene (LT) and hydroxy eicosatetraenoic acid (HETE) synthesis in human leukocytes *in vitro:* effect of arachidonic acid concentration, *Prostaglandins Leukotrienes Med.*, 13, 47, 1984.
39. **Nadeau, M., Fruteau de Laclos, B., Picard, S., Braquet, P., and Borgeat, P.,** Studies on leukotriene B_4 Ω-oxidation in human leukocytes, *Can. J. Biochem. Cell. Biol.*, 62, 1321, 1984.
40. **Gresele, P., Arnout, J., Coene, M. C., Deckmyn, H., and Vermylen, J.,** Leukotriene B_4 production by stimulated whole blood: comparative studies with isolated polymorphonuclear cells, *Biochem. Biophys. Res. Commun.*, 137, 334, 1986.
41. **Williams, J. D., Lee, T. H., Lewis, R. A., and Austen, F.,** Intracellular retention of the 5-lipoxygenase pathway product, leukotriene B_4, by human neutrophils activated with unopsonized zymosan, *J. Immunol.*, 34, 2624, 1985.
42. **Narumiya, S., Salmon, J. A., Cottee, F. H., Weatherley, B. C., and Flower, R. J.,** Arachidonic acid 15-lipoxygenase from rabbit peritoneal polymorphonuclear leukocytes, *J. Biol. Chem.*, 256, 9583, 1981.
43. **Turk, J., Maas, R. L., Brash, A. R., Jackson Roberts II, L., and Oates, J. A.,** Arachidonic acid 15-lipoxygenase products from human eosinophils, *J. Biol. Chem.*, 257, 7068, 1982.
44. **Maas, R. L., Brash, A. R., and Oates, J. A.,** A second pathway of leukotriene biosynthesis in porcine leukocytes, *Proc. Natl. Acad. Sci. U.S.A.*, 78, 5523, 1981.
45. **Lundberg, U., Radmark, O., Malmsten, C., and Samuelsson, B.,** Transformation of 15-hydroperoxy-5,9,11,13-eicosatetraenoic acid into novel leukotrienes, *FEBS Lett.*, 126, 127, 1981.
46. **Jubiz, W., Radmark, O., Lindgren, J. A., Malmsten, C., and Samuelsson, B.,** Novel leukotrienes: products formed by initial oxygenation of arachidonic acid at C-15, *Biochem. Biophys. Res. Commun.*, 99, 976, 1981.
47. **Sok, D. E., Han, C. O., Shieh, W. R., Zhou, B. N., and Sih, C. J.,** Enzymatic formation of 14,15-leukotriene A and C(14)-sulfur-linked peptides, *Biochem. Biophys. Res. Commun.*, 104, 1363, 1982.
48. **Hamberg, M. and Samuelsson, B.,** Prostaglandin endoperoxides. Novel transformation of arachidonic acid in human platelets, *Proc. Natl. Acad. Sci. U.S.A.*, 71, 3400, 1974.
49. **Borgeat, P., Fruteau de Laclos, B., Picard, S., Drapeau, J., Vallerand, P., and Corey, E. J.,** Studies on the mechanism of formation of the 5*S*,12*S*-dihydroxy-6,8,10,14(*E,Z,E,Z,*)-icosatetraenoic acid in leukocytes, *Prostaglandins,* 23, 713, 1982.
50. **Fruteau de Laclos, B., Maclouf, J., Poubelle, P., and Borgeat, P.,** Conversion of arachidonic acid into 12-oxo derivatives in human platelets. A pathway possibly involving the heme-catalysed transformation of 12-hydroperoxy-eicosatetraenoic acid, *Prostaglandins,* 33, 315, 1987.

51. **Glasgow, W. C., Harris, T. M., and Brash, A. R.,** A short-chain aldehyde is a major lipoxygenase product in arachidonic acid-stimulated porcine leukocytes, *J. Biol. Chem.*, 261, 200, 1986.
52. **Walker, I. C., Jones, R. L., and Wilson, N. H.,** The identification of an epoxy-hydroxy acid as a product from the incubation of arachidonic acid with washed blood platelets, *Prostaglandins,* 18, 173, 1979.
53. **Verhagen, J., Bruynzeel, L. B., Koedam, J. A., Wassink, G. A., de Boer, M., Terpstra, G. K., Kreukniet, J., Veldink, G. A., and Vliegenthart, J. F. G.,** Specific leukotriene formation by purified human eosinophils and neutrophils, *FEBS Lett.*, 168, 23, 1984.
54. **Shaw, R. J., Cromwell, O., and Kay, A. B.,** Preferential generation of leukotriene C_4 by human eosinophils, *Clin. Exp. Immunol.*, 56, 716, 1984.
55. **Laviolette, M., Picard, S., Braquet, P., and Borgeat, P.,** Comparison of 5- and 15-lipoxygenase activities in blood and alveolar leukocyte preparations from normal subjects and patients with eosinophilia, *Prostaglandins Leukotrienes Med.*, 23, 191, 1986.
56. **Fox, C. C., Kagey-Sobotka, A., Schleimer, R. P., Peters, S. P., MacGlashan, D. W., Jr., and Lichtenstein, L. M.,** Mediator release from human basophils and mast cells from lung and intestinal mucosa, *Int. Arch. Allergy Appl. Immunol.*, 77, 130, 1985.
57. **MacGlashan, D. W., Jr., Peters, S. P., Warner, J., and Lichtenstein, L. M.,** Characteristics of human basophil sulfidopeptide leukotriene release: releasability defined as the ability of the basophil to respond to dimeric cross-links, *J. Immunol.*, 136, 2231, 1986.
58. **Ziboh, V. A., Wong, T., Ming-Chi, W., and Yunis, A. A.,** Lipoxygenation of arachidonic acid by differentiated and undifferentiated human promyelocytic HL-60 cells, *J. Lab. Clin. Med.*, 108, 161, 1986.
59. **Bonser, R. W., Siegel, M. I., McConnell, R. T., and Cuatrecasas, P.,** The appearance of phospholipase and cyclo-oxygenase activities in the human promyelocytic leukemia cell line HL60 during dimethyl sulfoxide-induced differentiation, *Biochem. Biophys. Res. Commun.*, 98, 614, 1981.
60. **Lundberg, U., Serhan, C. N., and Samuelsson, B.,** Appearance of an arachidonic acid 15-lipoxygenase pathway upon differentiation of the human promyelocytic cell-line HL-60, *FEBS Lett.*, 185, 14, 1985.
61. **Naccache, P. H., Molski, T. F. P., Spinelli, B., Borgeat, P., and Abboud, C. N.,** Development of calcium and secretory responses in the human promyelocytic leukemia cell line HL60, *J. Cell. Physiol.*, 119, 241, 1984.
62. **Jakschik, B. A. and Kuo, C. G.,** Characterization of leukotriene A_4 and B_4 biosynthesis, *Prostaglandins,* 25, 767, 1983.
63. **Yoshimoto, T., Miyamoto, Y., Ochi, K., and Yamamoto, S.,** Arachidonate 12-lipoxygenase of porcine leukocyte with activity for 5-hydroxyeicosatetraenoic acid, *Biochim. Biophys. Acta,* 713, 638, 1982.
64. **Maas, R. L., Turk, J., Oates, J. A., and Brash, A. R.,** Formation of a novel dihydroxy acid from arachidonic acid by lipoxygenase- catalyzed double oxygenation in rat mononuclear cells and human leukocytes, *J. Biol. Chem.*, 257, 7056, 1982.
65. **Borgeat, P., Picard, S., Drapeau, J., and Vallerand, P.,** Metabolism of arachidonic acid in leukocytes: isolation of a 5,15-dihydroxy-icosatetraenoic acid, *Lipids,* 17, 676, 1982.
66. **Serhan, C. N., Hamberg, M., Samuelsson, B., Morris, J., and Wishka, D. G.,** On the stereochemistry and biosynthesis of lipoxin B, *Proc. Natl. Acad. Sci. U.S.A.*, 83, 1983, 1986.
67. **Fitzsimmons, B. J., Adams, J., Evans, J. F., Leblanc, Y., and Rokach, J.,** The Lipoxins. Stereochemical identification and determination of their biosynthesis, *J. Biol. Chem.*, 260, 13, 008, 1985.
68. **Corey, E. J., Mehrotra, M. M., and Su, W.,** On the synthesis and structure of lipoxin B, *Tetrahedron Lett.*, 26, 1919, 1985.
69. **Knapp, H. R., Oelz, O., Roberts, J., Sweetman, B. J., Oates, J. A., and Reed, P. W.,** Ionophores stimulate prostaglandin and thromboxane biosynthesis, *Proc. Natl. Acad. Sci. U.S.A.*, 74, 4251, 1977.
70. **Goldyne, M. E., Burrish, G. F., Poubelle, P., and Borgeat, P.,** Arachidonic acid metabolism among human mononuclear leukocytes. Lipoxygenase-related pathways, *J. Biol. Chem.*, 259, 8815, 1984.
71. **Salari, H., Borgeat, P., Fournier, M., Hebert, J., and Pelletier, G.,** Studies on the release of leukotrienes and histamine by human lung parenchymal and bronchial fragments upon immunologic and nonimmunologic stimulation, *J. Exp. Med.*, 162, 1904, 1985.
72. **Vanderhoek, J. Y., Bryant, R. W., and Bailey, J. M.,** Inhibition of leukotriene biosynthesis by the leukocyte product 15-hydroxy-5,8,11,13-eicosatetraenoic acid, *J. Biol. Chem.*, 255, 10,064, 1980.
73. **Maclouf, J., Fruteau de Laclos, B., and Borgeat, P.,** Stimulation of leukotriene biosynthesis in human blood leukocytes by platelet-derived 12-hydroperoxy-icosatetraenoic acid, *Proc. Natl. Acad. Sci. U.S.A.*, 79, 6042, 1982.
74. **Kanaji, K., Okuma, M., and Uchino, H.,** Deficient induction of leukotriene synthesis in human neutrophils by lipoxygenase-deficient platelets, *Blood,* 67, 903, 1986.
75. **Prescott, S. M., Zimmerman, G. A., and Morrison, A. R.,** The effects of a diet rich in fish oil on human neutrophils: identification of leukotriene B_5 as a metabolite, *Prostaglandins,* 30, 209, 1985.
76. **Stenson, W. F., Prescott, S. M., and Sprecher, H.,** Leukotriene B formation by neutrophils from essential fatty acid-deficient rats, *J. Biol. Chem.*, 259, 11,784, 1984.

77. **Stenson, W. F. and Parker, C. W.,** Metabolism of arachidonic acid in ionophore-stimulated neutrophils. Esterification of a hydroxylated metabolite into phospholipids, *J. Clin. Invest.*, 64, 1457, 1979.
78. **Hansson, G., Lindgren, J. A., Dahlen, S. E., Hedqvist, P., and Samuelsson, B.,** Identification and biological activity of novel Ω-oxidized metabolites of leukotriene B_4 from human leukocytes, *FEBS Lett.*, 130, 107, 1981.
79. **Shak, S. and Goldstein, I. M.,** Ω-oxidation is the major pathway for the catabolism of leukotriene B_4 in human polymorphonuclear leukocytes, *J. Biol. Chem.*, 259, 10,181, 1984.
80. **Powell, W. S.,** Properties of leukotriene B_4 20-hydroxylase from polymorphonuclear leukocytes, *J. Biol. Chem.*, 259, 3082, 1984.
81. **Shak, S. and Goldstein, I. M.,** Leukotriene B_4 Ω-hydroxylase in human polymorphonuclear leukocytes. Partial purification and identification as a cytochrome P-450, *J. Clin. Invest.*, 76, 1218, 1985.
82. **O'Flaherty, J., Kosfeld, S., and Nishihira, J.,** Binding and metabolism of leukotriene B_4 by neutrophils and their subcellular organelles, *J. Cell. Physiol.*, 126, 359, 1986.
83. **Henderson, W. R., Jörg, A., and Klebanoff, S. J.,** Eosinophil peroxidase-mediated inactivation of leukotrienes B_4, C_4, and D_4, *J. Immunol.*, 128, 2609, 1982.
84. **Henderson, W. R. and Klebanoff, S. J.,** Leukotriene B_4, C_4, D_4 and E_4 inactivation by hydroxyl radicals, *Biochem. Biophys. Res. Commun.*, 110, 266, 1983.
85. **Goetzl, E. J.,** The conversion of leukotriene C_4 to isomers of leukotriene B_4 by human eosinophil peroxidase, *Biochem. Biophys. Res. Commun.*, 106, 270, 1982.
86. **Lee, C. W., Lewis, R. A., Corey, E. J., Barton, A., Oh, H., Tauber, A. I., and Austen, K. F.,** Oxidative inactivation of leukotriene C_4 by stimulated human polymorphonuclear leukocytes, *Proc. Natl. Acad. Sci. U.S.A.*, 79, 4166, 1982.
87. **Raulf, M., Stüning, M., and König, W.,** Effect of cations on leukotriene release: requirements for the metabolism of peptido-leukotrienes (leukotrienes C_4, D_4) by human polymorphonuclear granulocytes, *Immunology*, 58, 479, 1986.
88. **Raulf, M., Stüning, M., and König, W.,** Metabolism of leukotrienes by gamma-glutamyl-transpeptidase and dipeptidase from human polymorphonuclear granulocytes, *Immunology*, 55, 135, 1985.
89. **Lee, C. W., Lewis, R. A., Corey, E. J., and Austen, K. F.,** Conversion of leukotriene D_4 to leukotriene E_4 by a dipeptidase released from the specific granule of human polymorphonuclear leucocytes, *Immunology*, 48, 27, 1983.
90. **Parker, C. W., Koch, D., Huber, M. M., and Falkenhein, S. F.,** Formation of the cysteinyl from of slow reacting substance (leukotriene E_4) in human plasma, *Biochem. Biophys. Res. Commun.*, 97, 1038, 1980.
91. **Köller, M., König, W., Brom, J., Bremm, K. D., Schönfeld, W., and Knöller, J.,** Functional characteristics of leukotriene C_4- and D_4-metabolizing enzymes (gamma-glutamyl transpeptidase, dipeptidase) within human plasma, *Biochem. Biophys. Acta*, 836, 56, 1985.
92. **Hagmann, W., Denzlinger, C., Rapp, S., Weckbecker, G., and Keppler, D.,** Identification of the major endogenous leukotriene metabolite in the bile of rats as *N*-acetyl leukotriene B_4, *Prostaglandins*, 31, 239, 1986.
93. **Orning, L., Norin, E., Gustafsson, B., and Hammarström, S.,** *In vivo* metabolism of leukotriene C_4 in germ-free and conventional rats, *J. Biol. Chem.*, 261, 766, 1986.
94. **Denzlinger, C., Guhlmann, A., Scheuber, P. H., Wilker, D., Hammer, D. K., and Keppler, D.,** Metabolism and analysis of cysteinyl leukotrienes in the monkey, *J. Biol. Chem.*, 261, 15,601, 1986.
95. **Serafin, W. E., Oates, J. A., and Hubbard, W. C.,** Metabolism of leukotriene B_4 in the monkey. Identification of the principal nonvolatile metabolite in the urine, *Prostaglandins*, 27, 899, 1984.
96. **Harper, T. W., Garrity, M. J., and Murphy, R. C.,** Metabolism of leukotriene B_4 in isolated rat hepatocytes, *J. Biol. Chem.*, 261, 5414, 1986.
97. **Rouzer, C. A. and Samuelsson, B.,** On the nature of the 5-lipoxygenase reaction in human leukocytes: enzyme purification and requirement for multiple stimulatory factors, *Proc. Natl. Acad. Sci. U.S.A.*, 82, 6040, 1985.
98. **Rouzer, C. A., Matsumoto, T., and Samuelsson, B.,** Single protein from human leukocytes possesses 5-lipoxygenase and leukotriene A_4 synthase activities, *Proc. Natl. Acad. Sci. U.S.A.*, 83, 857, 1986.
99. **Ueda, N., Kaneko, S., Yoshimoto, T., and Yamamoto, S.,** Purification of arachidonate 5-lipoxygenase from porcine leukocytes and its reactivity with hydroperoxyeicosatetraenoic acids, *J. Biol. Chem.*, 261, 7982, 1986.
100. **Shimizu, T., Izumi, T., Seyama, Y., Tadokoro, K., Radmark, O., and Samuelsson, B.,** Characterization of leukotriene A_4 synthase from murine mast cells: evidence for its identity to arachidonate 5-lipoxygenase, *Proc. Natl. Acad. Sci. U.S.A.*, 83, 4175, 1986.
101. **Hogaboom, G. K., Cook, M., Newton, J. F., Varrichio, A., Shorr, R. G. L., Sarau, H. M., and Crooke, S. T.,** Purification, characterization, and structural properties of a single protein from rat basophilic leukemia (RBL-1) cells possessing 5-lipoxygenase and leukotriene A_4 synthetase activities, *Mol. Pharmacol.*, 30, 510, 1986.

102. **Stüning, M., Raulf, M., and König, W.,** Localization of 5-lipoxygenase within human polymorphonuclear leukocytes, *Biochem. Pharmacol.*, 34, 3943, 1985.
103. **Rouzer, C. A., Shimizu, T., and Samuelsson, B.,** On the nature of the 5-lipoxygenase reaction in human leukocytes: characterization of a membrane-associated stimulatory factor, *Proc. Natl. Acad. Sci. U.S.A.*, 82, 7505, 1985.
104. **Borgeat, P., Nadeau, M., Salari, H., Poubelle, P., and Fruteau de Laclos, B.,** Leukotrienes: biosynthesis, metabolism, and analysis, *Adv. Lipid Res.*, 21, 47, 1985.
105. **Radmark, O., Shimizu, T., Jörnvall, H., and Samuelsson, B.,** Leukotriene A_4 hydrolase in human leukocytes, *J. Biol. Chem.*, 259, 12,339, 1984.
106. **Evans, J. F., Dupuis, P., and Ford-Hutchinson, A. W.,** Purification and characterisation of leukotriene A_4 hydrolase from rat neutrophils, *Biochim. Biophys. Acta*, 840, 43, 1985.
107. **Nathaniel, D. J., Evans, J. F., Leblanc, Y., Léveillé, C., Fitzsimmons, B. J., and Ford-Hutchinson, A. W.,** Leukotriene A_4 is a substrate and an inhibitor of rat and human neutrophil LTA_4 hydrolase, *Biochem. Biophys. Res. Comm.*, 131, 827, 1985.
108. **Evans, J. F., Nathaniel, D. J., Zamboni, R. J., and Ford-Hutchinson, A. W.,** Leukotriene A_3. A poor substrate but a potent inhibitor of rat and human neutrophil leukotriene A_4 hydrolase, *J. Biol. Chem.*, 260, 10,966, 1985.
109. **McGee, J. and Fitzpatrick, F.,** Enzymatic hydration of leukotriene A_4. Purification and characterization of a novel epoxide hydrolase from human erythrocytes, *J. Biol. Chem.*, 260, 12,832, 1985.
110. **Haeggström, J., Radmark, O., and Fitzpatrick, F. A.,** Leukotriene A_4-hydrolase activity in guinea pig and human liver, *Biochim. Biophys. Acta*, 835, 378, 1985.
111. **Izumi, T., Shimizu, T., Seyama, Y., Ohishi, N., and Takaku, F.,** Tissue distribution of leukotriene A_4 hydrolase activity in guinea pig, *Biochem. Biophys. Res. Comm.*, 135, 139, 1986.
112. **Yoshimoto, T., Soberman, R. J., Lewis, R. A., and Austen, K. F.,** Isolation and characterization of leukotriene C_4 synthetase of rat basophilic leukemia cells, *Proc. Natl. Acad. Sci. U.S.A.*, 82, 8399, 1985.
113. **Bach, M. K., Brashler, J. R., and Morton, D. R., Jr.,** Solubilization and characterization of the leukotriene C_4 synthetase of rat basophil leukemia cells: a novel, particulate glutathione *S*- transferase, *Arch. Biochem. Biophys.*, 230, 455, 1984.
114. **Söderström, M., Mannervik, B., Orning, L., and Hammarström, S.,** Leukotriene C_4 formation catalyzed by three distinct forms of human cytosolic glutathione transferase, *Biochem. Biophys. Res. Comm.*, 128, 265, 1985.
115. **Bach, M. K. and Brashler, J. R.,** A comparison of the leukotriene synthesizing ability of subfractions of rat liver glutathione S- transferases, *Prostaglandins, Leukotriene Med.*, 17, 125, 1985.
116. **Soberman, R. J., Harper, T. W., Betteridge, D., Lewis, R. A., and Austen, K. F.,** Characterization and separation of the arachidonic acid 5-lipoxygenase and linoleic acid Ω-6 lipoxygenase (arachidonic acid 15-lipoxygenase) of human polymorphonuclear leukocytes, *J. Biol. Chem.*, 260, 4508, 1985.
117. **Naccache, P. H., Showell, H. J., Becker, E. L., and Sha'afi, R. I.,** Arachidonic acid induced degranulation of rabbit peritoneal neutrophils, *Biochem. Biophys. Res. Commun.*, 87, 292, 1979.
118. **O'Flaherty, J. T., Showell, H. J., Becker, E. L., and Ward, P. A.,** Neutrophil aggregation and degranulation. Effect of arachidonic acid, *Am. J. Pathol.*, 95, 435, 1979.
119. **Volpi, M., Naccache, P. H., and Sha'afi, R. I.,** Arachidonic acid metabolites increase the plasma membrane permeability of the neutrophils to calcium, *Biochem. Biophys. Res. Commun.*, 92, 1231, 1980.
120. **Sha'afi, R. I., Naccache, P. H., Alobaidi, T., Molski, T. F. P., and Volpi, M.,** Effect of arachidonic acid and the chemotactic factor f-Met-Leu-Phe on cation transport in rabbit neutrophils, *J. Cell. Physiol.*, 106, 215, 1980.
121. **Molski, T. F. P., Naccache, P. H., Borgeat, P., and Sha'afi, R. I.,** Similarities in the mechanisms by which formyl-methionyl-leucyl-phenylalanine, arachidonic acid and leukotriene B_4 increase calcium and sodium fluxes in rabbit neutrophils, *Biochem. Biophys. Res. Commun.*, 103, 227, 1981.
122. **McPhail, L. C., Clayton, C. C., and Snyderman, R.,** A potential second messenger role for unsaturated fatty acids: activation of Ca^{2+}-dependent protein kinase, *Science*, 224, 622, 1984.
123. **Bromberg, Y. and Pick, E.,** Unsaturated fatty acids as second messengers of superoxide generation by macrophages, *Cell. Immunol.*, 79, 240, 1983.
124. **Ford-Hutchinson, A. W., Bray, M. A., Doig, M. V., Shipley, M. E., and Smith, M. J. H.,** Leukotriene B_4, a potent chemokinetic and aggregating substance released from polymorphonuclear leukocytes, *Nature (London)*, 286, 264, 1980.
125. **Palmer, R. M. J., Stepney, R. J., Higgs, G. A., and Eakins, K. E.,** Chemokinetic activity of arachidonic acid lipoxygenase products on leukocytes of different species, *Prostaglandins*, 20, 411, 1980.
126. **Kreisle, R. A., Parker, C. W., Griffin, G. L., Senior, R. M., and Stenson, W. F.,** Studies of leukotriene B_4 specific binding and function in rat polymorphonuclear leukocytes: absence of a chemotactic response, *J. Immunol.*, 134, 3356, 1985.
127. **Goetzl, E. J. and Pickett, W. C.,** Novel structural determinants of the human neutrophil chemotactic activity of leukotriene B_4, *J. Exp. Med.*, 153, 482, 1981.

128. **Hoffstein, S. T., Manzi, R. M., Razzartis, K. A., Bender, P. E., and Gleason, J.,** Structural requirements for chemotactic activity of leukotriene B_4, *Prostaglandins*, 31, 205, 1986.
129. **Goetzl, E. J. and Pickett, W. C.,** The human PMN leukocyte chemotactic activity of complex hydroxyeicosatetraenoic acids (HETEs), *J. Immunol.*, 125, 1789, 1980.
130. **Dahinden, C. A., Clancy, R. M., and Hugli, T. E.,** Stereospecificity of leukotriene B_4 and structure-function relationships for chemotaxis of human neutrophils, *J. Immunol.*, 133, 1477, 1984.
131. **Showell, H. J., Naccache, P. H., Borgeat, P., Picard, S., Vallerand, P., Becker, E. L., and Sha'afi, R. I.,** Characterization of the secretory activity of leukotriene B_4 towards rabbit neutrophils, *J. Immunol.*, 128, 811, 1982.
132. **Beckman, J. K., Gay, J. C., Brash, A. R., Lukens, J. N., and Oates, J. A.,** Investigations of the biological activity of leukotriene A_4 in human polymorphonuclear leukocytes, *Biochem. Biophys. Res. Commun.*, 133, 23, 1985.
133. **Lewis, R. A., Goetzl, E. J., Drazen, J. M., Soter, N. A., Austen, K. F., and Corey, E. J.,** Functional characterization of synthetic leukotriene B_4 and its stereochemical isomers, *J. Exp. Med.*, 154, 1243, 1981.
134. **Palmblad, J., Uden, A. M., Lindgren, J. A., Radmark, O., Hansson, G., and Malmsten, C. L.,** Effects of novel leukotrienes on neutrophil migration, *FEBS Lett.*, 144, 81, 1982.
135. **Shak, S., Perez, H. D., and Goldstein, I. M.,** A novel dioxygenation product of arachidonic acid possesses potent chemotactic activity for human polymorphonuclear leukocytes, *J. Biol. Chem.*, 258, 4948, 1983.
136. **Lee, T. H., Mencia-Huerta, J. M., Shih, C., Corey, E. J., Lewis, R. A., and Austen, K. F.,** Characterization and biologic properties of 5,12- dihydroxy derivatives of eicosapentanoic acid, including leukotriene B_5 and the double lipoxygenation product, *J. Biol. Chem.*, 259, 2383, 1986.
137. **Palmblad, J., Malmsten, C. L., Uden, A.-M., Radmark, O., Engstedt, L., and Samuelsson, B.,** Leukotriene B_4 is a potent and stereospecific stimulator of neutrophil chemotaxis and adherence, *Blood*, 58, 658, 1981.
138. **Goetzl, E. J., Brindley, L. L., and Goldman, D. W.,** Enhancement of human neutrophil adherence by synthetic leukotriene constituents of the slow-reacting substance of anaphylaxis, *Immunology*, 50, 35, 1983.
139. **Gimbrone, M. A., Brock, A. F., and Schafer, A. I.,** Leukotriene B_4 stimulates polymorphnuclear leukocyte adhesion to cultured endothelial cells, *J. Clin. Invest.*, 74, 1552, 1984.
140. **Hoover, R. L., Karnovsky, M. J., Austen, K. F., Corey, E. J., and Lewis, R. A.,** Leukotriene B_4 action of endothelium mediates augmented neutrophil/endothelial adhesion, *Proc. Natl. Acad. Sci. U.S.A.*, 81, 2191, 1984.
141. **Buchanan, M. R., Vazquez, M. J., and Gimbrone, M. A.,** Arachidonic acid metabolism and the adhesion of human polymorphonuclear leukocytes to cultured vascular endothelial cells, *Blood*, 62, 889, 1983.
142. **Fehr, J. and Huber, A.,** Complement-induced granulocyte adhesion and aggregation are mediated by different factors: evidence for non-equivalence of the two cell functions, *Immunology*, 53, 583, 1984.
143. **Damerau, B. and Keller, H. U.,** Comparison of adhesion to glass and aggregation of human leucocytes, *Agents Actions*, 13, 53, 1983.
144. **Craddock, P. R., Hammerschmidt, D., White, J. G., Dalmaso, A. P., and Jacob, H. S.,** Complement induced granulocyte aggregation in vitro. A possible mechanism of complement mediated leukostasis and leukopenia, *J. Clin. Invest.*, 60, 260, 1977.
145. **Fehr, J. and Jacob, H. S.,** In vitro granulocyte adherence and in vivo margination: two associated complement dependent functions. Studies based on the acute neutropenia of filtration leukophoresis, *J. Exp. Med.*, 146, 641, 1977.
146. **O'Flaherty, J. T., Thomas, M. J., Lees, C. J., and McCall, C. E.,** Neutrophil-aggregating activity of monohydroxyeicosatetraenoic acids, *Am. J. Pathol.*, 104, 55, 1981.
147. **Ford-Hutchinson, A. W., Bray, M. A., and Smith, M. J. H.,** The aggregation of rat neutrophils by arachidonic acid: a possible bioassay for lipoxygenase activity, *J. Pharm. Pharmacol.*, 31, 368, 1979.
148. **O'Flaherty, J. T., Showell, H. J., Becker, E. L., and Ward, P. A.,** Neutrophil aggregation and degranulation. Effect of arachidonic acid, *Am. J. Pathol.*, 95, 435, 1979.
149. **O'Flaherty, J. T.,** Involvement of bivalent cations and arachidonic acid in neutrophil aggregation, *Inflammation*, 4, 181, 1980.
150. **O'Flaherty, J. T., Bass, D., DeChatelet, T. L., Waite, M., and McCall, C.,** Role of arachidonic acid in neutrophil aggregation, *Prog. Lipid Res.*, 20, 671, 1981.
151. **Bokoch, G. M. and Reed, P. W.,** Effect of various lipoxygenase metabolites of arachidonic acid on degranulation of polymorphonuclear leukocytes, *J. Biol. Chem.*, 256, 5317, 1981.
152. **Serhan, C. N., Fahlstadius, P., Dahlen, S. E., Hamberg, M., and Samuelsson, B.,** Biosynthesis and biological activities of lipoxins, in *Advances in Prostaglandin Thromboxane Leukotriene Research*, Hayaishi, O. and Yamamoto, S., Eds., Raven Press, New York, 1985, 163.
153. **Palmblad, J., Gyllenhammar, H., Lindgren, J. A., and Malmsten, C. L.,** Effects of leukotrienes and f-Met-Leu-Phe on oxidative metabolism of neutrophils and eosinophils, *J. Immunol.*, 132, 3041, 1984.
154. **Gay, J. G., Beckman, J. K., Brash, A. R., Oates, J. A., and Lukens, J. N.,** Enhancement of chemotactic factor stimulated neutrophil oxidative metabolism by leukotriene B_4, *Blood*, 64, 780, 1984.

155. **Goetzl, E. J., Brash, A. R., Tauber, A. J., and Hubbard, W. C.,** Modulation of human neutrophil function by mono-hydroxy-eicosatetraenoic acids, *J. Immunol.*, 39, 491, 1980.
156. **Tak, H. L., Hoover, R. L., Williams, J. D., Sperling, R. I., Ravalese, J., Spur, B. W., Robinson, D. R., Corey, E. J., Lewis, R. A., and Austen, K. F.,** Effect of dietary enrichment with eicosapentaenoic and docohexaenoic acids on in vitro neutrophil and monocyte leukotriene generation and neutrophil function, *N. Engl. J. Med.*, 312, 1217, 1985.
157. **Lammi-Keefe, C. J., Hammerschmidt, D. E., and Weisdorf, D. J.,** Influence of dietary omega-3 fatty acids on granulocyte function, *Inflammation*, 6, 227, 1982.
158. **Kreisle, R. A. and Parker, C. W.,** Specific binding of leukotriene B_4 to a receptor on human polymorphonuclear leukocytes, *J. Exp. Med.*, 157, 628, 1983.
159. **Goldman, D. W., Gifford, L. A., Marotti, T., Koo, C. H., and Goetzl, E. J.,** Molecular and cellular properties of human polymorphonuclear leukocyte receptors for leukotriene B_4, *Fed. Proc.*, 46, 200, 1987.
160. **Gorman, R. R., Ruppel, P. L., and Lin, A. H.,** Evidence for leukotriene B_4 receptors in human neutrophils, in *Advances in Prostaglandin Thromboxane Leukotriene Research*, Hayaishi, O. and Yamamoto, S., Eds., Raven Press, New York, 1985, 661.
161. **Goldman, D. W. and Goetzl, E. J.,** Specific binding of leukotriene B_4 to receptors on human polymorphonuclear leukocytes, *J. Immunol.*, 129, 1600, 1982.
162. **Goldman, D. W. and Goetzl, E. J.,** Heterogeneity of human polymorphonuclear leukocyte receptors for leukotriene B_4. Identification of a subset of high affinity receptors that transduce the chemotactic response, *J. Exp. Med.*, 159, 1027, 1984.
163. **Lin, A. H., Ruppel, P. L., and Gorman, R. R.,** Leukotriene B_4 binding to human neutrophils, *Prostaglandins*, 28, 837, 1984.
164. **Clancy, R. M., Dahinden, C. A., and Hugli, T. E.,** Oxidation of leukotrienes at the Ω end: demonstration of a receptor for the 20-hydroxy derivative of leukotriene B_4 on human neutrophils and implications for the analysis of leukotriene receptors, *Proc. Natl. Acad. Sci. U.S.A.*, 81, 5729, 1984.
165. **O'Flaherty, J. T., Redman, J. F., and Jacobson, D. P.,** Protein kinase C regulates leukotriene B_4 receptors in human neutrophils, *FEBS Lett.*, 206, 279, 1986.
166. **Goldman, D. W., Olson, D. M., Payan, D. G., Gifford, L. A., and Goetzl, E. J.,** Development of receptors for leukotriene B_4 on HL-60 cells induced to differentiate by 1-alpha, 25-dihydroxyvitamin D_3, *J. Immunol.*, 136, 4631, 1986.
167. **Naccache, P. H., Molski, T. F. P., Borgeat, P., and Sha'afi, R. I.,** Association of leukotriene B_4 with the cytoskeleton of rabbit neutrophils. Effect of chemotactic factor and phorbol esters, *Biochem. Biophys. Res. Commun.*, 124, 963, 1984.
168. **Becker, E. L., Kermode, J. C., Naccache, P. H., Yassin, R., Marsh, M. L., Munoz, J. J., and Sha'afi, R. I.,** The inhibition of neutrophil granule enzyme secretion and chemotaxis by pertussis toxin, *J. Cell Biol.*, 100, 1641, 1985.
169. **Molski, T. F. P., Naccache, P. H., Marsh, M. L., Kermode, J., Becker, E. L., and Sha'afi, R. I.,** Pertussis toxin inhibits the rise in the intracellular concentration of free calcium that is induced by chemotactic factors in rabbit neutrophils. Possible role of the "G proteins" in calcium mobilization, *Biochem. Biophys. Res. Commun.*, 124, 644, 1984.
170. **Shefcyk, J., Yassin, R., Volpi, M., Molski, T. F. P., Naccache, P. H., Munoz, J. J., Becker, E. L., Feinstein, M. B., and Sha'afi, R. I.,** Pertussis toxin but not cholera toxin inhibits the stimulated increase in actin association with the cytoskeleton in rabbit neutrophils: role of the "G proteins" in stimulus-response coupling, *Biochem. Biophys. Res. Commun.*, 126, 1174, 1985.
171. **Bradford, P. G. and Rubin, R. P.,** Pertussis toxin inhibits chemotactic factor induced phospholipase C stimulation and lysosomal enzyme release in rabbit neutrophils, *FEBS Lett.*, 183, 317, 1985.
172. **Holian, A.,** Leukotriene B_4 stimulation of phosphatidylinositol turnover in macrophages and inhibition by pertussis toxin, *FEBS Lett.*, 201, 15, 1986.
173. **Ui, M.,** Islet-activating protein, pertussis toxin: a probe for function of the inhibitory guanine nucleotide regulatory component of adenylate cyclase, *Trends Pharmacol. Sci.*, 4, 272, 1984.
174. **Volpi, M., Yassin, R., Tao, W., Molski, T. F. P., Naccache, P. H., and Sha'afi, R. I.,** Leukotriene B_4 mobilizes calcium without the breakdown of polyphosphoinisitides and the production of phosphatidic acid in rabbit neutrophils, *Proc. Natl. Acad. Sci. U.S.A.*, 81, 5966, 1984.
175. **Bradford, P. G. and Rubin, R. P.,** Leukotriene B_4 stimulates phospholipase C mediated formation of inositol trisphosphate in neutrophils: implications for convergence of the phospholipase A_2 and C pathways, in *Advances in Prostaglandin Thromboxane Leukotriene Research*, Hayashi, O. and Yamamoto, S., Eds., Raven Press, New York, 1985, 105.
176. **Taylor, C. W., Merritt, J. E., Putney, J. W., and Rubin, R. P.,** Effects of Ca^{2+} on phosphoinositide breakdown in exocrine pancreas, *Biochem. J.*, 238, 765, 1986.
177. **Andersson, T., Schlegel, W., Monod, A., Krause, K. H., Stendahl, O., and Lew, P. D.,** Leukotriene B_4 stimulation of phagocytes results in the formation of inositol 1,4,5-trisphosphate. A possible second messenger for Ca^{2+} mobilization, *Biochem. J.*, 240, 333, 1986.

178. **Bradford, P. G. and Rubin, R. P.**, Characterization of formylmethionyl-leucyl-phenylalanine stimulation of inositol trisphosphate accumulation in rabbit neutrophils, *Mol. Pharmacol.*, 27, 1985.
179. **Naccache, P. H., Sha'afi, R. I., Borgeat, P., and Goetzl, E. J.**, Mono- and dihydroxy-eicosatetraenoic acids alter calcium homeostasis in rabbit neutrophils, *J. Clin. Invest.*, 64, 1584, 1981.
180. **Sha'afi, R. I., Naccache, P. H., Molski, T. F. P., Borgeat, P., and Goetzl, E. J.**, Cellular regulatory role of leukotriene B_4: its effects on cation homeostasis in rabbit neutrophils, *J. Cell. Physiol.*, 108, 401, 1981.
181. **White, J. R., Naccache, P. H., Molski, T. F. P., Borgeat, P., and Sha'afi, R. I.**, Direct demonstration of increased intracellular concentration of free calcium in rabbit and human neutrophils following stimulation by chemotactic factors, *Biochem. Biophys. Res. Commun.*, 113, 44, 1983.
182. **Naccache, P. H., Molski, T. F. P., Borgeat, P., White, J. R., and Sha'afi, R. I.**, Phorbol esters inhibit the f-Met-Leu-Phe and leukotriene B_4 stimulated calcium mobilization and enzyme secretion in rabbit neutrophils, *J. Biol. Chem.*, 260, 2125, 1985.
183. **Lew, P. D., Dayer, J. M., Wollheim, C. B., and Pozzan, T.**, Effect of leukotriene B_4, prostaglandin E_2 and arachidonic acid on cytosolic calcium in human neutrophils, *FEBS Lett.*, 166, 44, 1984.
184. **Dyett, D. E., Malawista, S. E., Naccache, P. H., and Sha'afi, R. I.**, Stimulated cytokineplasts from human polymorphonuclear leukocytes mobilize calcium and polymerize actin. Cytoplasts made in cytochalasin B retain a defect in actin polymerization, *J. Clin. Invest.*, 77, 34, 1986.
185. **Naccache, P. H., Molski, T. F. P., Borgeat, P., and Sha'afi, R. I.**, Mechanism of action of leukotriene B_4: intracellular calcium redistribution in rabbit neutrophils, *J. Cell. Physiol.*, 118, 13, 1984.
186. **Naccache, P. H., Molski, T. F. P., Borgeat, P., and Sha'afi, R. I.**, Intracellular calcium redistribution and its relationship to f-Met-Leu-Phe, leukotriene B_4 and phorbol ester induced rabbit neutrophil degranulation, *J. Cell. Physiol.*, 122, 273, 1985.
187. **Yassin, R., Shefcyk, J., White, J. R., Tao, W., Volpi, M., Molski, T. F. P., Naccache, P. H., Feinstein, M. B., and Sha'afi, R. I.**, Effects of chemotactic factors and other agents on the amount of actin and a 65,000-mol-wt protein associated with the cytoskeleton of rabbit and human neutrophils, *J. Cell. Biol.*, 101, 182, 1985.
188. **Sklar, L. A., Omann, G. M., and Painter, R. G.**, Relationship of actin polymerization and depolymerization to light scattering in human neutrophils: dependence on receptor occupancy and intracellular Ca^{2+}, *J. Cell Biol.*, 101, 1161, 1985.
189. **Sha'afi, R. I., Shefcyk, J., Yassin, R., Molski, T. F. P., Volpi, M., Naccache, P. H., White, J. R., Feinstein, M. B., and Becker, E. L.**, Is a rise in intracellular concentration of free calcium necessary of sufficient for actin polymerization, *J. Cell Biol.*, 102, 1459, 1986.
190. **Sha'afi, R. I., Naccache, P. H., Molski, T. F. P., and Volpi, M.**, Chemotactic stimuli-induced changes in the pHi of rabbit neutrophils, in *Intracellular pH: Its Measurement, Regulation and Utilization in Cellular Function,* Nuccitelli, R. and Deamer, D. W., Eds., Alan R. Liss, New York, 1982, 513.
191. **Molski, T. F. P., Ford, C., Weisman, S. J., and Sha'afi, R. I.**, Cell alkalinization is not necessary and increased sodium influx is not sufficient for stimulated superoxide production, *FEBS Lett.*, 203, 267, 1986.
192. **Grinstein, S. and Furuya, W.**, Cytoplasmic pH regulation in phorbol ester activated human neutrophils, *Am. J. Physiol.*, 251, C55, 1986.
193. **Grinstein, S., Furuya, W., and Biggar, W. D.**, Cytoplasmic pH regulation in normal and abnormal neutrophils. Role of superoxide generation and Na^+/H^+ exchange, *J. Biol. Chem.*, 261, 512, 1986.
194. **Simchowitz, L. and Cragoe, E. J.**, Regulation of human neutrophil chemotaxis by intracellular pH, *J. Biol. Chem.*, 261, 6492, 1986.
195. **Nishizuka, Y.**, Studies and perspectives on protein kinase C, *Science,* 233, 305, 1986.
196. **Hansson, A., Serhan, C. N., Haeggstrom, J., Ingelman-Sundberg, M., and Samuelsson, B.**, Activation of protein kinase C by lipoxin A and other eicosanoids. Intracellular action of oxygenated products of arachidonic acid, *Biochem. Biophys. Res. Commun.*, 13, 1215, 1986.
197. **Naccache, P. H., Molski, M. M., and Sha'afi, R. I.**, Polymyxin B inhibits phorbol 12,myristate 13,acetate, but not chemotactic factor, induced effects in rabbit neutrophils, *FEBS Lett.*, 193, 227, 1985.
198. **Sha'afi, R. I., Molski, T. F. P., Huang, C. K., and Naccache, P. H.**, The inhibition of neutrophil responsiveness caused by phorbol esters is blocked by the protein kinase inhibitor H-7, *Biochem. Biophys. Res. Commun.*, 137, 50, 1986.
199. **Gerard, C., McPhail, L., Marfat, A., Stimler-Gerard, N., Bass, D. A., and McCall, C. E.**, Role of protein kinases in stimulation of human polymorphonuclear leukocyte oxidative metabolism by various agonists. Differential effects of a novel protein kinase inhibitor., *J. Clin. Invest.*, 77, 61, 1986.
200. **Schell-Frederick, E.**, A comparison of the effects of soluble stimuli on free cytoplasmic and membrane bound calcium in human neutrophils, *Cell Calcium,* 5, 237, 1984.
201. **Lagast, H., Pozzan, T., Wadvogel, F. A., and Lew, P. D.**, Phorbol myristate acetate stimulates ATP dependent calcium transport by the plasma membrane of neutrophils, *J. Clin. Invest.*, 73, 878, 1984.

202. **Naccache, P. H., Molski, T. F. P., Borgeat, P., and Sha'afi, R. I.,** Phorbol esters inhibit the f-Met-Leu-Phe and leukotriene B_4 stimulated calcium mobilization and enzyme secretion in rabbit neutrophils, *J. Biol. Chem.*, 260, 2125, 1985.
203. **Kato, H., Ishitoya, J., and Takenawa, T.,** Inhibition of inositol phospholipid metabolism and calcium mobilization by cyclic AMP-increasing agents and phorbol ester in neutrophils, *Biochem. Biophys. Res. Commun.*, 139, 1272, 1986.
204. **Claesson, H. E.,** Leukotriene A_4 and B_4 stimulate the formation of cyclic AMP in human leukocytes, *FEBS Lett.*, 139, 305, 1982.
205. **Gorman, R. R., Lin, A. H., and Hopkins, N. K.,** Acetylglycerylether phosphorylcholine (AGEPC) and leukotriene B_4 stimulated cyclic AMP levels in human polymorphonuclear leukocytes, *Adv. Cyclic Nucl. Prot. Phosphor. Res.*, 17, 631, 1984.
206. **Jackowski, S. and Sha'afi, R. I.,** Response of adenosine cyclic 3′, 5′monophosphate level in rabbit neutrophils to the chemotactic peptide formyl-methionyl-leucyl-phenylalanine, *Mol. Pharmacol.*, 16, 473, 1979.
207. **Simchowitz, L., Spilberg, I., and Atkinson, J. P.,** Evidence that the functional responses of human neutrophils occur independently of transient elevations in cyclic AMP levels, *J. Cyclic Nucl. Prot. Phosphor. Res.*, 9, 35, 1983.
208. **Showell, H. J., Otterness, I. G., Marfat, A., and Corey, E. J.,** Inhibition of leukotriene B_4 induced neutrophil degranulation by leukotriene B_4-dimethylamide, *Biochem. Biophys. Res. Commun.*, 106, 741, 1982.
209. **Sha'afi, R. I., Molski, T. F. P., Borgeat, P., and Naccache, P. H.,** Deactivation of the effects of f-Met-Leu-Phe and leukotriene B_4 on calcium homeostasis in rabbit neutrophils, *Biochem. Biophys. Res. Commun.*, 103, 766, 1981.
210. **Yuli, I. and Oplatka, A.,** Cytosolic acidification as an early transduction signal of human neutrophil chemotaxis, *Science*, 235, 340, 1987.
211. **Borgeat, P.,** unpublished observations, 1986.
212. **Borgeat, P.,** unpublished data, 1987.
213. **Beaumier, L., Faucher, N., and Naccache, P. H.,** Arachidonic acid-induced release of calcium in permeabilized human neutrophils, *FEBS Lett.*, 221, 289, 1987.
214. **Faucher, N. and Naccache, P. H.,** Relationship between pH, sodium and shape changes in chemotactic-factor-stimulated human neutrophils, *J. Cell. Physiol.*, 132, 483, 1987.
215. **Naccache, P. H., Sha'afi, R. I., and Borgeat, P.,** unpublished observations, 1986.

I. INTRODUCTION

The average lifespan of neutrophil leucocytes released from bone marrow in healthy individuals is believed to be on the order of hours. During this short lifespan the average neutrophil is thought to spend its initial hours shared about equally between residence in a marginated pool, reversibly attached to the endothelium of small blood vessels, and a circulating pool.[1] Marginated cells adjacent to an inflammatory locus can migrate between the endothelial cells and, through their basal lamina, into the underlying connective tissue.[2] Migration through the blood vessel wall and towards the focus of inflammation appears to occur by a taxic mechanism sensitive to the release of chemoattractant factors at that site.[3] One of the most potent chemoattractants released is C5a, generated via the complement pathway,[4] but diffusible peptides or denatured proteins resulting from proteolytic action at the site might also contribute to the concentration of neutrophils at these loci.[5] The major role of neutrophils at the inflammatory site appears to be phagocytosis and killing of microorganisms and the general destruction and clearance of cellular material. Therefore, neutrophils are able to adhere to endothelial surfaces, penetrate the blood vessel wall, locomote towards the site of inflammation, and thence recognize and internalize invading microorganisms. For these eventually sacrificial functions, the neutrophil is specialized.

Neutrophil leucocytes are perhaps the most obviously motile mammalian cell. Human cells can locomote on flat surfaces at average velocities up to 20 $\mu m\ min^{-1}$. Locomotion of cells can be regulated by a number of substances which might have significance for motility and directed migration in vivo. Denatured proteins, possibly derived from general proteolytic damage of cellular materials,[6] lipids of bacterial origin[7] and derivatives related to arachidonic acid,[8] *N*-formylated hydrophobic peptides[9] presumably derived from bacterial damage, and soluble complement fragments, particularly C5a,[4] will all promote general locomotory activity (chemokinesis), and many will promote orientation and directional migration up a concentration gradient (chemotaxis). Cell-surface binding characteristics for both C5a[4] and *N*-formylmethionyl peptides[10,11] suggest specific cell-surface receptors, but neither has yet been isolated and characterized. There is somewhat less certainty as to whether lipid or denatured protein chemoattractants operate via stereospecific receptors, and their mode of action remains largely unresolved. For receptor-mediated chemotaxis it is assumed that receptor occupation, albeit by entirely mysterious mechanisms, regulates the motile machinery of cells to result in net upgradient movement. Cells, remarkably, appear to be able to detect differences in concentration at opposite ends of the cell in the order of only 1%.[12] The nature of chemotactic mechanisms, particularly as it relates to the behavior of neutrophils, is discussed in an excellent review by Zigmond;[13] however, recent observations by Vicker et al.[14] suggest that neutrophils cannot detect a preexistent gradient but rather respond to temporal (and directional) changes in chemoattractant concentration.

The locomotion of neutrophils, as with locomotion of other crawling cells like fibroblasts, is assumed to be due to the activity of actin filament meshworks and other actin filament structures in conjunction with myosin. However, this assumption is essentially deductive rather than rigorously established by experiment. The superficial morphology of cell locomotion appears to be somewhat different in neutrophils and fibroblasts, although both types of locomotion appear to result from continuous cycles of protrusion, adhesion of the protruded part, and retraction of the rear portion of the cell (see Figure 6). The nature of the molecular mechanisms underlying these arbitrary phases of locomotion are at present merely a matter for speculation, although some pertinent observations have been made. In particular, Svitkina et al.[15] have shown that in fibroblasts whose actin filament architecture has been disassembled by depletion of ATP by metabolic inhibitors, removal of inhibition initiates reassembly which follows a defined sequence of assembly events presumably analogous to the constructive half of the actin filament cycle in locomoting cells. These observations are consistent with assembly of actin filament meshworks in the leading lamellae followed by organization and rearward contraction of this meshwork as a dorsal sheath. This sheath will presumably eventually disassemble at the rear and provide subunit material for continuation of the cycle. These observations are also consistent with those of Grebecki[16] on amebae, which show that fluid endoplasm containing construction materials for growing pseudopodia flows towards the tip of the pseudopodium and contributes to the growing gelled exoplasmic sheath, which then contracts slowly rearwards (towards the cell center) relative to the substratum. Interestingly, this contracting sheath carries surface membrane binding sites rearwards with it which may be analogous to the rearward movement of actin-containing arcs on the dorsal surface of fibroblasts.[17] Thus, a common sequence of molecular events might underlie both ameboid and fibroblast-like locomotion despite the superficial morphological differences. Neutrophil locomotion in vitro on planar substrata has an ameboid quality which may well be appropriate in vivo where locomotion predominantly occurs within a 3-dimensional matrix of collagen fibers and other complex connective tissue components, and where the adhesive phase of the locomotory cycle might be simply jamming of the leading pseudopodium in the interstices of the connective tissue matrix.[18]

At the site of inflammation, neutrophils will phagocytose material derived from both damaged cells in the area and any microorganisms which have been recognized (opsonized) by the humoral arm of the immune system. Particles coated with immunoglobulin and/or with covalently attached C3b from hydrolysis of C3 by C.142 are bound by neutrophils and phagocytosed.[19] A range of other molecules when coated onto surfaces will also initiate phagocytosis, notably the plant lectins.[20] Neutrophils will also indulge in cannibalism.[21] Phagocytosis and killing of live microrganisms is clearly an important facet of the overall immune response to invading cells, as is indicated by the prevalence of infection when neutrophil function is compromised.

Phagocytosis involves both the recognition and internalization of particles, essentially separate events which can be independently inhibited. The internalization phase in some respects resembles aspects of locomotion in that the cell effectively moves over the surface of the particle until it has surrounded and engulfed it. To what extent these events involve common mechanisms remains to be evaluated. The organization of actin filaments in neutrophils, their reorganization during phagocytosis and locomotion, and the mechanism by which receptors might regulate actin filament function to mediate these motile phenomena are the subject of this essay.

II. THE CORTICAL ACTIN FILAMENT MESHWORK

It was established early on in the study of actin filament function in nonmuscle cells that when nonmuscle tissues were homogenized, at least half of the actin was recovered in an unpolymerized state.[22] This was in contrast to experience with skeletal muscle, which at the time provided (and perhaps unfortunately still does) the model system for actin filament function, and where the majority of the actin is polymerized into filaments. Furthermore, the unpolymerized half of the tissue actin would not readily polymerize under conditions in which skeletal muscle actin would. These observations were interpreted at the time to indicate that nonmuscle actins were different from muscle actin with regard to polymerization. It was subsequently discovered that there were indeed specific isoforms of actin which were present in nonmuscle cells (as well as in different muscle tissues),[23] which from their position on isoelectric focusing gels were termed β-and γ-actin. Nonmuscle cells express only these isoforms (with molar ratios of about 5:1). Both the primary sequence and the characteristics of polymerization of these nonmuscle isoforms are very similar to those of skeletal muscle actin. The crucial differences observed in the tissue extracts were subsequently revealed by Carlsson et al.[24] to be due to the stoichiometric association of actin monomers with a small

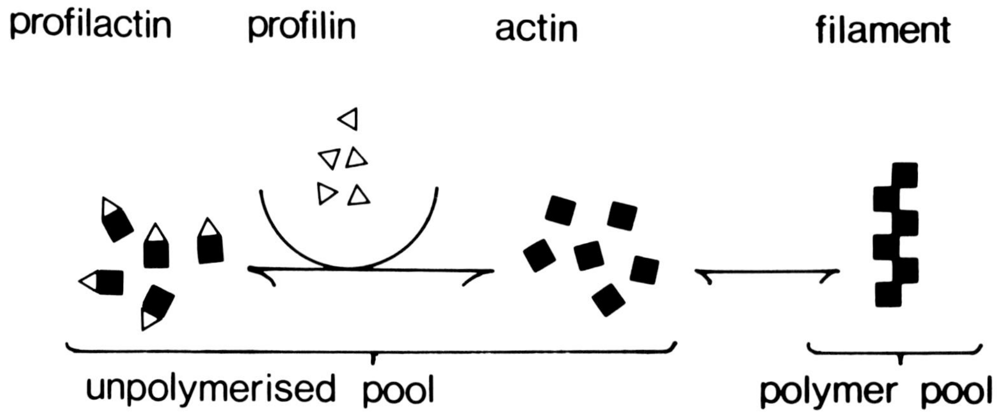

FIGURE 1. Buffering of the actin monomer pool by profilin. The proportion of total actin sequestered by profilin is a function of the critical concentration for polymer assembly (the free actin monomer concentration) and the affinity of profilin for actin.

basic protein called profilin, which essentially prevents self-nucleation of actin filaments in vitro and accounts for the observed differences in polymerization. The wide distribution of profilin, the approximate 1:1 stoichiometry of profilin to actin in tissues, and the isolation of soluble unpolymerized actin from tissues as the stoichiometric complex, profilactin, has led to the proposal that profilin buffers a pool of unpolymerized actin which can serve for filament assembly as and when required by the cell (Figure 1). The prediction from this hypothesis that net increases in the proportion of polymerized actin in cells should be paralleled by increases in the concentration of free profilin has been experimentally verified in thrombin-treated platelets.[25] The presence of profilin-like proteins in eukaryotic cells in widely divergent phyla[26] has engendered the view that maintenance of an actin monomer pool in cells by profilin is an essentially general mechanism for control of actin filament behavior in cells, although little direct evidence exists for this mechanism in neutrophils.

The proportion of actin in polymerized and unpolymerized pools is difficult to measure, since it is not clear in what form unpolymerized actin is present, i.e., as monomers, oligomers, profilactin, or other small complexes, nor, for that matter, is it known whether different forms of polymer exist in cells other than those observed and studied in vitro. The only currently available measure of the proportion of actin assembled in cells, and of the total cellular content, comes from the use of a DNAase I inhibition assay originally developed by Blikstad et al.,[27] which exploits the observation that DNAase I is inhibited by monomeric actin (or profilactin) but not by polymeric actin. Thus, it is possible to measure total and unpolymerized actin in cell extracts, at least as discriminated by the criteria of this assay system. In general and in experiments reported from our laboratory, there is precise correlation between the results of this assay and other visual or biochemical analyses. The localization of polymerized actin can be selectively achieved by using fluorescent derivatives of phalloidin, a cyclic peptide isolated from the toadstool *Amanita phalloides*. This peptide selectively, and with high affinity, binds to polymerized actin.[28] Thus, with the two approaches it is possible both to analyze net changes in the extent of actin polymerization and to localize polymers. More sophisticated techniques using microinjection of hapten- or fluorophore-labeled actin, as have proved useful in studying dynamic microtubule behavior in cells,[29] will be required to follow the flux of assembly, disassembly, and reorganization of actin filaments in living cells.

Both the DNAase I inhibition assay[30-34] and labeling of polymerized actin with fluorescent phalloidin derivatives[30-32,34-36] have been used to study actin filament behavior in neutrophils; in addition, association of actin with a sedimentable nonionic detergent-extracted cell residue has been used as an indicator of filament network assembly.[36-40] Semiquantitative analysis of phalloidin fluorescence has been achieved by using a fluorescence-activated cell sorter (FACS).[34-36] In resting cells, that is peripheral blood cells in suspension, the proportion of total actin in the monomer pool has been shown to be between 70 and 80% for both pig[31] and human[34] neutrophils. Thus, relatively little of the total actin is polymerized in resting cells, as determined by the DNAase I inhibition assay, compared with, for example, blood platelets, where some 50% of total actin is present as polymer. In pig cells, total actin is some 5 to 8% of total cell protein. The approximate monomer concentration in pig cells can be calculated if morphometric estimates for the proportional volume of the cytoplanic component (43%) and the size ($\phi = 6.8$ μm) of these cells are used, and assuming that actin monomer is restricted to and evenly distributed within this compartment. The content of unpolymerized actin is 3.52 ± 0.18 (11)[31] pg per cell in a cytoplasmic volume of 65 μm^3 which gives an unpolymerized actin concentration of about 5 mg/mℓ, some three orders of magnitude higher than the critical concentration for actin assembly in vitro under estimated cytoplasmic ionic conditions.[41] It is assumed that profilin and possibly other actin-binding species are responsible for sequestration of these large amounts of actin in a monomer pool, and that free actin monomer exists at whatever the effective critical concentration is within the cell. The advantage for cells in maintaining a large monomer pool is presumably that assembly of polymer can occur rapidly in any region of the cell independently of diffusion and that cells have a significant margin of monomer to satisfy any physiological requirements for assembly which may occur on adhesion to a substration, during phagocytosis, or during chemotactic factor-stimulated locomotion.

That polymerized actin exists in resting cells is indicated by the binding of fluorescent phalloidin to the cells[30-32,34-36] and by the DNAase I inhibition assay.[30-34] In addition, values obtained for the cellular content of polymerized actin by using the DNAase I inhibition assay correspond satisfactorily with estimates for the amount of actin sedimenting with detergent-insoluble cell residues.[31] Nevertheless, these estimates can only be considered as simplistic indicators for what are clearly complex processes. The polymerized actin content of pig neutrophils as determined by the DNAase I inhibition assay is about 1.2 pg/cell (note that pig neutrophils are small compared to human and rabbit cells). According to the localization of phalloidin fluorescence in resting cells, essentially all the polymerized actin is present in a thin cortical submembranous layer[30,32] (Figure 2), although others have noted a diffuse cytoplasmic fluorescence.[35] If neutrophils are treated with nonionic detergents, then a detergent-insoluble residue (cytoskeleton) is left which, remarkably, retains the overall shape, size, and nuclear localization of the whole cell and in which a thin cortical filament meshwork is retained (Figure 3).[42] The absence of any obvious actin filament-like structures between the cortex and the nucleus confirms the view dictated by the rhodamine-phalloidin images that very few actin filaments exist within the bulk of the cytoplasm. The cortical meshwork contains a feltwork of filaments some of which have the morphological characteristics of actin filaments. From transmission electron microscopy of such cortical meshworks (Figure 4) an approximate thickness of 0.1 to 0.2 μm can be obtained, a value similar to that reported for other cells (see review[43]). This cortical meshwork seen in detergent-extracted cells has similar dimensions to the cortical rim of cytoplasm which excludes organelles in intact cells, and is also characteristic of the appearance of lamellipodia, which engulf phagocytic particles and which occur at the leading edge of locomoting cells. If the assumption is made from the subjective appearance of this meshwork that filaments are oriented at angles between about 10 and 90° to the plasma membrane, then the actin filaments are going to have average lengths of only a fraction of a micrometer, that is, only some 30 to 100 subunits. This, together with the widespread presence of fodrin-like proteins, the most widely distributed isoform of spectrin in vertebrate cells,[44,45] and of other proteins thought at one time to be characteristic of erythrocytes, including band 4.1[46] and the SITS-sensitive anion transporter

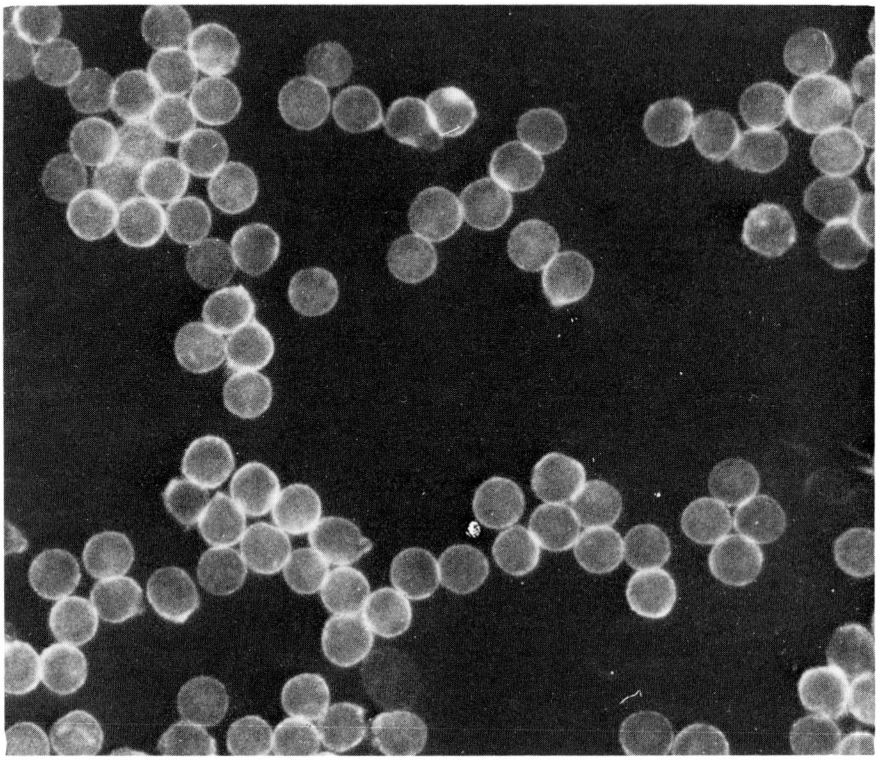

FIGURE 2. Localization of polymerized actin in suspended neutrophils using rhodamine phalloidin. Suspended cells were fixed, permeabilized then incubated with rhodamine phalloidin. Note the characteristic location of fluorescence within a thin cortical rim.

(band III)[47] (see also review[48]) suggests that the basic structural principles revealed in anucleate erythrocyte skeletons may extend to most other cells. The unperturbed actin filament cortex of neutrophils might thus be a complex of relatively short actin polymers cross-linked by a spectrin isoform and attached to the plasma membrane via ankyrin-like and band III-like proteins. The complete insensitivity of this polymerized actin pool to saturating concentrations of cytochalasin D[32] suggests that it may be protected by end-binding proteins and, in resting cells at least, be metabolically inactive. It has been previously established that cell-surface glycoproteins, particularly an 81-kdalton protein, exhibit properties consistent with their attachment to the cortical filament meshwork.[42] The involvement of this cortical meshwork in the generation of lamellipodia during locomotion and phagocytosis will be discussed in subsequent sections.

The presence of other actin filament-binding proteins in the cell cortex of phagocytic cells has been established by Stossel and colleagues over a number of years. Predominant among these are a filamin isoform which appears to allow flexible cross-links to occur between isotropic actin filaments[49] and which may make a major contribution to the gelling of actin filaments in cytoplasm. In addition, two Ca^{2+}-sensitive proteins have been identified and characterized: gelsolin, a protein which caps the barbed ends of actin filaments and, in the presence of micromolar Ca^{2+}, severs filaments[50] and α-actinin which can cross-link parallel actin filaments in the absence of Ca^{2+}.[51] Both proteins are widely distributed in eukaryotic cells. They are all probably located within the cortical filament meshwork of resting cells. There are, however, clearly additional relevant proteins in filament gels, which have yet to be characterized or even identified.[52]

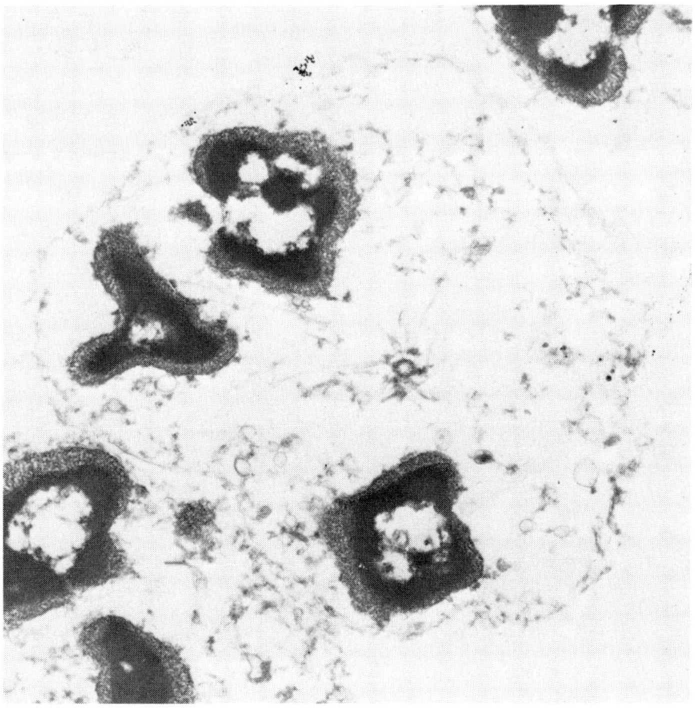

FIGURE 3. Utrastructural morpholoy of a detergent-insoluble residue of a neutrophil. Cells were treated with a polymer-stabilizing buffer containing nonionic detergent and protease inhibitors[42] prior to fixation. Note retention of the general organization of the cell, particularly the position of the cell periphery and the association of centrioles and their microtubules with the nuclear complex.

The structure and function of the cortex can be more directly studied by using functional cellular fragments of neutrophils isolated by Ficoll step density gradient centrifugation of cytochalasin B-treated neutrophils.[30,53] These fragments are spherical, sealed vesicles of cytoplasm surrounded by a portion (~25%) of the original plasma membrane, but which exclude nearly all membrane-bound intracellular organelles. These so-called cytoplasts survive for days in appropriate media and are functional as regards Fc-receptor-dependent phagocytosis,[30] although given the surface area: volume restrictions of spheres, these fragments find both phagocytosis and spreading somewhat difficult. The appearance and composition of the cortical filament meshwork can be studied in the essential absence of the lysosomal proteolytic enzymes which make much of the biochemistry with neutrophils difficult. The appearance of these structures and their cortical meshworks are shown in Figure 5. These structures have not been widely exploited to study cell cortex function, but it seems likely that they might yield answers to questions which are hard to ask (experimentally) in the whole cell.

III. GENERATION OF 3-DIMENSIONAL MOVEMENT

Phagocytosis and cell locomotion are excellent examples of motile activities which involve complex 3-dimensional shape changes in the cytoplasm during the generation of locomotory or engulfing lamellipodia. Both phenomena are by appearances, at least, protrusive events. An associated phenomenon particularly relevant to cell locomotion in subsequent contractile activity, which, after mechanical establishment of the protrusion at some site forward of the

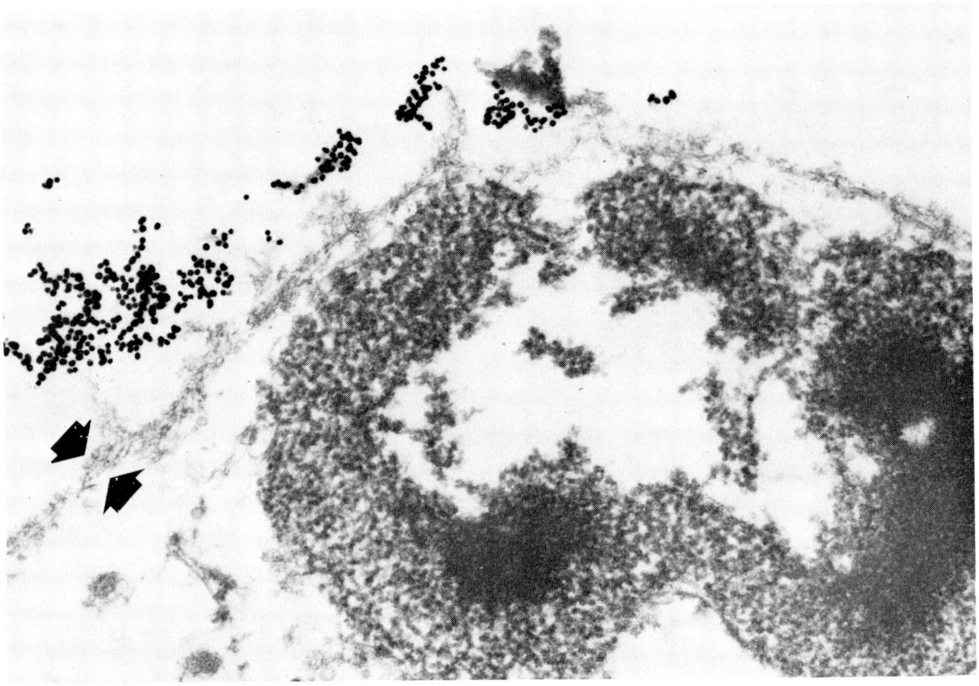

FIGURE 4. Structure of the cortical filament meshwork in detergent-insoluble cell residues. Neutrophils were treated with detergent as described in the legend to Figure 3, except that prior to extraction cells were exposed to Fc-coated colloidal gold aggregates which the cells phagocytose. Note the fine filamentous meshwork (arrows) which corresponds in position to the fluorescence shown in Figure 2. Linkages between phagocytic substrate and the cortical meshwork, presumably via Fc receptors, clearly survive detergent extraction of the lipid bilayer.

current position of the cell, is used to pull the rear of the cell towards this point. During phagocytosis, a similar mechanism might be used to draw a particle into the cell (Figure 6). A working hypothesis for both protrusion and contraction is that these processes are generated by the dynamic behavior of actin filaments or of actin filament interaction with myosin. This hypothesis is based predominantly on three classes of observation: first, on the sensitivity of at least some of these processes to cytochalasins, a class of fungal metabolite which interferes with actin filament function;[54] second, on the dramatic reorganization of actin filaments which occurs during these processes; and third, on the close association of actin and other 'contractile' proteins with sites of movement generation. Of course, the lessons learned on actin and myosin function in skeletal muscle have been influential.

Immunocytochemical evidence implicates a number of contractile proteins at the sites of movement generation. Some like filamin,[55] fimbrin,[56] actin,[32] and gelsolin[57] are concentrated in protrusive structures (note potential problems with the localization of gelsolin[57]) while others, including actin, myosin, myosin light chain kinase, γ-actinin, and tropomyosin,[59,60] are concentrated in contractile structures like stress fibers, the dorsal sheath, and the cleavage furrow. There are, however, with the exception of obvious specialized structures, difficulties in discriminating between protrusive regions and contractile regions, and it is possible that both activities can take place within the same actin filament meshwork by the dynamic interchange of different regulatory proteins, either in response to appropriate signals or to the immediate history of the gel.

Several models for the generation of 3-dimensional movement in nonmuscle cells have been proposed, most notably by Stossel and colleagues[61] in relation to locomotion and phagocytosis, and by Oster[63] using computer simulations of available kinetic data. The

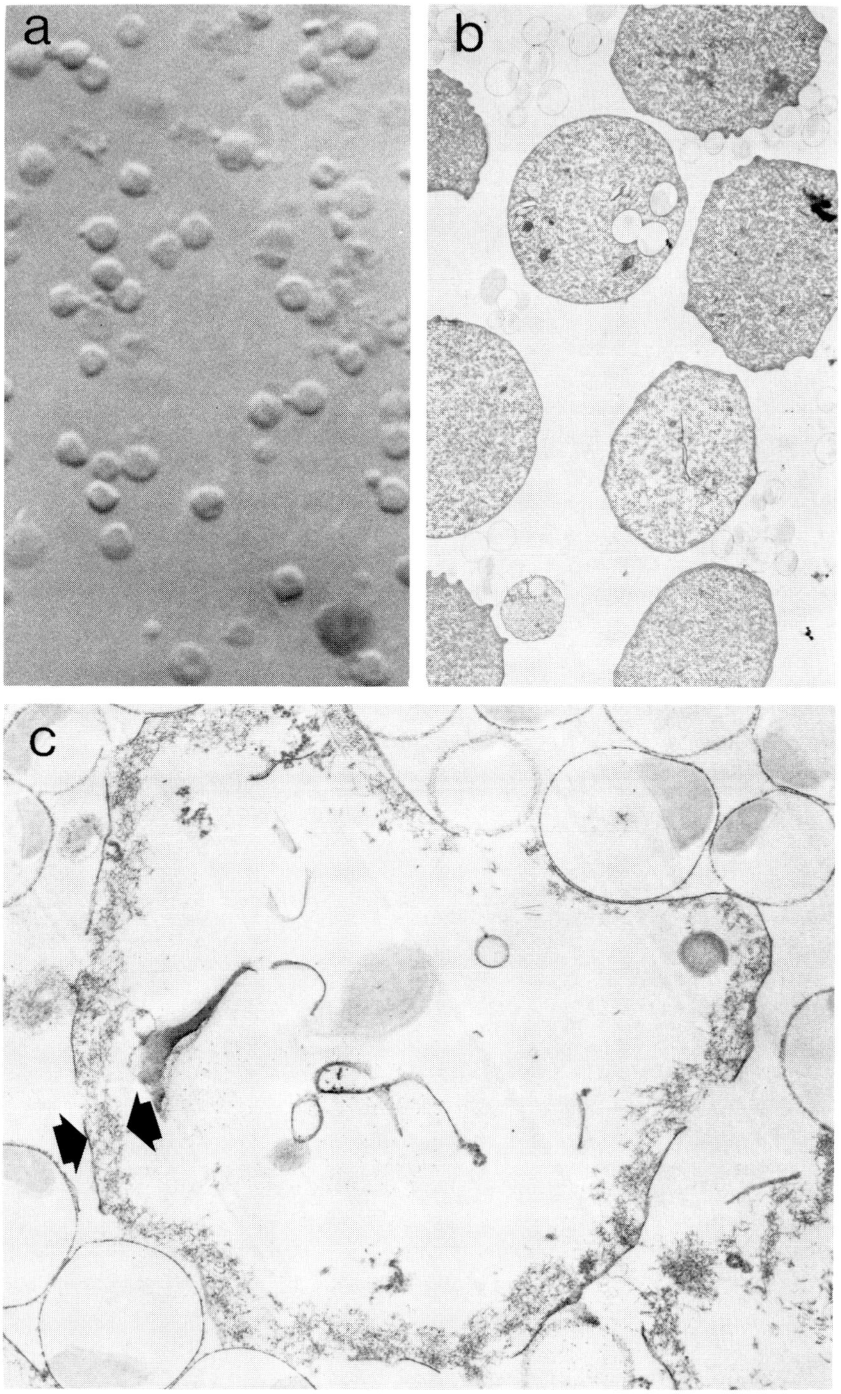

FIGURE 5. The cortical filamentous network of functional granule-free cytoplasts from neutrophils. Light- (a) and electron- (b) micrographs of granule-free cytoplasts prepared by Ficoll step density gradient centrifugation of cytochalasin-B-treated neutrophils.[30] Cytoplasts in (b) exposed to Fc-coated polystyrene spheres prior to fixation. Cortical meshworks (c) were prepared by detergent treatment of cytoplasts as indicated in the legend to Figure 3 for cells, except that cytoplasts had been exposed to Fc-coated polystyrene spheres prior to lysis. Note the even cortical meshwork (arrows) and the incomplete removal of lipid bilayer.

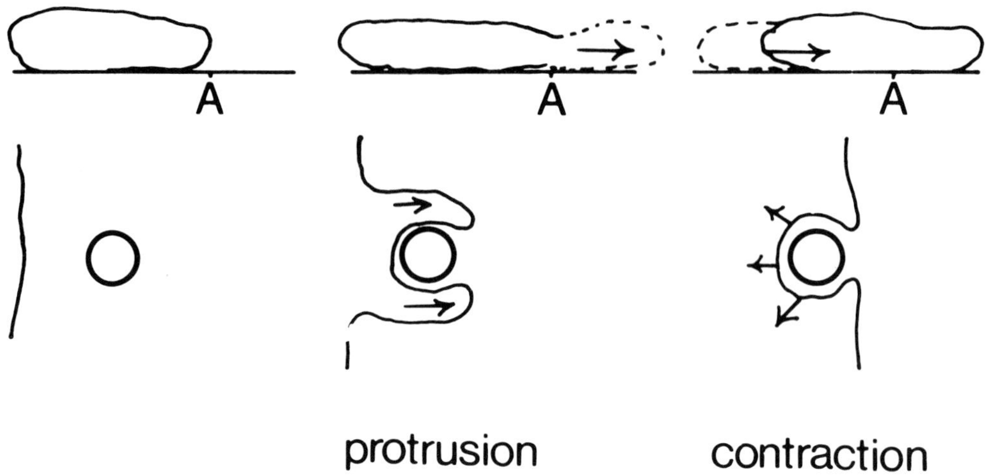

FIGURE 6. Possible relationship of protrusive and contractile events to phagocytosis and cell locomotion. Equivalent events during locomotion of a cell past a fixed point "A" and during phagocytosis of a spherical particle. Cell protrusions move over the surface of the substratum or particle, contraction of cell pulls either rear of cell to substratum attachment or attached particle to cell.

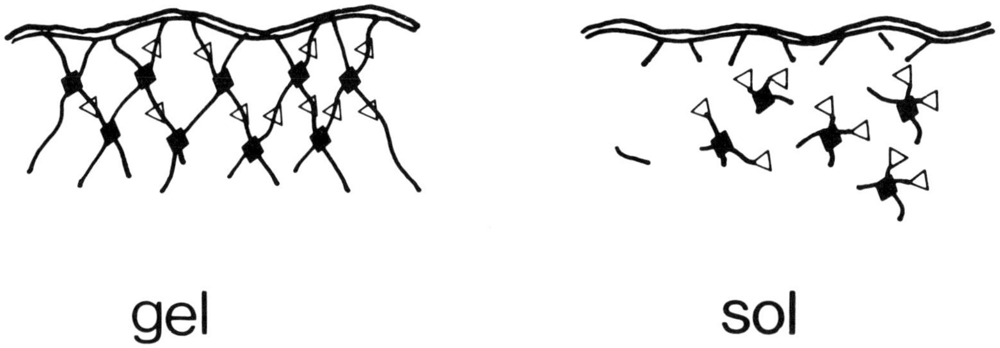

FIGURE 7. Possible role of filamin and gelsolin in controlling the rigidity of cortical cytoplasm. Actin filaments attached to the plasma membrane are cross-linked by filamin (♦) and bound by gelsolin (△). When gelsolin is activated (increase in $[Ca^{2+}]_i$?), actin filaments are severed, reducing rigidity. The less rigid gel would be deformable by myosin-mediated sliding (see text).

models of Stossel and colleagues have evolved from the study of the composition and behavior of actin filament gels formed from extracts of macrophages. These gels are predominantly actin filaments cross-linked by filamin in an isotropic array and associated with gelsolin and myosin (and probably several other relevant proteins).[61] The relative stoichiometries of these proteins are similar in both cell extracts and in vitro gels with both filamin and gelsolin being present at about 1:50 to 1:100 actin subunits. Stossel has proposed that an extensively cross-linked (by filamin) actin gel would be essentially rigid and that myosin, if activated in such a gel, would be working against the "elasticity" of the gel, rather than producing net movement or shape change. However, if the gel were made less rigid, that is, partially solated, myosin could then generate net relative movement in this loosened filament meshwork (Figure 7). Regulation in this model is provided by the Ca^{2+}-sensitivity of gelsolin which cleaves actin filaments and thus solates the gel in the presence of Ca^{2+}.[50] Shape change is also modulated by Ca^{2+}, via calmodulin and myosin light chain kinase,[62] controlling myosin activity. Thus, Ca^{2+} both reduces the rigidity of the gel and activates myosin

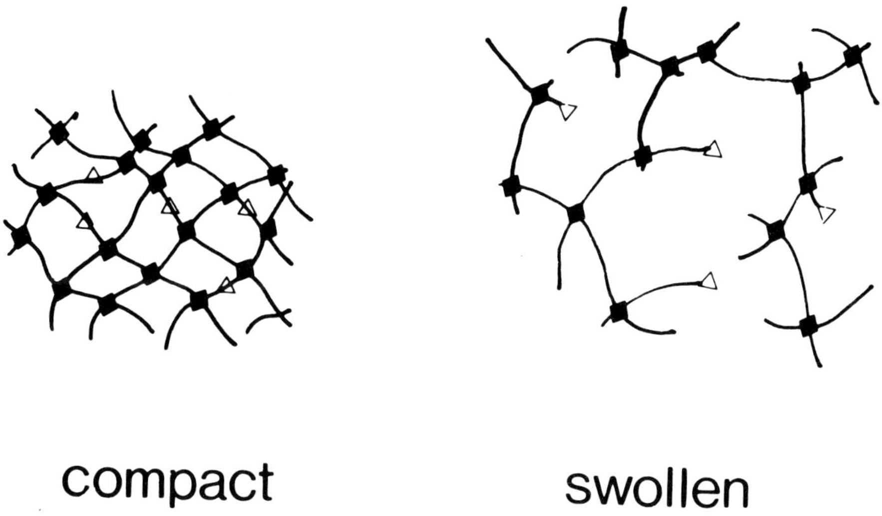

FIGURE 8. Osmotic swelling of cross-linked polymer meshworks. Actin filaments are cross-linked by filamin (♦) and cleaved by gelsolin (△). See text for description.

interaction with actin. This model essentially provides for complex contractile shape changes, but does not address the mechanism of protrusion.

A different perspective of the behavior of actin filament meshworks during complex 3-dimensional shape changes has been presented by Oster who, rather than considering the individual biochemical properties of the filament meshworks, has considered the overall physicochemical behavior of the system.[63] The actin filament meshwork is essentially a charged (net negatively; actin has a pI of ~ 5.3) cross-linked (filamin) polymer network within an ionic aqueous environment. Under these conditions, the meshwork can be considered in the same way as charged polymers trapped within a semipermeable membrane, since filaments cannot escape from a confined location. Thus, the charged mass is subject to a Donnan influx of ions which raises the effective osmotic pressure within the vicinity of the filaments. The filament mass thus has a tendency to swell. The net negative charges on the polymers may also contribute to interfilament repulsion in dense meshworks, as will the tendency of filaments to disperse to simply increase the degree of disorder of the system. Since the filament meshwork is restrained by its filamin cross-links, the meshwork will adopt an equilibrium volume under a particular set of conditions where the dispersive forces are equal and opposite to the elastic energy stored in the cross-linked meshwork. If cross-links or polymers are then severed, some of the elastic energy is released and the system will adopt a new (larger) equilibrium volume where greater distortion is conferred on a less highly cross-linked meshwork. Thus, in this model the action of gelsolin in the presence of Ca^{2+} could lead to an increase in volume of an actin filament network (Figure 8). The additional influence of myosin on such a meshwork would in principle be similar to that described by Stossel. Thus, subtle 3-dimensional Ca^{2+} gradients from, for example, interactions of the cell surface with external structures, could lead to complex shape change.

Although osmotic swelling might provide for a certain amount of volume increase under certain circumstances, growth of new actin filament meshwork might provide for major protrusive activity of cells. In principle, this would appear to require the net polymerization of new filaments and their organization with actin-binding proteins into higher order structures. It is possible to speculate that for actin polymers against a flexible membrane surface, under conditions where net assembly prevailed, random movements of the membrane away from the filament end would allow subunit exchange. Since under net polymerizing conditions

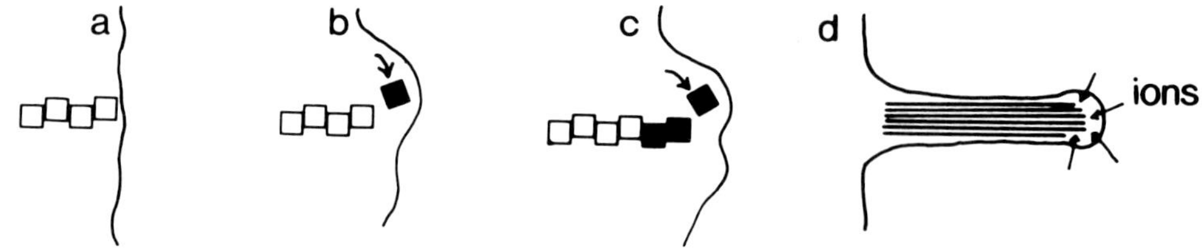

FIGURE 9. Consolidation of random thermal movements by polymerization. (a) Diagram showing how random movements of a flexible surface can be given direction by net polar assembly of actin filaments. (a to c) Random thermal fluctuations of the membrane allow the addition of monomers at the growing (barbed) end. Generation of an ion influx at the tip of a growing filopodium (d) could make "forward" protrusion more efficient.

the probability is that a subunit addition (rather than loss) would take place, then the random thermal movement of the membrane would have been consolidated by assembly of actin (Figure 9). In essence, random thermal movement is given direction by linear polar assembly. This process would be somewhat more efficient if a bundle of actin filaments participated since cooperation between filaments would occur. It is also possible that net ion pumping and its associated osmotic consequences by membrane transport mechanisms could generate "forward" movement of the advancing membrane. It is likely that assembly against a surface occurs in the rapidly assembling axonemal process of *Thyone* sperm.[634]

Contractile activity in nonmuscle cells is inferred from the presence of both myosin and actin in the same region of the cell. Although in most cases the inference is barely substantiated by additional evidence, in stress fibers there is reasonable evidence for actin and myosin-based contraction. The arrangement of proteins in stress fibers is clearly reminiscent of that in striated muscle where electron-dense cross-banding regions exist with linear periodicity, with alternating dense and less-dense regions occupying 0.2 to 0.4 μm and 0.2 to 0.7 μm, respectively. The dense regions contain myosin,[65] tropomyosin,[60] and myosin light chain kinase[59] and alternate with the less-dense regions containing α-actinin.[59] Similar ultrastructure exists in the cleavage furrow[66] of dividing cells.

The organization of myosin in nonmuscle cells remains largely unknown except within stress fibers where ultrastructural immunolocalization of both myosin head and myosin rod suggest that myosin exists in small bipolar aggregates.[67] Since myosin is also apparently present in the cortical meshwork of cells, which is only some 0.1 to 0.2 μm thick, polymers of myosin, which thus far have remained unobserved, must be relatively small. Stress fibers have also been dissected out of cultured cells and been shown to contract in the presence of ATP and Ca^{2+}.[68] Bundled actin filaments have not, to the author's knowledge, been observed in neutrophil leucocytes. However, stress fibers are rare in cells in vivo and may represent an extreme level of organization induced by adhesion to flat substrata, but one which nonetheless represents a common set of structural principles for actin filaments and their accessory proteins in nonmuscle cells. The untested assumption remains that interaction of myosin with actin filaments generates contractile motility during cell locomotion and phagocytosis.

IV. ACTIN FILAMENT BEHAVIOR DURING PHAGOCYTOSIS

Phagocytosis appears to be a continuous process where, on adhesion of an appropriate particle at the cell surface, lamellipodia progressively migrate over the particle surface until they meet, where they apparently fuse, releasing the (plasma) membrane-bound particle, the phagosome, into the cytoplasm (Figure 10). The process is arbitrarily broken down into stages which include recognition, binding, engulfment, release into the cytoplasm, and finally fusion with lysosomes. Whether the particle is pulled into the cytoplasm or is surrounded

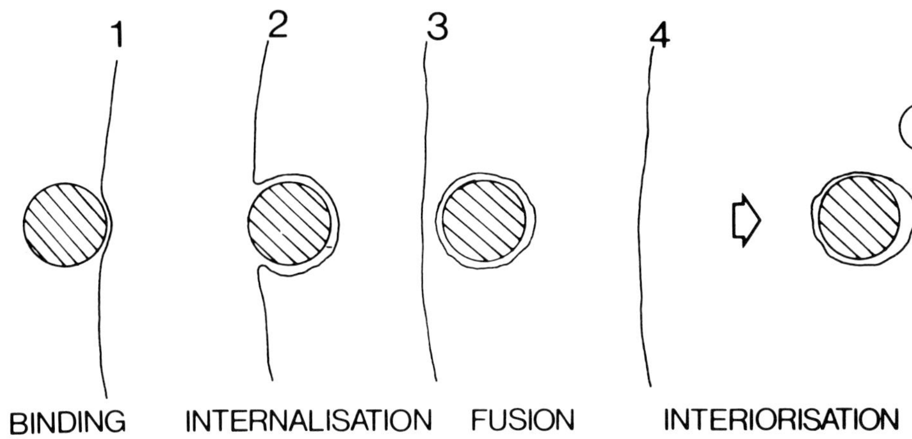

FIGURE 10. The phagocytic pathway. See text for explanation.

by cytoplasmic protrusions has mechanistic implications, but discrimination is not easy from morphological observation, particularly since most phagocytosis has been closely observed for cells locomoting on a flat substratum. One of the major barriers to the study of sequential or transient events during phagocytosis is that within a phagocytosing cell population, individual phagocytic events occur stochastically. Within the population, therefore, there are not only cells with phagocytic particles at different "stages" of the phagocytic process, but individual cells may also be simultaneously engaged in phagocytosis of several particles, each at a different stage.

The involvement of actin filaments during phagocytosis became a widespread expectation after the establishment of the involvement of actin in muscle contraction and the verification of its presence in nonmuscle cells. Early experiments to test the hypothesis simply involved the application of cytochalasins, metabolites of *Helminthosporium* species, which appeared by their effects on all things motile (see Reference 69), to interfere with actin filament function. It has been subsequently established that cytochalasins bind with high affinity ($K_d \sim 10^{-7}$ to $10^{-8}M$) to the barbed ends of actin polymers and at lower affinity ($K_d \sim 10^{-5}$ M) to the sides of filaments or to monomers.[70,78] A summary of the observations of the several papers[72-79] describing the effects of cytochalasins on phagocytosis is given in Table 1. Since the possible mechanisms of shape change involving actin filaments might be complex and involve both protrusive and contractile mechanisms (Figure 11), it is not easy to interpret those inhibition studies despite the likelihood that some of the cytochalasins (notably cytochalasin D) are relatively specific in their effects on cells. Cytochalasin B, the most frequently used cytochalasin in these studies, is also a potent inhibitor of hexose and nucleoside transport.[80] The concentration range, over which 50% effects on phagocytosis are observed, are rather higher than estimated K_d values for binding to filament ends and are closer to those expected for fragmentation of filaments.[71] Thus, insofar as the data can be interpreted, internalization of phagocytic particles does not involve net assembly at barbed ends of actin filaments, but requires their tensile or structural properties. More detailed studies on cytochalasin effects will be described below.

In addition to evidence from the effects of cytochalasins, the involvement of contractile proteins during phagocytosis has been inferred from the preferential localization of contractile proteins at sites of phagocytosis, as determined by immunocytochemical techniques and ultrastructural analysis. Actin filaments are preferentially associated with sites of particle ingestion as determined by transmission electron microscopy of stained sections[79,81] by immunnocytochemical analysis,[79,82] by use of fluorescent derivatives of phalloidin,[30,31] and by association with isolated phagocytic membrane.[83,84] Myosin and filamin have also been located preferentially at sites of particle ingestion by immunocytochemical techniques.[55,85,86]

Table 1
REPORTED EFFECTS OF CYTOCHALASIN B ON PHAGOCYTOSIS

Cell type	Phagocytic substrate	Concentration (μM)	% Inhibition	Ref.
Macrophages	Ig-coated RBC	2	80	26
		20	98	
Macrophages L-cells	Latex spheres	10	50	18
Macrophages	PVP[a]	2—5	50	16
Neutrophils	Bacteria	2	30	24
		40	60	
Macrophages	Bacteria	20	50	15
Neutrophils	Bacteria	10	50	73
Macrophages				

[a] Polyvinyl pyrollidone.

Stained thin sections at the site of particle engulfment show fibrillar organelle-free cytoplasm within the engulfing pseudopodia (Figure 12). This appearance is characteristic of the cortical rim of cytoplasm and is assumed to be that of regions of the cytoplasm rich in actin filaments. The filamentous cytoplasm does not remain associated with the engulfed particle, but rather the particle after fusion of the plasma membrane appears to pass through the cortical meshwork into the deep cytoplasm.

To study the increase in filamentous actin at this site and to determine whether reversible assembly of actin polymers occurs or whether filaments are redistributed from elsewhere, a temperature pulse-chase protocol was developed[31] which allowed synchronization of phagocytosing cells in a population at a particular ''stage'' of phagocytosis. This protocol revealed that transient net assembly followed by disassembly of actin filaments occurs at the site of particle engulfment.[31] Net assembly occurs on binding of the particle to the cell surface at a pulse temperature of 20°C as determined by using a deoxyribonuclease I inhibition assay. On following the pulse with 3 min ''chases'' at 26 and 37°C, then particles progress further along the phagocytic pathway (see Figure 10) and, concurrently, the actin assembled during the pulse is progressively disassembled[31] (Figure 13). These observations appear to confirm previous speculation that actin filament structures are assembled and take part in engulfment, but are disassembled on completion of engulfment, which presumably returns that portion of the cell surface back into its receptive resting phase. These results obtained by using the DNAase assay are amply confirmed by the localization of polymeric actin by rhodamine phalloidin (See Figure 15a).

Interestingly, net assembly of actin induced by phagocytic stimuli is not inhibited by saturating concentrations of cytochalasin D. This observation has been made by using both the deoxyribonuclease I inhibition assay (Figure 14) and by localization of polymeric actin by rhodamine phalloidin (Figure 15). Since cytochalasins block barbed-end assembly of actin, these observations are most simply interpreted by invoking pointed-end assembly. This interpretation would have been inconsistent with in vitro studies on actin assembly which have consistently shown a higher critical concentration for assembly at the pointed end of actin polymers.[87] However, recent studies[41] have shown that, in the presence of cytoplasmic concentrations of inorganic phosphate, the critical concentrations at opposite ends are the same. Thus, the mechanism of assembly of actin filament meshworks associated with phagocytosis may be different from that involved in protrusion at ruffling borders of locomoting cells or in filopodia protrusion by thrombin-treated platelets, which are highly sensitive to cytochalasins.[88]

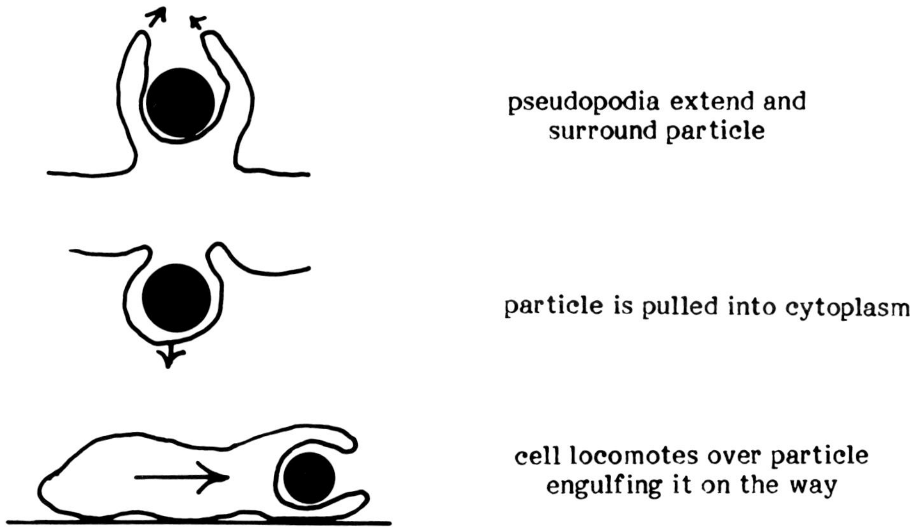

FIGURE 11. Does phagocytosis involve protrusive and/or contractile events? Schematic diagram indicating the possible modes of particle engulfment.

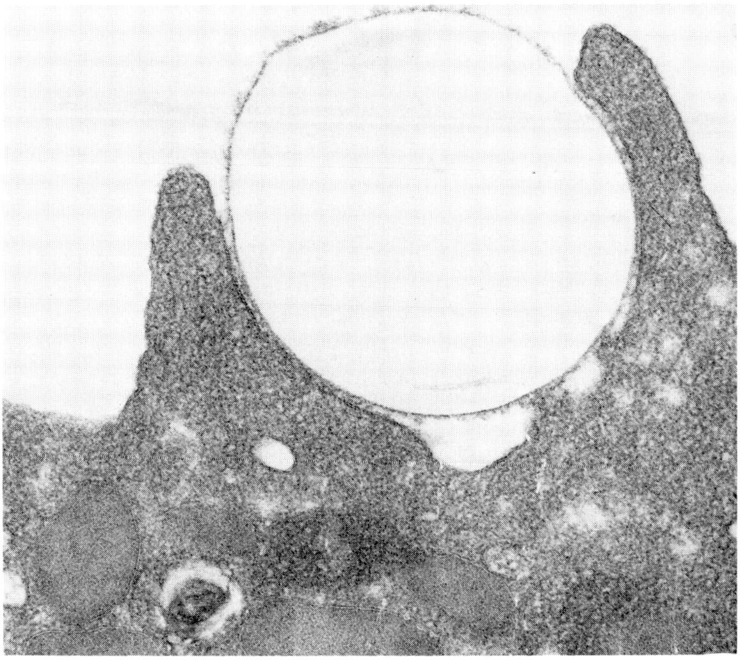

FIGURE 12. Electron micrograph of phagocytosis of an Fc-coated polystyrene sphere by a neutrophil. Note the extension of pseudopodia to surround the particle, the close opposition of the plasma membrane to the particle surface and the characteristic organelle-free ''motile'' cytoplasm. (Electron micrograph courtesy of Dr. R. Clive Richards and P. S.)

Control of actin filament assembly and reorganization presumably occurs in response to occupation of Fc-receptors. Quite how the signal is transduced across the plasma membrane

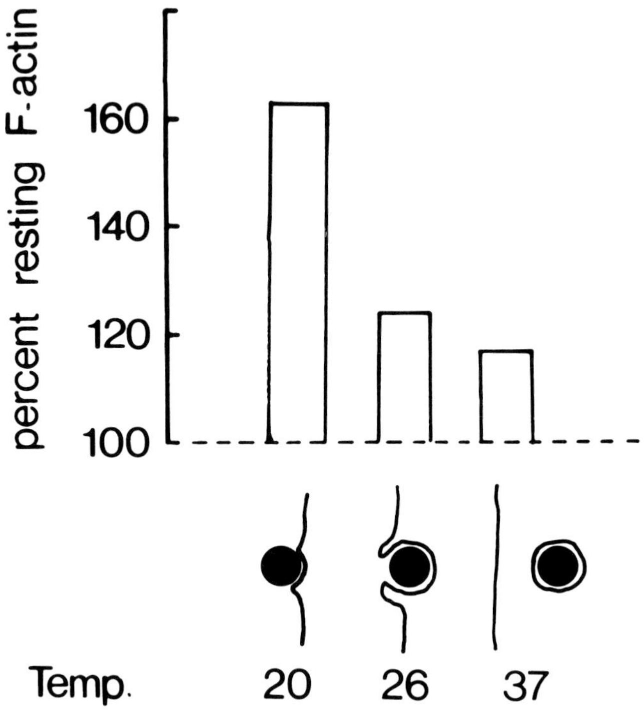

FIGURE 13. Transient actin assembly associated with particle engulfment. Cells were "pulsed" with Fc receptor-diverted phagocytic substrate for 3 min at 20 or 37°C. The proportion of polymerized actin in these cells was estimated directly after the "chase" incubation by using the DNAase I inhibition assay. The phagocytic substrate appears to be perdominantly trapped at various stages of the pathway as indicated in Figure 10. This technique does not absolutely discriminate "phases" of the pathway which probably accounts for residual excess F-actin after the 37°C "chase".

remains essentially unknown. Redistribution of Ca^{2+} from outside to inside occurs in the presence of Fc-receptor-directed ligands (Figure 16) as judged by the fluorescence signal of intracellular quin2 in the presence and absence of external calcium. Other reports document changes in the apparent localization of Ca^{2+} on the inner surface of the plasma membrane.[89] In experiments designed to test the effect of severe intracellular Ca^{2+} depletion[90] on the ability of phagocytic stimuli to induce localized actin assembly, little effect was seen despite the reduction of $[Ca^{2+}]_i$ to 50% of its resting level and the abolition of ligand (Concanavalin A)-induced increases in $[Ca^{2+}]_i$ (data not shown). Similar observations were made for cells in which pH had been experimentally raised or lowered by addition of weak bases or acids, respectively.[90] These data tend to discourage the view that actin reorganization and assembly induced by binding of phagocytic stimuli to Fc receptors is regulated via changes in $[Ca^{2+}]_i$ or pH_i. This conclusion has been reached by Sha'afi et al. by a different experimental route.[40] An alternative mechanism by which Fc-receptor occupation could be recognized on the cytoplasmic side of the plasma membrane might involve the clustering of Fc-receptors at the site of particle adhesion. The observed requirements for a continuous coating of Fc on the surface of the phagocytic stimulus and for a threshold density[91,92] are consistent with such a model.

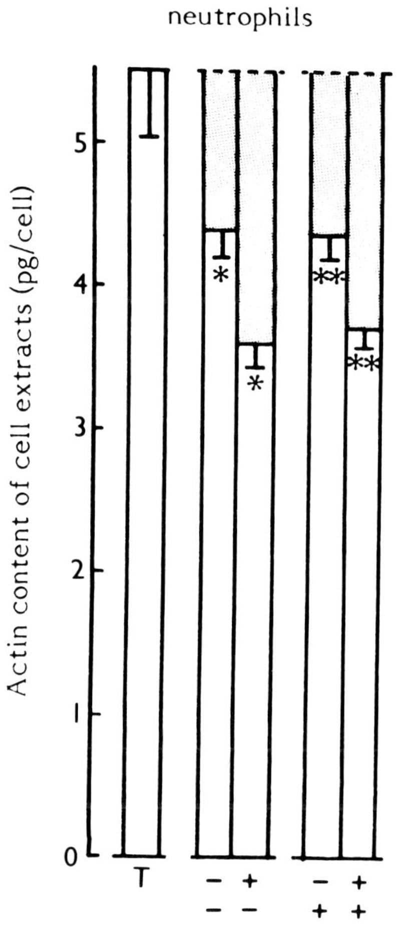

FIGURE 14. Insensitivity of Fc-receptor induced actin assembly to cytochalasin D. Neutrophils were incubated with homologous Fc-bearing phagocytic substrate for 3 min at 20°C to induce net actin assembly (see Figure 13). The proportion of polymerized actin in these cells incubated as above in either the absence or presence of saturating (10^{-6} *M*) concentrations of cytochalasin D was estimated. Cells destined for cytochalasin D were also preincubated in cytochalasin for 3 min prior to addition of phagocytic substrate. Note the lack of inhibition by cytochalasin of the increase in the polymerized actin pool.

V. ACTIN FILAMENT BEHAVIOR DURING CHEMOTAXIS AND LOCOMOTION

A variety of techniques have been used to demonstrate reorganization of actin in response to treatment of neutrophils with chemotactic stimuli, usually the formylated peptide f-Met-Leu-Phe (fMLP). Increased actin polymerization in response to fMLP treatment is indicated by using the DNAase I inhibition assay,[33,34] quantitative assessment of the binding of fluorescent phalloidin by cells,[34-36] or by association of actin with sedimentable detergent-

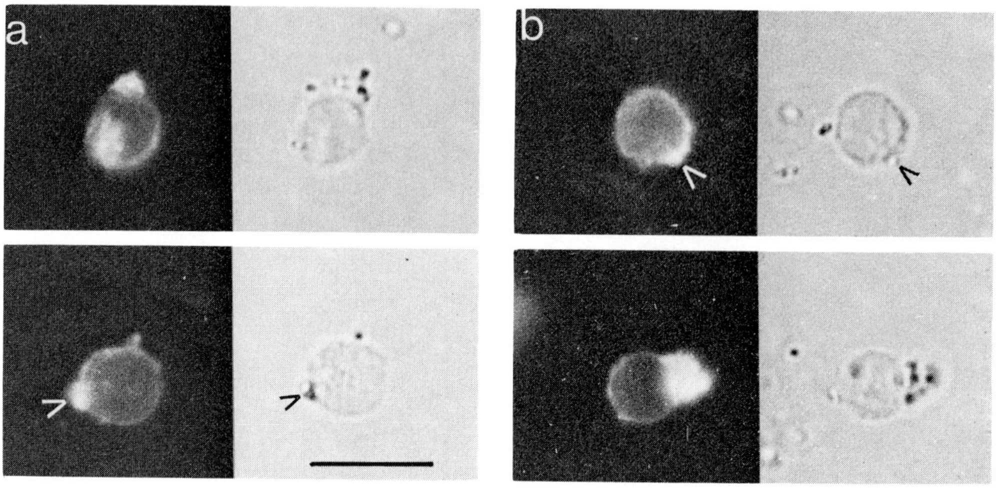

FIGURE 15. Localization of actin filaments in phagocytosing neutrophils using rhodamine-phalloidin. Cells were incubated with Fc-coated polystyrene spheres as phagocytic substrate in the absence (a) or presence (b) of cytochalasin D as indicated in the legend to Figure 14. Cells were then fixed and permeabilized prior to localization of polymerized actin using 10^{-8} *M* rhodamine-phalloidin. Images are fluoresence/phase contrast pairs of the same cell. The position of some of the polystyrene spheres is indicated by arrowheads. Note the distinct localization of fluorescence at the site of particle attachment and the absence of any effect of saturating (10^{-6} *M*) concentrations of cytochalasin D.

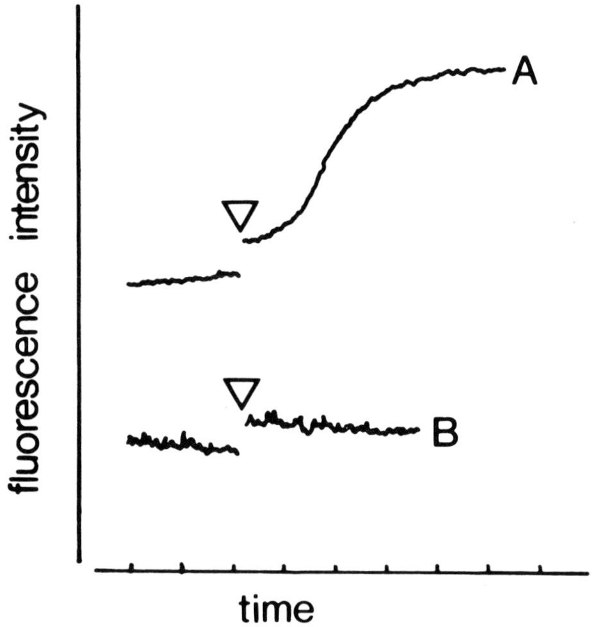

FIGURE 16. Fc-induced $[Ca^{2+}]_i$ increase in quin-2-loaded neutrophils. Cells were loaded with quin-2 and the fluorescence signal recorded as described previously.[90] Fluorescence intensity is given in arbitrary units and the time markers are at 2-min intervals. Cells were stimulated with Fc-bearing phagocytic stimuli in the presence (A) or absence (B) of extracellular Ca^{2+}. Note that while the absence of extracellular Ca^{2+} does not prevent phagocytosis, it completely abolishes the Fc-induced increase in $[Ca^{2+}]_i$.

insoluble residues.[37-40] Although these experiments were performed in zero-gradient conditions, it seems reasonable to suppose that increased actin assembly is one of the cellular responses involved in cell alignment within a chemotactic gradient. However, since neither the mechanistic principles underlying gradient detection nor the mechanisms by which detection is translated into orientation towards the source of chemoattractant are understood, it remains merely an attractive supposition. Increased actin assembly would be consistent with increased protrusive activity at the leading edge of cells which may accompany both the chemokinetic and chemotactic behavioral responses.

The intracellular concentration of both Ca^{2+} and H^+ (pH) are altered in the presence of chemotactic stimuli.[93-101] Blockage of Na^+/H^+ exchange or modulation pH_i in neutrophils, while allowing random locomotion, prevents orientation of cells upgradient and chemokinetic responses.[101] These observations suggest that while locomotion per se is insensitive to intracellular pH changes, pH_i may regulate orientation of the locomotory response. Simultaneous changes in $[Ca^{2+}]_i$ and pH_i are characteristic of responses to the receptor-mediated breakdown of phosphatidylinositol-4,5-bisphosphate,[102] which also occurs in neutrophils in response to chemotactic factors.[103-105] The major route of phosphoinositide breakdown in response to FMLP (~98%) occurs via the action of a phospholipase D[105] in contrast to the presumed signal pathway which utilizes a phospholipase C.[102] Nevertheless, sufficient inositol trisphosphate is produced to potentially release Ca^{2+} from intracellular stores.[105] The coordinate production of diacylglycerol is presumed to activate protein kinase C[106] which may then phosphorylate target proteins in the cell. Protein kinase C is also the major target enzyme for phorbol esters. These compounds induce a variety of physiological responses in neutrophils normally regulated via cell surface receptors[107,108] as well as promoting changes in pH_i and $[Ca^{2+}]_i$.[90]

Phorbol esters initiate rapid and dramatic changes in cell morphology in neutrophils (Figure 17) giving them a distinct rose-bud morphology. The morphological change appears to result from protrusion by the cell of petal-shaped lamellae. Protrusion is completely inhibited by $10^{-6}M$-cytochalasin D[32] which indicates that protrusion results from *de novo* actin filament assembly, presumably at the barbed end. This assumption appears to be confirmed by the labeling pattern of cells by rhodamine phalloidin (Figure 18). Since actin assembly is presumed to occur at the protrusive leading edge of locomoting cells, it is possible that diacylglycerol, the endogenous homologue of phorbol esters, is involved in regulation of actin assembly in cell protrustion. This supposition gains support from studies on cultured fibroblasts, where both phorbol esters and growth factors which induce phosphatidylinositol turnover, promote rapid actin filament assembly at the cell cortex.[109]

To distinguish whether phorbol esters stimulated actin assembly in neutrophils via changes in pH_i or $[Ca^{2+}]_i$ cells were treated with phorbol ester after severe Ca^{2+} depletion or after experimental manipulation of pH_i using weak acids and bases. Neither constraint on $[Ca^{2+}]_i$ or pH_i interfered with phorbol ester-induced lamellae protrusion,[32] which implies mechanisms of action independent of these ions. The effect of phorbol esters on protrusion at subthreshold concentrations (30 to 100 p*M*) is synergized by $10^{-9}M$-A23187 (data not shown) which is consistent with regulation of assembly via protein kinase C. It thus appears that the "steering" mechanism may operate via changes in pH_i,[102] whereas the locomotory machinery functions essentially independently of pH_i but may be modulated by diacyglycerol.[32]

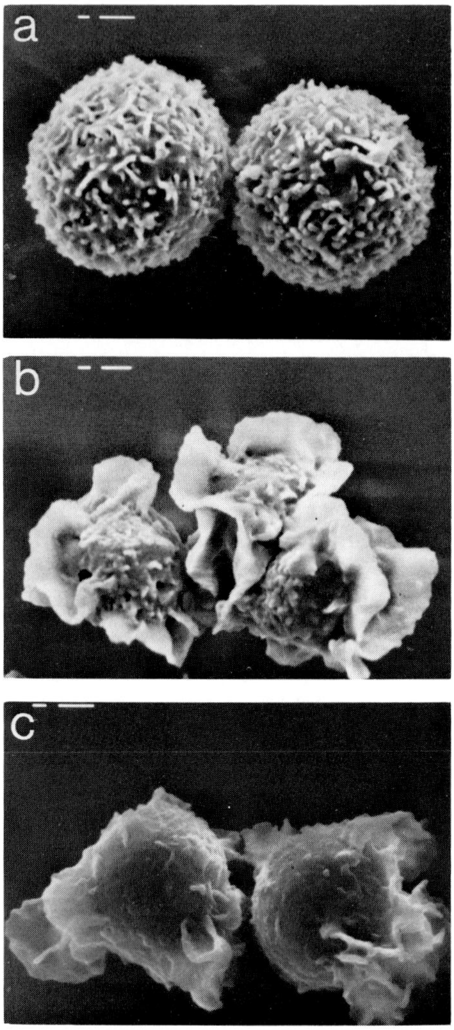

FIGURE 17. Phorbol ester-induced lamellae protrusion in neutrophils. Neutrophils were treated with 30 n*M*-12-*O*-tetradecanoyl-13-phorbol ester (b) or 50 μg/mℓ-1-oleoyl-2-acetyl glycerol (c) for 3 min at 37°C prior to fixation and preparation for scanning electron microscopy. The dramatic morphology change can be compared with control cells in (a).

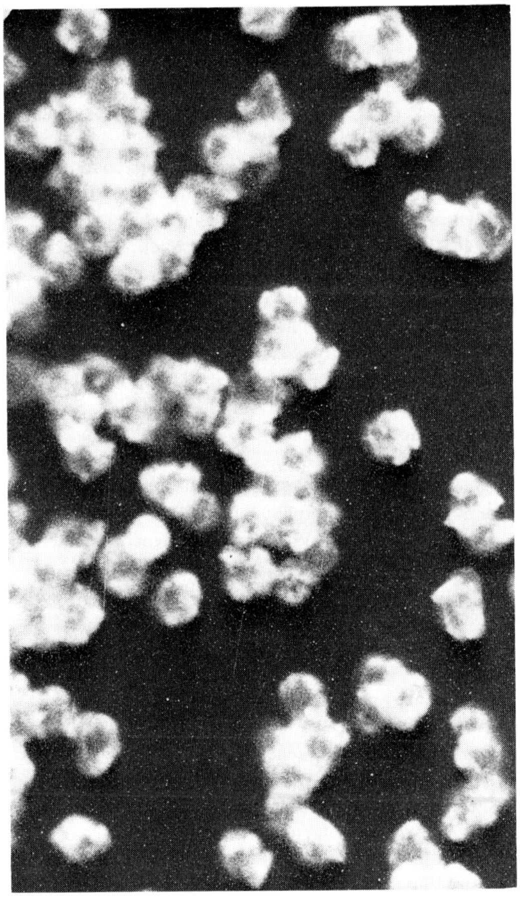

FIGURE 18. Phorbol ester-induced actin assembly. Cells were treated with phorbol ester as described in the legend to Figure 17. After a 3-min incubation, cells were fixed, permeabilized and labeled with rhodamine phalloidin. Note that almost every cell has produced petal-shaped lamellae and that these lamellae are rich in actin filaments.

REFERENCES

1. **Athens, J. W., Raabs, S. O., Haab, O. P., Mauer, A. M., Ashenbrucker, M., Cartwright, G. E., and Wintrobe, M. M.,** Leukokinetic studies. IV. The total blood, circulating and marginal granulocyte pools and the granulocyte turnover rate in normal subjects, *J. Clin. Invest.*, 40, 989, 1961.
2. **Allison, F., Smith, M. R., and Wood, W. B.,** Studies on the pathogenesis of acute inflammation. I. The inflammatory reaction to thermal injury as observed in the rabbit ear chamber, *J. Exp. Med.*, 102, 655, 1955.
3. **Buckley, I. K.,** Delayed secondary damage and leucocyte chemotaxis following local aseptic heat injury *in vivo, Exp. Mol. Pathol.*, 2, 402, 1963.

$I(1,4,5)P_3$ is released into the cytosol and can be metabolized by two separate routes.[17-21] The first is dependent on a cascade of dephosphorylation steps while the second utilizes an $I(1,4,5)P_3$ kinase to form $I(1,3,4,5)P_4$, which is then sequentially dephosphorylated. Both $I(1,4,5)P_3$ and $I(1,3,4,5)P_4$ can be dephosphorylated by a membrane-bound phosphatase, which specifically removes the 5′-phosphate to yield $I(1,4)P_2$ and $I(1,3,4)P_3$, respectively. Sequential dephosphorylation by soluble phosphatases leads to the release of the myo-inositol which can be reutilized for the resynthesis of PI.

In parotid homogenates only 15% of added $I(1,4,5)P_3$ is converted into $I(1,3,4,5)P_4$ and 75% is converted into IP_2, IP_1, and P_i.[19] However, the principal pathway for removing $I(1,4,5)P_3$, during the early stages of cell stimulation under conditions where cytosol Ca^{2+} is elevated, may involve the $I(1,4,5)P_3$ kinase rather than the phosphatase. This stems from the observation that the activity of the $I(1,4,5)P_3$ kinase appears to be controlled by the level of cytosolic Ca^{2+}, i.e., an increase of cytosol Ca^{2+} stimulates $I(1,4,5)P_3$ kinase activity.[20,22] Thus, if the receptor-mediated rise in cytosol Ca^{2+} is curtailed, then the $I(1,4,5)P_3$ is maintained in the cell for a longer time, and so the rise in cytosol Ca^{2+} may act as a negative feedback control by limiting the duration of the $I(1,4,5)P_3$ signal.

II. RECEPTOR-MEDIATED INOSITOL LIPID METABOLISM IN ACTIVATED NEUTROPHILS

Until the mid 1970s, most studies (including those on the neutrophil) on stimulated inositide metabolism utilized the enhanced incorporation of labeled isotopes into PA and PI. Sbarra and Karnovsky (1960)[23] were the first to show enhanced incorporation of $^{32}P_i$ into the total phospholipids, when guinea pig neutrophils were incubated with starch granules. That this enhanced incorporation during phagocytosis was specific to PA and PI was soon demonstrated.[24] Subsequent work on inositol lipid metabolism has mainly focused on using soluble agonists such as f-Met-Leu-Phe, since this enables the simultaneous measurement of functional responses as well. Most of the published work that will be discussed below concerns the effect of soluble agonists such as f-Met-Leu-Phe on neutrophils.

A. Methodological Considerations

The methodology employed to study stimulated inositide metabolism in neutrophils has changed over a period of time as our understanding of the inositide cycle has developed. The various techniques that have been employed are based on the measurement of:

1. Stimulated radiolabeling of PI with $^{32}P_i$ or ^3H-inositol and labeling of PA with $^{32}P_i$. This is the oldest method but caution needs to be exercised in interpreting results because PI labeling is separated from PIP_2 hydrolysis by at least 3 enzymatic steps (see Figure 1). In principle, formation of DG from any other source (e.g., triglycerides) can also lead to increased labeling of PA and PI.
2. Loss of label from PI and PPI in pulse-labeled neutrophils—^3H-glycerol, ^3H-arachidonic acid, and $^{32}P_i$ have been used for this. ^3H-glycerol can only label the lipids by the *de novo* synthesis route and, thus, a brief exposure will only label the newly synthesized pool. ^3H-arachidonic acid gets incorporated into inositol lipids with high specific activity since they are highly enriched in arachidonic acid. $^{32}P_i$, on the other hand, one can label the monoester phosphates of PPI to isotopic equilibrium because they are turning over very rapidly. The diester phosphate on the inositol lipid turns over slowly and will therefore be poorly labeled. This technique does not provide any information regarding the route of loss of the lipid.
3. Formation of DG and PA in pulse-labeled neutrophils using ^3H-glycerol and ^3H-arachidonic acid. This provides supportive evidence that a phospholipase C or D action probably takes place, but does not provide confirmatory evidence regarding its source.

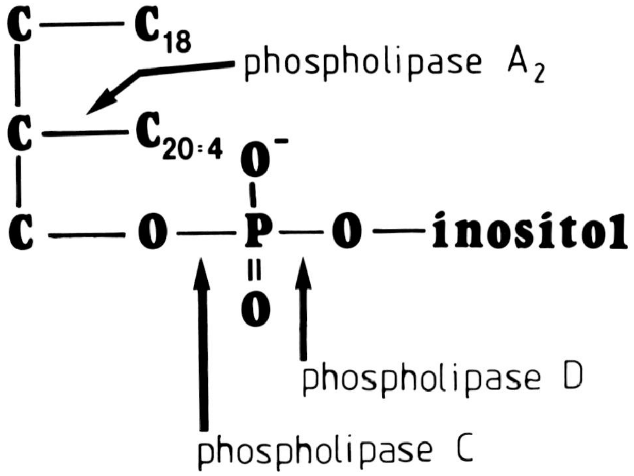

FIGURE 2. A schematic illustration of the site of action of different phospholipases. Phospholipids can be cleaved by at least three types of phospholipases. A phosphodiesteratic attack on an inositol phospholipid can occur at two sites: (1) Phospholipase C cleaves between the phosphate molecule and the glycero-backbone forming diacyglycerol and the water-soluble derivatives of inositol. I(1)P, $I(1,4)P_2$ and $I(1,4,5)P_3$ are the respective products of PI, PIP and PIP_2; (2) Phospholipase D cleaves between the phosphate and the inositol ring forming phosphatidate and the water-soluble inositol derivatives. Inositol, I(4)P and $I(4,5)P_2$ are the corresponding products of PI, PIP and PIP_2. Phospholipase A_2 hydrolyzes the fatty acid at the second position on the glycero-backbone generating arachidonic acid ($C_{20:4}$) and the corresponding lyso-derivative. (From Cockcroft, S., *Receptor Biochemistry and Methodology,* Vol. 7, *Phosphoinositides and Receptor Mechanisms,* Putney, J. W., Jr., Ed., Alan R. Liss, New York, 1986, 287. With permission.)

4. Mass determinations of inositol lipids, PA and DG. This requires large amounts of material since most of these lipids are minor components.
5. Accumulation, often in the presence of Li^+, of ^{3}H-inositol-labeled IP_1, IP_2, and IP_3. The short half life of the neutrophil and because ^{3}H-inositol is relatively impermeant in these cells, makes this method costly. However, the differentiated human leukemic cell-line, HL-60, can be grown in the presence of ^{3}H-inositol and it gets incorporated into the inositol lipids well. This neutrophil-like cell-line has been used in many studies with results that are similar to those obtained from neutrophils.

Of all these methods, the formation, of inositol phosphates, particularly $I(1,4,5)P_3$, provides an unambiguous measure of a phospholipase C attack on PIP_2. Other methods of analysis cannot always distinguish between a phospholipase C-type action, a phospholipase A_2 or a phospholipase D attack (Figure 2).

B. Source of Neutrophils

Neutrophils from a variety of sources have been used in the study of receptor-mediated inositol lipid metabolism. Most studies have utilized either human neutrophils from peripheral blood or rabbit and guinea pig neutrophils that have been elicited by injection of glycogen or casein into the peritoneum. Exudates from the peritoneal cavity provide a nearly pure source of neutrophils after 4 to 18 hr of injection. However, such cells have been partially activated, in that they have responded to a chemotactic stimulus. This response primes the

Since the increase in cytosol Ca^{2+} caused by ionophores has not been quantitated accurately, it is not possible to know precisely what level of Ca^{2+} is required to activate the PPI-pde in intact cells. The Ca^{2+}-activated PPI-pde will be discussed in Section VII.A.

D. Leukocidin as an Agent for Increasing Plasma Membrane Permeability to Ca^{2+}

Leukocidin is a cytotoxic protein produced by bacteria, e.g., *Pseudomonas aeruginosa*[132,133] and *Staphylococcus aureus*.[134,135] It consists of two components which act synergistically to induce cytotoxic changes in neutrophils, macrophages, and lymphocytes, but not erythrocytes or platelets.[132,135] The mode of action of leukocidin is unknown, but it has been reported that it leads to an accumulation of Ca^{2+} which is accompanied by secretion.[136] In a more recent study, it was shown that secretion and cytotoxicity could be separated in a time-dependent manner: secretion occurred within the first 10 min of leukocidin treatment without any release of lactate dehydrogenase. Cytotoxicity was observed after 15 min and maximized at 20 min.[137]

Leukocidin also causes an increase in labeling of PA and degradation of PPI from ^{32}P-labeled neutrophils.[137,138] It is not known whether it stimulates the formation of inositol phosphates. However, if the action of leukocidin is to destroy the membrane permeability to Ca^{2+}, then the resulting elevation of cytosol Ca^{2+} must be triggering inositide metabolism in neutrophils, an effect not dissimilar to that of Ca^{2+}-ionophores.

VI. INHIBITORS OF INOSITIDE METABOLISM IN NEUTROPHILS

A. Phorbol Esters

Phorbol esters have multiple effects on neutrophils. Thus, PMA by itself is able to stimulate secretion and the respiratory burst, and when added with Ca^{2+}-ionophores (to raise cytosol Ca^{2+}), it synergizes to trigger cellular activation.[48,49,67,104] In addition to its stimulatory activity, it is also a potent inhibitor of neutrophil function when triggered by receptor-directed agonists. Thus, chemotaxis, secretion, and a rise in cytosol Ca^{2+} due to either f-Met-Leu-Phe, PAF, or leukotriene B_4 and con A-induced capping are all inhibited by prior treatment of neutrophils with PMA.[39,76,104,139,140]

DG phosphorylation due to f-Met-Leu-Phe and IP_3 formation stimulated by LTB_4 is also inhibited by PMA,[58,76] suggesting that activation of protein kinase C may transmit a negative signal to inhibit the receptor from coupling to its effector, the PPI-pde. Inhibition may be achieved by protein kinase C-induced phosphorylation of either the receptor or the guanine nucleotide regulatory protein (G_p) which couples the receptor to the PPI-pde (see Section VII). PMA leads to phosphorylation of a membrane-associated protein corresponding to the molecular weight of the α subunit of G_p and also a 64 kdalton protein whose molecular weight corresponds to that of the f-Met-Leu-Phe receptor.[76] The G-protein is possibly the more likely target because inhibition by PMA is not confined to the f-Met-Leu-Phe receptor, but is also found for other receptors on neutrophils as well as other cell types including smooth muscle[141] and platelets.[142]

B. Cyclic AMP

Elevation of cAMP levels by forskolin, dibutyryl cAMP, and prostaglandin E_2 in neutrophils generally leads to inhibition of physiological functions stimulated by f-Met-Leu-Phe, such as arachidonic acid release, secretion, and respiratory burst.[143,144] The effect of cAMP is to inhibit the formation of IP_3, implying that the activity of PPI-pde is inhibited.[143] Thus, it is possible that, like protein kinase C, cAMP-dependent protein kinase may regulate the activity of PPI-pde via the G-protein.

VII. REQUIREMENT FOR A GUANINE NUCLEOTIDE REGULATORY PROTEIN IN COUPLING THE f-MET-LEU-PHE RECEPTOR TO THE CATALYTIC UNIT, PPI-PDE

The neutrophil was the first system in which it was possible to control the activity of the enzyme responsible for hydrolyzing the inositol lipids via a guanine nucleotide regulatory protein. This became possible when it was shown that a membrane-bound enzyme is responsible for the hydrolysis of the PPI.[66]

A. Discovery of the Membrane Bound PPI-pde

The enzyme that hydrolyzes the inositol lipids has been called by various names, e.g., phospholipase C,(poly)phosphoinositide phosphodiesterase or phosphoinositidase C. A soluble phospholipase C, able to hydrolyze PI to generate the water-soluble inositol phosphates and DG, was initially discovered in pancreas[145] and such an enzyme has been found in all mammalian tissues where it has been looked for.[146,147] The water-soluble products formed are inositol-1-phosphate and inositol-1,2-cyclic phosphate and these are produced in roughly equal amounts.[148] In two separate reviews,[146,147] it was concluded that there was no evidence for the presence of a plasma membrane-bound PI-specific phospholipase.

More recently, it has been shown that this ubiquitous soluble phospholipase C is also able to hydrolyze PIP and PIP_2.[149,150] The products of the reaction have been identified as the corresponding inositol bis- and tris-phosphates present, both in cyclic and noncyclic forms.[151] Several properties of the cytosolic enzyme should be noted. It is able to attack its substrate only when the latter is presented in a nonbilayer form. Thus, substrate normally present in a biological membrane is resistant to attack by the soluble enzyme. It is thus envisaged that the receptor-agonist interaction leads to a perturbation of the membrane that permits the cytosolic enzyme to act upon its substrate.[152]

Despite the widespread presence of the soluble enzyme in the cytosol of many cells including neutrophils, the possibility of a membrane-bound phospholipase C that could couple to its receptor was considered. In parallel with another well-defined transmembrane signaling system, the formation of cyclic AMP by adenylate cyclase,[153] the presence of a membrane-bound species of a phospholipase C was envisaged.

It has been demonstrated that plasma membranes prepared from rabbit or human neutrophils do indeed contain a membrane-bound enzyme that is specific for PPI. This is often referred to as PPI-pde because the hydrolysis involves a phosphodiesterase cleavage of the phospholipase C type.[66] The first point to remember is that the enzyme should be able to attack its substrate within the membrane, and so one has to use the endogenous substrate present in the membranes. It had previously been demonstrated that the red cell plasma membrane contains such an enzyme and it can be activated by Ca^{2+}.[126,154,155] In the intact cell the concentration of cytosol Ca^{2+} is maintained at 100 n*M* and this level of Ca^{2+} does not activate the enzyme. However, by exposing red cell membranes to Ca^{2+} concentrations in the range of 100 μ*M* and above, the enzyme can hydrolyze both PIP and PIP_2 (but not PI). Approximately 60 to 70% of the endogenous substrate is available for hydrolysis. The water-soluble products of the reaction have been identified as $I(1,4)P_2$ and $I(1,4,5)P_3$ and the membrane-soluble product is DG. The fatty acid composition of the DG formed is predominantly stearic acid and arachidonic acid and thus resembles its parent phospholipid.[156] The Ca^{2+} requirement for the red cell enzyme is dependent on the presence of Mg^{2+} and ionic strength. Thus, if the enzyme is assayed in the absence of Mg^{2+} and low ionic strength, then the Ca^{2+} requirement is in the region of 1 to 10 μ*M*; however, this is shifted to much higher concentrations (100 to 200 μ*M*) when Mg^{2+} is present. This is similar to conditions that pertain in the intact erythrocyte.[155]

A similar enzyme is also present in the plasma membrane of neutrophils.[66] As in the red cell, the enzyme can be activated by Ca^{2+}: 50 to 100 μM Ca^{2+} induces maximal hydrolysis. PPI-pde activity was monitored by adding Ca^{2+} to plasma membranes prepared from neutrophils prelabeled with $^{32}P_i$ for 90 min (at which time the monoester phosphates of the PPIs are labeled to equilibrium with ^{32}P-ATP). Very little change in the levels of the PPIs occur in the absence of Ca^{2+}, but in its presence approximately 40 to 60% of the label in the PPI is lost. The counts lost from the PPI are quantitatively recovered in the aqueous phase as I(1,4)P_2 and I(1,4,5)P_3. Both PIP and PIP_2 are hydrolyzed with equal facility, that is, the enzyme shows no specificity for the two inositides. No hydrolysis of PI nor of any of the other lipids occur. The radiochemical hydrolysis represents a mass loss of the PPIs. Hydrolysis is complete by 2 min.

The other product of hydrolysis is DG whose fatty acid composition resembles that of the parent PPIs, i.e., it is highly enriched in stearic acid and arachidonic acid. Thus, in the plasma membranes of neutrophils there is present an enzyme that can hydrolyze PPIs, and this could be the potential candidate for coupling to the receptor. Ca^{2+} itself cannot be the physiological activator, since the levels of Ca^{2+} in receptor-stimulated cells do not increase beyond 1 to 2 μM, and this level of Ca^{2+} is too low to activate the enzyme. However, Ca^{2+} activation of PPI-pde does occur in intact neutrophils when A23187 or leukocidin is used to raise cytosol Ca^{2+} (as discussed in Section V).

Addition of f-Met-Leu-Phe directly to membranes has no effect on enzyme activity.[66] By analogy with the adenylate cyclase system, the possibility that the membrane-bound PPI-pde may be regulated by a guanine nucleotide binding protein (G-protein) was considered. The ability of a nonhydrolyzable guanine nucleotide analogue, GTPγS, was tested for its ability to induce hydrolysis of PPI in ^{32}P-prelabeled plasma membranes. Enzyme activity was monitored under resting physiological conditions, i.e., in the presence of 100 nM Ca^{2+} and high ionic strength (150 mM-KCl). Similarly to Ca^{2+}, GTPγS is an effective activator of the PPI-pde; PIP and PIP_2 are hydrolyzed with equal facility and the products of the reaction are identified as IP_3 and IP_2.[157]

The ratio of PIP to PIP_2 present in isolated membranes is 2:1.[66] The concentrations of PPIs are maintained by Mg·ATP.[158] Two inositol lipid kinases, PI kinase and PIP kinase, phosphorylate PI sequentially to form PIP and PIP_2. The concentration of the inositol lipids can then be manipulated by the addition of Mg·ATP. Indeed, it was found that if neutrophil membranes (prepared from unlabeled cells) are incubated with ^{32}P-labeled ATP then the PPIs can be phosphorylated to yield a ratio of PIP to PIP_2 of 1:2.[223] In effect, the presence of Mg·ATP enables the regeneration of membrane PIP_2, and this probably mimics the situation present in the intact cells.

In membrane preparations where radioactive ATP is used to label the PPIs, PIP_2 predominates as the major substrate for hydrolysis. Both Ca^{2+} and GTPγS are effective in inducing hydrolysis of the PPIs (Figure 4). The plasma membrane PPI-pde can only be activated by the f-Met-Leu-Phe receptor in the presence of a guanine nucleotide, GTPγS (Figure 5) or GTP.[159,161]

The model (see Figure 6) proposed for neutrophils is that the occupied receptor (with agonists such as PAF, leukotriene B_4, C5a, and f-Met-Leu-Phe) interacts with a guanine nucleotide binding protein (termed G_p or N_p) which regulates the activity of the PPI-pde. The activated PPI-pde generates two intracellular signals, IP_3, to mobilize intracellular Ca^{2+} stores, and DG, to activate protein kinase C. The increase in cytosol Ca^{2+} activates Ca^{2+}-regulated enzymes, such as the calmodulin-dependent protein kinase[162] or calmodulin-activated NAD kinase,[163] while protein kinase C phosphorylates other key proteins, such as those responsible for switching on NADPH oxidase.[164] Both pathways can operate separately, but it is most likely that these act synergistically in intact cells. A major unanswered question concerns the role of cytochalasin B in neutrophil activation. From the available data it is not clear whether or not this compound enhances the hydrolysis of PIP_2, perhaps by increasing the coupling between G_p and PPI-pde.

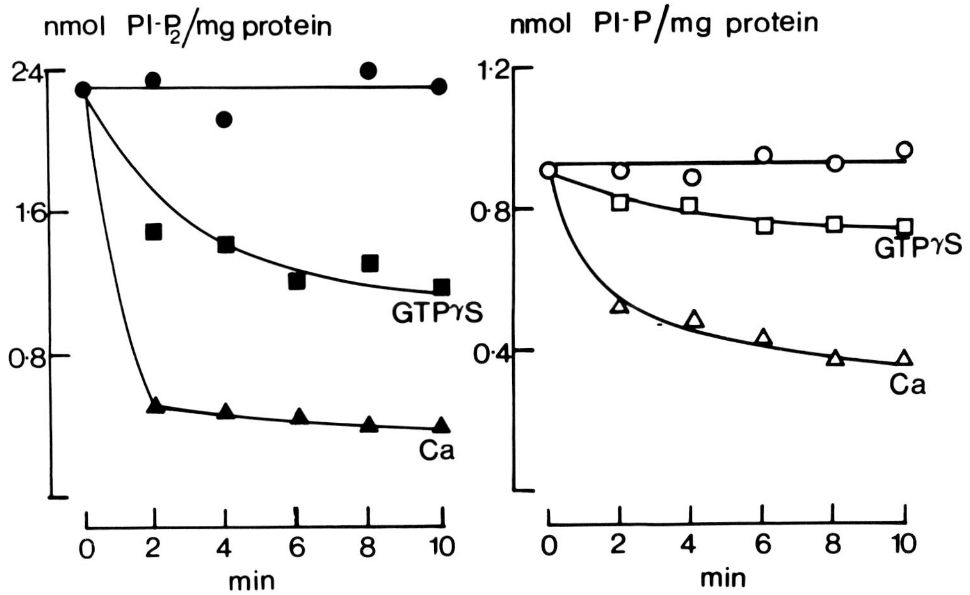

FIGURE 4. Time course of hydrolysis of PIP and PIP_2 by Ca^{2+} and GTPγS. Plasma membranes from human neutrophils were incubated with ^{32}P-labeled ATP (2 m*M*) in the presence of Mg^2 (10 m*M*) and Ca^{2+} (100 n*M*). By 5 min the PPI were labeled to a steady state and these membranes were transferred to tubes containing either Ca^{2+} (0.5 m*M*) or GTPγS (100 μ*M*). At the indicated times, samples were withdrawn, quenched, and the lipids extracted for the analysis of PIP and PIP_2. The amounts of PIP and PIP_2 were calculated from the specific activity of ATP.

This model of activation is not specific for neutrophils, but is a general mechanism for cell activation in systems where the receptor controls the hydrolysis of inositol lipids. Thus, there have been many reports confirming the requirement for guanine nucleotides in hormonal regulation of PPI-pde in various cell types including liver,[165] the GH_3 pituitary cell line,[166] pancreas,[167] blowfly salivary glands,[168] and brain cortex.[169]

B. Ca^{2+} Regulation of PPI-pde

Studies with intact cells have suggested that Ca^{2+} is important for PPI-pde activity, even though the rise in cytosol Ca^{2+} is a consequence of PIP_2 hydrolysis. This paradox is resolved by studying the Ca^{2+} requirement for GTPγS-stimulated activation. Figure 7 shows that PPI-pde shows optimal activity when Ca^{2+} is held at 100 n*M*. Increasing the level to 10 μ*M* does not increase or potentiate PPI-pde activity with either a maximal or submaximal concentration of GTPγS. Reducing the Ca^{2+} level to 1 n*M* inhibits GTPγS-activated PPI-pde.

Thus Ca^{2+} is essential for the expression of GTPγS activity but its requirements are fulfilled by the presence of resting levels (100 n*M*) of Ca^{2+}. Depletion below these basal levels leads to inhibition of receptor stimulated PPI-pde activity.

The model illustrated in Figure 6 shows a site for Ca^{2+} on PPI-pde. Ca^{2+} in the μmolar alone may interact with the site to activate the PPI-pde directly but in the presence of GTPγS, the Ca^{2+} requirement may be reduced to the nmolar range. Thus, the effect of the subunit, α_p can be seen to shift the Ca^{2+} requirement to lower concentrations and this may occur by a change in affinity for Ca^{2+}. This model can only be substantiated when the enzyme, PPI-pde, is purified and a change in Ca^{2+} affinity in the presence and absence of α_p is demonstrated.

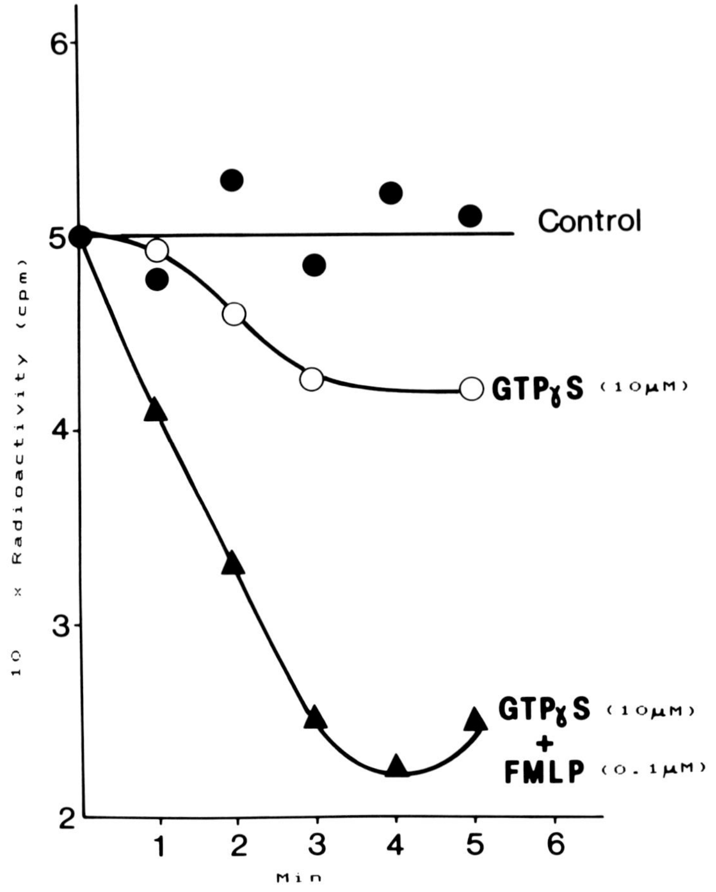

FIGURE 5. Time course of PIP_2 hydrolysis by f-Met-Leu-Phe and GTPγS. Human neutrophil plasma membranes were labeled as described in Figure 4. The labeled membranes were transferred into tubes containing buffer, GTPγS (10 μ*M*) or GTPγS (10 μ*M*) plus f-Met-Leu-Phe (100 n*M*). f-Met-Leu-Phe alone had no effect on PIP_2 hydrolysis. (From Cockcroft, S., *Receptor Biochemistry and Methodology,* Vol. 7, *Phosphoinositides and Receptor Mechanisms,* Putney, J. W., Jr., Ed., Alan R. Liss, New York, 1986, 287. With permission.)

C. Identification of a Novel Guanine Nucleotide Regulatory Protein, G_p

The use of bacterial toxins, such as those derived from cholera and pertussis, has greatly facilitated our understanding of G-protein function. It is well established that G_s (the G-protein that activates adenylate cyclase), exists as a heterotrimer αβγ, [170] and can be ADP-ribosylated by cholera toxin releasing the α subunit to activate the adenylate cyclase. Similarly, modification of G_i activity (the G-protein which mediates inhibition of adenylate cyclase) is achieved by pertussis toxin-induced ADP-ribosylation.

Pretreatment with pertussis toxin leads to inhibition of all the receptor-directed neutrophil responses, such as the rise in cytosol Ca^{2+}, inositol lipid metabolism (measured as IP_3 generation or by any other parameter), secretion, oxidative burst, chemotaxis, and arachidonic acid release.[171-181] Pertussis toxin catalyses ADP ribosylation of a protein with a molecular weight of 41 kdalton.[172-175,182] Since G_i also has a molecular weight of 41 kdalton and is ADP-ribosylated by pertussis toxin, it was originally thought that G_i coupled the PPI-pde to its receptor.[172,173,176] However, as the receptors controlling the inhibitory activity of adenylate cyclase are distinct from those coupling to PPI-pde, it was highly unlikely that

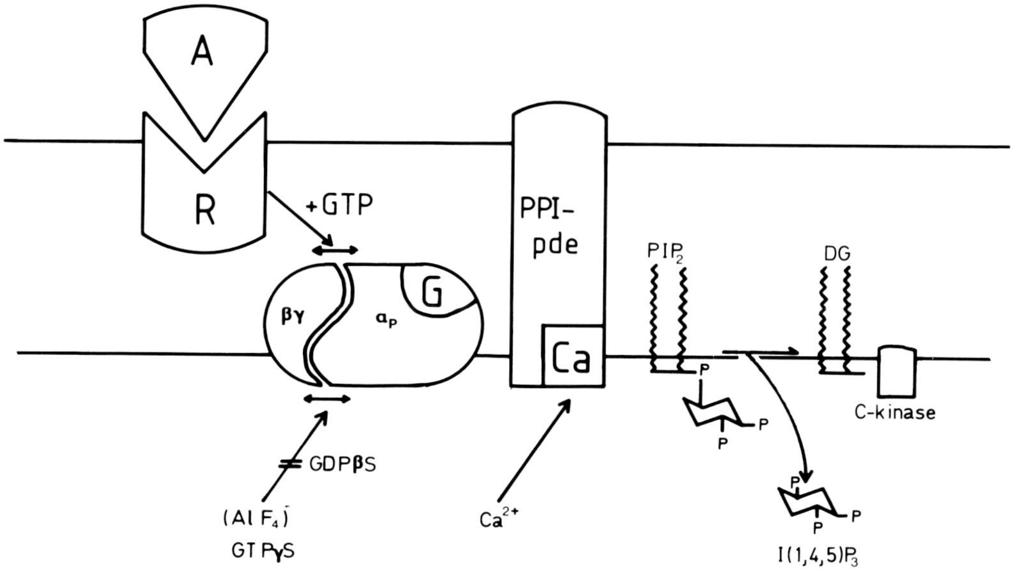

FIGURE 6. Model depicting the control of PPI-PDE by agonist plus GTP, GTPγS, fluoride, and Ca^{2+}. PPI-pde catalyses the hydrolysis of PIP_2 to yield 1,2-DG and $I(1,4,5)P_3$. The catalytic unit can be activated by the receptor via the guanine nucleotide regulatory protein (G_p) which is depicted as a heterotrimer. The occupied receptor catalyses the exchange of GDP to GTP on the α-subunit. This leads to subunit dissociation of the heterotrimer to α_p and βγ subunits. α_p can now activate the PPI-pde. Subunit dissociation can also occur directly by fluoride or GTPγS. GDPβS is an inhibitor of fluoride, GTPγS and GTP plus agonist. Ca^{2+} can directly activate the enzyme without the intervention of the G-protein. Ca^{2+} activation is not affected by GDPβS.

G_i could be the coupling factor. Although G_p has not yet been purified, it is clear that it is not the same as G_i.[182] This is based on immunological evidence that the pertussis toxin substrate in neutrophils is distinct from the well-characterized subunit of G_i and also other previously identified pertussis-toxin substrates, G_0 and transducin. However, it has been reported that when G_p is deactivated (by pertussis toxin), then G_i or G_0 is able to couple functionally the f-Met-Leu-Phe receptor to the PPI-pde. This does not prove that G_p is identical to G_i or G_0, but provides support that the various G-proteins (G_0, G_i, G_p) are all related products bearing both structural and functional similarities[183] and can therefore substitute for one another in an in vitro system.

Although the inositol lipid signaling pathway in neutrophils is sensitive to pertussis toxin inhibition, this is not generally true for all cell types. Thus, cells fall into two categories. Mast cells,[184,185] neutrophils,[171-181] macrophages,[186] lymphocytes,[181] fibroblasts,[187] and renal mesangial cells[188] are systems where receptors that are known to couple to inositol lipid metabolism are inhibited by pertussis toxin, while brown adipocytes,[189] 1321N1 astrocytoma cells,[190] chick heart cells,[191] liver,[190] GH_3 cells,[192] and pancreas,[167] are systems where the toxin is without effect. The reason for this difference is not clear at the present time, but one implication is that there are subtypes of G_p. In analogy with α_s, [193] it is possible that α_p also exists in two forms. α_s is a product of one gene, but due to differential splicing of the mRNA, two α_s subunits are present which differ in size. One of the subunits has absent a 46 amino acid long region. If a similar situation is present for α_p, then it is envisaged that in those cells where G_p is pertussis-toxin insensitive, the α_p subunit has a deletion in the region where the pertussis toxin would be responsible for ADP-ribosylation.

Another candidate for G_p is the product of N-ras, p21, which is also a G-protein.[194] It has been shown that increasing the expression of p21 leads to an increase in receptor-stimulated PPI-pde activity in 3T3 fibroblasts.[195] Because of the complex nature of the

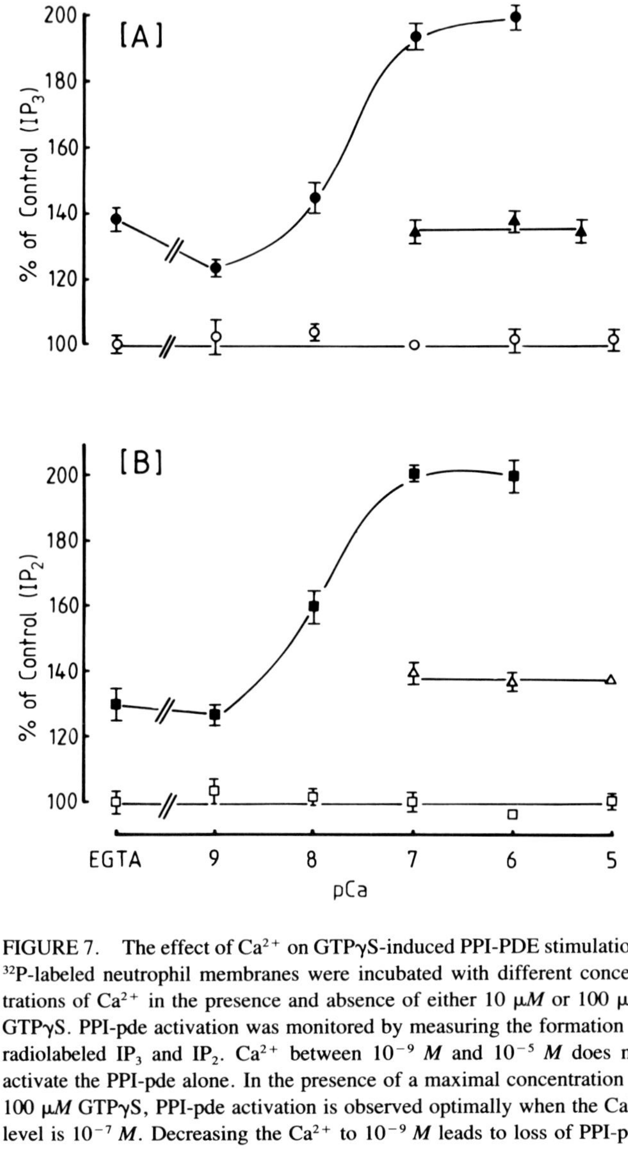

FIGURE 7. The effect of Ca^{2+} on GTPγS-induced PPI-PDE stimulation. ^{32}P-labeled neutrophil membranes were incubated with different concentrations of Ca^{2+} in the presence and absence of either 10 μ*M* or 100 μ*M* GTPγS. PPI-pde activation was monitored by measuring the formation of radiolabeled IP_3 and IP_2. Ca^{2+} between 10^{-9} *M* and 10^{-5} *M* does not activate the PPI-pde alone. In the presence of a maximal concentration of 100 μ*M* GTPγS, PPI-pde activation is observed optimally when the Ca^{2+} level is 10^{-7} *M*. Decreasing the Ca^{2+} to 10^{-9} *M* leads to loss of PPI-pde activity. A suboptimal concentration of GTPγS (10 μ*M*) also stimulates PPI-pde and this response is not potentiated by increasing the level of Ca^{2+} between 10^{-7} *M* and 10^{-5} *M* (From Cockcroft, S., *Biochem. J.*, 240, 503, 1986. With permission.)

experimental model, it is not entirely clear whether p21 directly couples to the PPI-pde, but it is clear from the molecular weight, however, that in the neutrophil, G_p is not N-ras.

Pertussis toxin inhibits the activity of PPI-pde in membrane preparations of neutrophils when activated by GTPγS alone,[196] or GTP plus f-Met-Leu-Phe.[159] The observation that Ca^{2+} activation of PPI-pde is not inhibited by pertussis toxin [159,196] lends further support to the argument that Ca-ionophores stimulate PPI-pde by virtue of increasing cytosol [Ca^{2+}], since ionophore-mediated IP_3 formation is also not blocked by pertussis toxin.[54,197]

It is interesting to note that pertussis toxin can block GTPγS-induced activation of PPI-pde.[196] It is well known that activation of G_i by nonhydrolyzable analogues of GTP, such as GTPγS or GppNHp, and by fluoride is resistant to pertussis-toxin treatment, compared to activation by GTP or GTP plus agonist.[198,199] Thus, G_p and G_i differ in this particular aspect.

D. Mechanism of Fluoride Activation in Neutrophils

It has been known since 1959 that fluoride can activate the respiratory burst in neutrophils,[200-203] and the mechanism of its action can only now be explained by the ability of fluoride to activate G-proteins. Fluoride is able to cause a rise in cytosolic Ca^{2+}, stimulate inositol phosphate formation, which is not inhibited by pertussis toxin,[204,205] and stimulate the phosphorylation of neutrophil proteins.[206] Fluoride can also activate PPI-pde in membranes from hepatocytes,[207] WRK-1 (a mammary cell line),[208] and GH_3 cells.[192] Therefore, it seems most likely that these are due to the ability of fluoride to stimulate G_p and, consequently, this ion should also stimulate PPI-pde in neutrophil membranes.

G proteins, in general, can be activated by fluoride,[209-211] and the mechanism involves the formation of a fluoroaluminate complex $[AlF_4]$.[209] This can interact with the bound GDP on the α subunit of G-proteins. The GDP-$[AlF_4]$ complex is analogous to GTP, and this leads to subunit dissociation, thereby activating the G-protein and stimulating its catalytic unit.[210,211] Thus, it is likely that G_p bears a structural similarity to the other well-characterized G-proteins and may also be a heterotrimer, sharing a common βγ subunit with the other G-proteins, the difference being in the α subunit. Even the α subunits must bear some structural resemblance since they all possess GTPase activity and a GTP binding site.[212]

E. Activation of GTPase activity in neutrophil membranes

Termination of the hormone-mediated stimulation of G-proteins is generally regulated by hydrolysis of bound GTP to the α-subunit by a high-affinity GTPase.[153] If the analogy with the adenylate cyclase system is carried through, then it predicts that termination of the receptor-mediated regulatory activation of G_p may occur by hydrolysis of bound GTP, by a high-affinity GTPase. The presence of a f-Met-Leu-Phe-stimulated GTPase which is inhibited by pertussis toxin has been demonstrated.[213,214] Other evidence supporting the concept of a G-protein regulation includes the effect of guanine nucleotides on the affinity states of the f-Met-Leu-Phe receptor, such that binding of f-Met-Leu-Phe in the presence of GTP converts the high-affinity state of the receptor to a low-affinity state.[215-217]

VIII. GUANINE NUCLEOTIDE REGULATION OF SECRETION

All the available information supports the concept that GTP couples the f-Met-Leu-Phe receptor to its catalytic unit, PPI-pde generating the two intracellular messengers. The receptor can be by-passed by using a guanine nucleotide analogue, GTPγS, or fluoride, both of which directly activate PPI-pde (see Figure 6). What happens when GTPγS is introduced inside cells? Addition of guanine nucleotide analogues into permeabilized neutrophils stimulates secretion[49] and therefore one can speculate that this is mediated by G_p.

However, several characteristics of GTPγS-induced secretion lead to the proposal that the effects of GTPγS on secretion are due to its ability to interact with a G-protein, which is distinct from G_p. It has been designated G_e; e for exocytosis. Activation of PPI-pde by GTPγS is dependent on the presence of 100 n*M* Ca^{2+} (Figure 7). Secretion due to GTPγS is insensitive to cytosol Ca^{2+} in that GTPγS can stimulate secretion from permeabilized neutrophils in the presence of 30 m*M* EGTA, a condition which leads to total inhibition of PPI-pde activity.[49,126] Also, pertussis toxin, which inhibits GTPγS-stimulated PPI-pde activity is totally without effect on GTPγS-induced secretion in permeabilized neutrophils.[49,196] Because guanine nucleotides can have multiple effects in cells, it should not be assumed that a response triggered by guanine nucleotides, such as secretion, is due to PPI-pde activation.

IX. CONCLUSIONS AND PERSPECTIVES

Inositol lipids play a very special role in cellular activation. In neutrophils, stimulated inositol lipid metabolism has two main functions. As in many other cell types, hydrolysis of PIP_2 leads to the formation of two intracellular messengers, $I(1,4,5)P_3$ and DG. The enzyme that catalyzes this reaction is present in plasma membranes of neutrophils and is coupled to the receptor via a guanine nucleotide-binding protein, G_p. The second function of inositol lipids is the hydrolysis of PI by phospholipase A_2 for the release of arachidonic acid. This is synthesized to generate further activators of cell function, including neutrophils, and thus should be regarded as signaling molecules that act as extracellular agonists.

Can IP_3 and DG be entirely responsible for activation of neutrophil responses such as secretion, chemotaxis, and the respiratory burst? Grzeskowiak et al.[77,218] have shown that the respiratory burst can be stimulated by f-Met-Leu-Phe under conditions that completely inhibit the activation of phosphoinositide metabolism. This implies that the effects of f-Met-Leu-Phe are not limited to its ability to activate PPI-pde, and this question has been addressed recently by us in a separate review.[219] It is conceivable that f-Met-Leu-Phe also couples to another signaling system (e.g., PC hydrolysis) and our current concepts on cellular activation may be incomplete.

Questions have also been raised regarding the role of protein kinase C in neutrophil function. Although provision of the two signals, Ca^{2+} and a protein kinase C activator, can synergize and give full expression of neutrophil responses, when f-Met-Leu-Phe is the stimulus, activation of protein kinase C appears to be minimal. This conclusion is based on the use of the protein kinase C inhibitors, H-7, H-9, polymyxin B, and retinal which all inhibit PMA-induced neutrophil function, but do not inhibit f-Met-Leu-Phe stimulation.[220-222] However, none of these inhibitors can be regarded as specific, so that a role for this arm of the pathway cannot be totally ruled out. In conclusion, inositol lipids by generating intracellular and extracellular signals play a key role in neutrophil activation.

ACKNOWLEDGMENT

I wish to thank the Medical Research Council for their support. S. C. is a Lister Institute Research Fellow.

REFERENCES

1. **Hokin, M. R. and Hokin, L. E.,** Enzyme secretion and the incorporation of ^{32}P into phospholipides of pancreas slices, *J. Biol. Chem.,* 203, 967, 1953.
2. **Nishizuka, Y.,** The role of protein kinase C in cell surface signal transduction and tumor promotion, *Nature (London),* 308, 693, 1984.
3. **Berridge, M. J. and Irvine, R. F.,** Inositol trisphosphate, a novel second messenger in cellular signal transduction, *Nature (London),* 312, 315, 1984.
4. **Cohen, P.,** The role of protein phosphorylation in neural and hormonal control of cellular activity, *Nature (London),* 296, 613, 1982.
5. **Irvine, R. F.,** How is the level of free arachidonic acid controlled in mammalian cells?, *Biochem. J.,* 204, 3, 1982.
6. **Downes, C. P. and Michell, R. H.,** Inositol phospholipid breakdown as a receptor-controlled generator of second messengers, in *Molecular Mechanisms of Transmembrane Signalling,* Vol. 4, Cohen, P. and Houslay, M. D., Eds., Elsvier, Amsterdam, 1985, 3.
7. **Berridge, M. J.,** Inositol trisphosphate and diacylglycerol as second messengers, *Biochem. J.,* 220, 345, 1984.
8. **Michell, R. H.,** Inositol phospholipids in cell surface receptor function, *Biochim. Biophys. Acta,* 415, 81, 1975.
9. **Michell, R., Kirk, C., Jones, L., Downes, C., and Creba, J.,** The stimulation of inositol lipid metabolism that accompanies calcium mobilisation in stimulated cells: defined characteristics and unanswered questions, *Philos. Trans. R. Soc. B.,* 296, 123, 1981.

10. **Hawthorne, J. N.**, Polyphosphoinositide metabolism in excitable cells, *Bioscience Rep.*, 3, 887, 1983.
11. **Abdel-Latif, A. A.**, Calcium-mobilizing receptors, polyphosphoinositides, and the generation of second messengers, *Pharmacol. Rev.*, 38, 227, 1986.
12. **Berridge, M. J.**, Intracellular signalling through inositol trisphosphate and diacylglycerol, *Biol. Chem. Hoppe-Seyler*, 367, 447, 1986.
13. **Fisher, S., van Rooijen, L., and Agranoff, B.**, Renewed interest in the polyphosphoinositides, *Trends Neurosciences*, 9, 53, 1984.
14. **Cockcroft , S.**, Phosphatidylinositol metabolism in mast cells and neutrophils, *Cell Calcium*, 3, 337, 1982.
15. **Cockcroft, S. and Allan, D.**, Loss of phosphatidylinositol and gain in phosphatidate in neutrophils stimulated with fMetLeuPhe, in *Inositol and Phosphoinositides: Metabolism and Regulation* Bleasdale, J. E., Eichberg, J., and Hauser, G., Eds., Humana Press, Clifton, N.J., 1985, 161.
16. **Cockcroft, S.**, Phosphoinositides and neutrophil activation, in *Phosphoinositides and Receptor Mechanisms*, Putney, J. W., Jr., Ed., *Receptor Biochemistry and Methodology*, Vol. 7, Allan R. Liss, New York, 1986, 287.
17. **Storey, D. J., Shears, S. B., Kirk, C. J., and Michell, R. H.**, Stepwise enzymatic dephosphorylation of inositol 1,4,5-trisphosphate to inositol in liver, *Nature (London)*, 312, 374, 1984.
18. **Irvine, R. F., Letcher, A. J., Heslop, J. P., and Berridge, M. J.**, The inositol tris/tetrakisphosphate pathway-demonstration of Ins(1,4,5)P_3 3-kinase activity in animal tissues, *Nature (London)*, 320, 631, 1986.
19. **Hawkins, P. T., Stephens, L., and Downes, C. P.**, Rapid formation of inositol 1,3,4,5-tetrakisphosphate and inositol 1,3,4-trisphosphate in rat parotid glands may both result indirectly from receptor-stimulated release of inositol 1,4,5-trisphosphate from phosphatidylinositol 4,5-bisphosphate, *Biochem. J.*, 238, 507, 1986.
20. **Biden, T. J. and Wollheim, C. B.**, Ca^{2+} regulates the inositol tris/tetrakisphosphate pathway in intact and broken preparations of insulin-secreting RINm5F cells, *J. Biol. Chem.*, 261, 11, 931, 1986.
21. **Downes, C. P., Hawkins, P. T., and Irvine, R. F.**, Inositol 1,3,4,5-tetrakisphosphate and not phosphatidylinositol 3,4-bisphosphate is the probable precursor of inositol 1,3,4-trisphosphate in agonist-stimulated parotid gland, *Biochem. J.*, 238, 501, 1986.
22. **Lew, P. D., Monod, A., Krause, K. H., Waldvogel, F. A., Biden, T. J., and Schlegel, W.**, The role of cytosolic free calcium in the generation of inositol 1,4,5-trisphosphate and inositol 1,3,4-trisphosphate in HL-60 cells, *J. Biol. Chem.*, 261, 13, 121, 1986.
23. **Sbarra, A. J. and Karnovsky, M. L.**, The biochemical basis of phagocytosis. Incorporation of ^{14}C-labelled building blocks into lipid, protein and glycogen of leukocytes during phagocytosis, *J. Biol. Chem.*, 235, 2224, 1960.
24. **Karnovsky, M. L. and Wallach, D. F. H.**, The metabolic basis of phagocytosis: incorporation of inorganic phosphate into various classes of phosphatides during phagocytosis, *J. Biol. Chem.*, 236, 1895, 1961.
25. **Cockcroft, S.**, Ca^{2+}-dependent conversion of phosphatidylinositol to phosphatidate in neutrophils stimulated with fMetLeuPhe or ionophore A23187, *Biochim. Biophys. Acta*, 795, 37, 1984.
26. **Zimmerli, W., Seligman, B., and Gallin, J. I.**, Exudation primes human and guinea pig neutrophils for subsequent responsiveness to the chemotactic peptide *N*-formylmethionylleucylphenylalanine and increase complement component C3bi receptor expression, *J. Clin. Invest.*, 77, 925, 1986.
27. **Cockcroft, S., Barrowman, M. M., and Gomperts, B. D.**, Breakdown and synthesis of polyphosphoinositides in fMetLeuPhe-stimulated neutrophils, *FEBS Lett.*, 181, 259, 1985.
28. **Zurier, R. B., Hoffstein, S., and Weissman, G.**, Cytochalasin B. Effect on lysosomal enzyme release from human leukocytes, *Proc. Natl. Acad. Sci. U.S.A.*, 70, 844, 1973.
29. **Bennett, J. P., Cockcroft, S., and Gomperts, B. D.**, Use of cytochalasin B to distinguish between early and late events in neutrophil activation, *Biochim. Biophys. Acta.*, 601, 584, 1980.
30. **Tou, J.-S. and Stjernholm, R.**, Cytochalasin B: effect on phospholipid metabolism and lysosomal enzyme release by leukocytes, *Biochim. Biophys. Acta*, 392, 1, 1975.
31. **Berger, R. R. and Karnovsky, M. L.**, Biochemical basis of phagocytosis. V. Effect of phagocytosis on cellular uptake of extracellular fluid, and on the intracellular pool of L-alpha-glycerophosphate, *Fed. Proc.*, 25, 840, 1966.
32. **Sastry, P. S. and Hokin, L. E.**, Studies on the role of phospholipids in phagocytosis, *J. Biol. Chem.*, 241, 3354, 1966.
33. **Tou, J.-S. and Stjernholm, R. L.**, Stimulation of the incorporation of $^{32}P_i$ and myo-[2-^{3}H]inositol into the phosphoinositides in polymorphonuclear leukocytes during phagocytosis, *Arch. Biochem. Biophys.*, 160, 487, 1974.
34. **Silverstein, S. C., Steinman, R. M., and Cohn, Z. A.**, Endocytosis, *Annu. Rev. Biochem.*, 46, 665, 1977.
35. **Lew, D. P., Andersson, T., Hed, J., Di Virgilio, F., Pozzan, T., and Stendahl, O.**, Ca^{2+}-dependent and Ca^{2+}-independent phagocytosis in human neutrophils, *Nature (London)*, 315, 509, 1985.

36. **Becker, E. L., Showell, H. S., Henson, P. M., and Hsu, L. S.,** The ability of chemotactic factors to induce lysosomal enzyme release: the characteristics of the release, the importance of surfaces and the relation of enzyme release to chemotactic responsiveness, *J. Immunol.*, 112, 2047, 1974.
37. **Cockcroft, S., Bennett, J. P., and Gomperts, B. D.,** The dependence on Ca^{2+} of phosphatidylinositol breakdown and enzyme secretion in rabbit neutrophils stimulated by formylmethionylleucylphenylalanine or ionomycin, *Biochem. J.*, 200, 501, 1981.
38. **Chenoweth, D. and Hugli, T.,** C5a and C5a analogs as probes of the neutrophil C5a receptor, *Mol. Immunol.*, 17, 151, 1980.
39. **Naccache, P. H., Molski, M. M., Volpi, M., Becker, E. L., and Sha'afi, R. I.,** Unique inhibitory profile of platelet activating factor induced calcium mobilization and granule enzyme secretion in rabbit neutrophils towards pertussis toxin and phorbol ester, *Biochem. Biophys. Res. Commun.*, 130, 677, 1985.
40. **Smith, R. J. and Bowman, B.,** Stimulation of human neutrophil degranulation with 1-*O*-octadecyl-2-*O*-acetyl-*SN*-glyceryl-3-phosphorylcholine: modulation by inhibitors of arachidonic acid metabolism, *Biochem. Biophys. Res. Commun.*, 104, 14,495, 1982.
41. **O'Flaherty, J. T.,** Neutrophil degranulation: evidence pertaining to its mediation by the combined effects of leukotriene B_4 platelet-activating factor, and 5-HETE, *J. Cell Physiol.*, 122, 229, 1985.
42. **O'Flaherty, J. T., Wykle, R. L., Thomas, M. J., and McCall, C. E.,** Neutrophil degranulation responses to combinations of arachidonate metabolites and platelet-activating factor, *Res. Commun. Chem. Pathol. Pharmacol.*, 43, 3, 1984.
43. **Smith, R. J. and Iden, S. S.,** Properties of concanavalin A-elicited granule exocytosis from human polymorphonuclear neutrophils, *Inflammation*, 4, 343, 1980.
44. **Smith, R. J., Speziale, S. C., and Bowman, B. J.,** Properties of interleukin-1 as a complete secretagogue for human neutrophils, *Biochem. Biophys. Res. Commun.*, 130, 1233, 1985.
45. **Volpi, M., Yassin, R., Tao, W., Molski, T. F. P., Naccache, P. H., and Sha'afi, R. I.,** Leukotriene B_4 mobilizes calcium without the breakdown of polyphosphoinositides and the production of phosphatidic acid in rabbit neutrophils, *Proc. Natl. Acad. Sci. U.S.A.*, 81, 5966, 1984.
46. **Prescott, S. M., Zimmerman, G. A., and Seeger, A. R.,** Leukotriene B_4 is an incomplete agonist for the activation of human neutrophils, *Biochem. Biophys. Res. Commun.*, 122, 535, 1984.
47. **Serhan, C. N., Radin, A., Smolen, J. E., Korchak, H., Samuelsson, B., and Weissmann, G.,** Leukotriene B_4 is a complete secretogogue in human neutrophils: a kinetic analysis, *Biochem. Biophys. Res. Commun.*, 107, 1006, 1982.
48. **Di Virgilio, F., Lew, D. P., and Pozzan, T.,** Protein kinase C activation of physiological processes in human neutrophils at vanishingly small cytosolic Ca^{2+} levels, *Nature (London)*, 310, 691, 1984.
49. **Barrowman, M. M., Cockcroft, S., and Gomperts, B. D.,** Two roles for guanine nucleotides in stimulus secretion sequence of neutrophils, *Nature (London)*, 319, 504, 1986.
50. **Volpi, M., Yassin, R., Naccache, P. H., and Sha'afi, R. I.,** Chemotactic factor causes rapid decreases in phosphatidylinositol 4,5-bisphosphate and phosphatidylinositol 4-monophosphate in rabbit neutrophils, *Biochem. Biophys. Res. Commun.*, 112, 957, 1983.
51. **Dougherty, R. W., Godfrey, P. P., Hoyle, P. C., Putney, J. W., Jr., and Freer, R. J.,** Secretagogue-induced phosphoinositide metabolism in human leukocytes, *Biochem. J.*, 222, 307, 1984.
52. **Bradford, P. and Rubin, R.,** Pertussis toxin inhibits chemotactic factor-induced phospholipase C stimulation and lysosomal enzyme secretion in rabbit neutrophils, *FEBS Lett.*, 183, 317, 1985.
53. **Volpi, M., Naccache, P. H., Molski, T. F. P., Shefcyk, J., Huang, C.-K., Marsh, M. L., Munoz, J., Becker, E. L., and Sha'afi, R. I.,** Pertussis toxin inhibits the formyl-methionyl-leucyl-phenylalanine but not the phorbol ester stimulated changes in ion fluxes, protein phosphorylation and phospholipid metabolism in rabbit neutrophils: role of the "G-proteins" in exitation response coupling, *Proc. Nat. Acad. Sci. U.S.A.*, 82, 2708, 1985.
54. **Ohta, H., Okajima, F., and Ui, M.,** Inhibition by islet activating protein of a chemotactic peptide-induced early breakdown of inositol phospholipids and Ca^{2+} mobilisation in guinea pig neutrophils, *J. Biol. Chem.*, 260, 15,771, 1985.
55. **Shefcyk, J., Yassin, R., Volpi, M., Molski, T. F. P., Naccache, P. H., Munoz, J. J., Becker, E. L., Feinstein, M. B., and Sha'afi, R. I.,** Pertussis but not cholera toxin inhibits the stimulated increase in actin association with the cytoskeleton in rabbit neutrophils: role of the "G proteins" in stimulus-response coupling, *Biochem. Biophys. Res. Commun.*, 126, 1174, 1985.
56. **Verghese, M. W., Smith, C. D., and Snyderman, R.,** Potential role for a guanine nucleotide regulatory protein in chemoattractant receptor mediated polyphosphoinositide metabolism, Ca^{++} mobilization and cellular responses by leukocytes, *Biochem. Biophys. Res. Commun.*, 127, 450, 1985.
57. **Andersson, T., Schlegel, W., Monod, A., Krause, K-H., Stendahl, O., and Lew, D. P.,** Leukotriene B_4 stimulation of phagocytes results in the formation of inositol 1,4,5-trisphosphate: a second messenger for Ca^{2+} mobilization, *Biochem. J.*, 240, 333, 1986.

58. **Mong, S., Chi-Rosso, G., Miller, J., Hoffman, K., Razgaitis, K. A., Bender, P., and Crooke, S. T.**, Leukotriene B_4 induces formation of inositol phosphates in rat peritoneal polymorphonuclear leukocytes, *Molec. Pharmacol.*, 30, 235, 1986.
59. **Lochner, J. E., Badwey, J. A., Horn, W., and Karnovsky, M. L.**, All-trans-retinal stimulates superoxide release and phospholipase C activity in neutrophils without significantly blocking protein kinase C, *Proc. Natl. Acad. Sci.* U.S.A., 83, 7673, 1986.
60. **Downes, C. P. and Wusteman, M. M.**, Breakdown of polyphosphoinositides and not phosphatidylinositol accounts for muscarinic agonist-stimulated inositol phospholipid metabolism in rat parotid glands, *Biochem. J.*, 216, 633, 1983.
61. **Bradford, P. G. and Rubin, R. P.**, Characterization of formylmethionyl-leucyl-phenylalanine stimulation of inositol trisphosphate accumulation in rabbit neutrophils, *Mol. Pharmacol.*, 27, 74, 1985.
62. **Di Virgilio, F., Vicentini, L. M., Treves, S., Riz, G., and Pozzan, T.**, Inositol phosphate formation in fMet-Leu-Phe-stimulated human neutrophils does not require an increase in the cytosolic free Ca^{2+} concentration, *Biochem. J.*, 229, 361, 1985.
63. **Burgess, G. M., McKinney, J. S., Irvine, R. F., and Putney, J. J. W.**, Inositol 1,4,5-trisphosphate and inositol 1,3,4-trisphosphate formation in Ca^{2+}-mobilizing-hormone-activated cells, *Biochem. J.*, 232, 237, 1985.
64. **Berridge, M. J., Downes, C. P., and Hanley, M. R.**, Lithium amplifies agonist-dependent phosphatidylinositol response in brain and salivary glands, *Biochem. J.*, 206, 587, 1982.
65. **Bennett, J. P., Cockcroft , S., Caswell, A. H., and Gomperts, B. D.**, Plasma membrane location of phosphatidylinositol hydrolysis in rabbit neutrophils stimulated with formylmethionylleucylphenylalanine, *Biochem. J.*, 208, 801, 1982.
66. **Cockcroft, S., Baldwin, J. M., and Allan, D.**, The Ca^{2+}-activated polyphosphoinositide phosphodiesterase of human and rabbit neutrophil membranes, *Biochem. J.*, 221, 477, 1984.
67. **Cockcroft, S., Bennett, J. P., and Gomperts, B. D.**, Stimulus-secretion coupling in rabbit neutrophils is not mediated by phosphatidylinositol breakdown, *Nature (London)*, 288, 275, 1980.
68. **Cockcroft, S. and Allan, D.**, The fatty acid composition of phosphatidylinositol, phosphatidate and 1,2-diacylglycerol in stimulated human neutrophils, *Biochem. J.*, 222, 557, 1984.
69. **Holub, B. J. and Kuksis, A.**, Metabolism of molecular species of diacylglycerophospholipids, *Adv. Lipid Res.*, 9, 1, 1978.
70. **Yano, K., Hattori, H., Imai, A., and Nozawa, Y.**, Modification of positional distribution of fatty acids in phosphatidylinositol of rabbit neutrophils stimulated with formylmethionyl-leucyl-phenylalanine, *Biochim. Biophys. Acta*, 752, 137, 1983.
71. **Cook, H. W., Clarke, J. T. R., and Spence, M. W.**, Inability of rabbit peritoneal polymorphonuclear leukocytes to synthesise arachidonic acid from linoleic acid, *Prostaglandins Leukotrienes Med.*, 10, 39, 1983.
72. **Takenawa, T., Ishitoya, J., Homma, Y., Kato, M., and Nagai, Y.**, Role of enhanced inositol phospholipid metabolism in neutrophil activation, *Biochem. Pharmacol.*, 34, 1931, 1985.
73. **Takenawa, T., Homma, Y., and Nagai, Y.**, Role of Ca^{2+} in phosphatidylinositol response and arachidonic acid release in formylated tripeptide- or Ca^{2+} ionophore A23187- stimulated guinea pig neutrophils, *J. Immunol.*, 130, 2849, 1983.
74. **Rossi, F., Della Bianca, V., Grzeskowiak, M., De Togni, P., and Cabrini, G.**, Relationships between phosphoinositide metabolism, Ca^{2+} changes and respiratory burst in formyl-methionyl-leucyl-phenylalanine-stimulated human neutrophils, *FEBS Lett.*, 181, 253, 1985.
75. **Yano, K., Nakashima, S., and Nozawa, Y.**, Coupling of polyphosphoinositide breakdown with calcium efflux in formylmethionyl-leucyl-phenylalanine-stimulated rabbit neutrophils, *FEBS Lett.*, 161, 296, 1983.
76. **Barrowman, M. M., Cockcroft, S., and Gomperts, B. D.**, Potentiation and inhibition of secretion from neutrophils by phorbol ester, *FEBS Lett.*, 201, 137, 1986.
77. **Grzeskowiak, M., Della Bianca, V., Cassatella, M., and Rossi, F.**, Complete dissociation between the activation of phosphoinositide turnover and of NADPH oxidase by formyl-methionyl-leucyl-phenylalanine in human neutrophils depleted of Ca^{2+} and primed by subthreshold doses of phorbol 12-myristate, *Biochem. Biophys. Res. Comm.*, 135, 785, 1986.
78. **Chock, S. P. and Schmauder-Chock, E. A.**, Evidence of *de novo* membrane generation in the mechanism of mast cell secretory granule activation, *Biochem. Biophys. Res. Commun.*, 132, 134, 1985.
79. **Dunlop, M. and Larkins, R.**, Pancreatic islets synthesize phospholipids *de novo* from glucose via acyl-dihydroacetone phosphate, *Biochem. Biophys. Res. Comm.*, 132, 467, 1985.
80. **Ishizuka, Y., Imai, A., Nakashima, S., and Nozawa, Y.**, Evidence for *de novo* synthesis of phosphatidylinositol coupled with histamine release in activated rat mast cells, *Biochem. Biophys. Res. Commun.*, 111, 581, 1983.
81. **Farese, R., Davis, J., Barnes, D., Standaert, M., Babischkin, J., Hock, R., Rosic, N., and Pollet, R.**, The *de novo* phospholipid effect of insulin is associated with increases in diacylglycerol, but not inositol phosphates or cytosolic Ca^{2+}, *Biochem. J.*, 231, 269, 1985.

82. **Sundler, R. and Papahadjopoulos, D.**, Control of membrane fusion by phospholipid head groups: phosphatidate/phosphatidylinositol specificity, *Biochem. Biophys. Acta,* 649, 743, 1981.
83. **Sundler, R., Duzgunes, N., and Papahadjopoulos, D.**, Control of membrane fusion by phospholipid head groups: the role of phosphatidylethanolamine in mixtures with phosphatidate and phosphatidylinositol, *Biochim. Biophys. Acta,* 649, 751, 1981.
84. **Sundler, R. and Wijkander, J.**, Protein-mediated intermembrane contact specifically enhances Ca^{2+}-induced fusion of phosphatidate-containing membranes, *Biochim. Biophys. Acta,* 730, 391, 1983.
85. **Sundler, R.**, Role of phospholipid head group structure and polarity in the control of membrane fusion, *Membr. Fluidity,* 563, 1984.
86. **Besterman, J. M., Duonio, V., and Cuatreacasas, P.**, Rapid formation of diacylglycerol from phosphatidylcholine: a pathway for generation of a second messenger, *Proc. Natl. Acad. Sci. U.S.A.*, 83, 6785, 1986.
87. **Daniel, L. W., Waite, M., and Wykle, R. L.**, A novel mechanism of diglyceride formation. 12-*O*-tetradeconylphorbol-13-acetate stimulates the cyclic breakdown and resynthesis of phosphatidylcholine, *J. Biol. Chem.*, 261, 9128, 1986.
88. **Serhan, C. N., Broekman, J., Korchak, H. M., Marcus, A. J., and Weissmann, G.**, Endogenous phospholipid metabolism in stimulated neutrophils: differential activation by FMLP and PMA, *Biochem. Biophys. Res. Commun.*, 107, 951, 1982.
89. **Serhan, C. N., Broekman, M. J., Korchak, H. M., Smolen, J. E., Marcus, A. J., and Weissmann, G.**, Changes in phosphatidylinositol and phosphatidic acid in stimulated human neutrophils. Relationship to calcium mobilization, aggregation and superoxide radical generation, *Biochim. Biophys. Acta,* 762, 420, 1983.
90. **Naccache, P., Showell, H., Becker, E., and Sha'afi, R.**, Arachidonic acid induced degranulation of rabbit peritoneal neutrophils, *Biochem. Biophys. Res. Commun.*, 87, 292, 1979.
91. **Naccache, P., Showell, H., Becker, E., and Sha'afi, R.**, Pharmacological differentiation between the chemotactic factor induced intracellular calcium redistribution and transmembrane calcium influx in rabbit neutrophils, *Biochem. Biophys. Res. Commun.*, 89, 1224, 1979.
92. **Bokoch, G. and Reed, P.**, Inhibition of the neutrophil oxidative response to a chemotactic peptide by inhibitors of arachidonic acid oxygenation, *Biochem. Biophys. Res. Commun.*, 90, 481, 1979.
93. **Showell, H. J., Naccache, P. H., Walenga, R. W., Dalecki, M., Feinstein, M. B., Sha'afi, R. I., and Becker, E. L.**, The effects of quercitin, 1-tosylamido-2-phenylethyl chloromethyl ketone, cytochalasin A and nordihydroguaiaretic acid on lysosomal enzyme secretion, arachidonic acid metabolism and Ca^{2+} fluxes in rabbit neutrophils, *J. Reticuloendothelial Soc.*, 30, 167, 1981.
94. **Sha'afi, R., Naccache, P., Alobaidi, T., Molski, T., and Volpi, M.**, Effect of arachidonic acid and the chemotactic factor F-Met-Leu-Phe on cation transport in rabbit neutrophils, *J. Cell Physiol.*, 106, 215, 1981.
95. **Smith, R. J., Sun, F. F., Bowman, B. J., Iden, S. S., Smith, H. W., and McGuire, J. C.**, Effect of 6,9-deepoxy-6,9-(phenylimino)-delta 6,8-prostaglandin I_1 (U-60,257), an inhibitor of leukotriene synthesis, on human neutrophil function, *Biochem. Biophys. Res. Commun.*, 109, 943, 1982.
96. **Lanni, C. and Becker, E. L.**, Inhibition of neutrophil phospholipase A_2 by *p*-bromophenacyl bromide, nordihydroguaiaretic acid, 5,8,11,14-eicosatetraenoic acid and quercitin, *Int. Arch. Allergy Appl. Immun.*, 76, 214, 1985.
97. **Bokoch, G. M. and Reed, P. W.**, Evidence for inhibition of leukotriene A_4 synthesis by 5,8,11,14-eicosatetranoic acid in guinea pig polymorphonuclear leukocytes, *J. Biol. Chem.*, 256, 4156, 1981.
98. **O'Flaherty, J. T., Showell, H. J., Ward, P. A., and Becker, E. L.**, A possible role of arachidonic acid in human neutrophil aggregation and degranulation, *Am. J. Pathol.*, 96, 799, 1979.
99. **Smith, R. J., Sun, F. F., Iden, S. S., Bowman, B. J., Sprecher, H., and McGuire, J. C.**, An evaluation of the relationship between arachidonic acid lipoxygenation and human neutrophil degranulation, *Clin. Immunol. Immunopathol.*, 20, 157, 1981.
100. **Palmer, R. M. J. and Salmon, J. A.**, Release of leukotriene B_4 from human neutrophils and its relationship to degranulation induced by *N*-formyl-methionyl-leucyl-phenylalanine, serum treated zymosan and the ionophore A23187, *Immunology,* 50, 65, 1983.
101. **Clancy, R., Dahinden, C., and Hugli, T.**, Arachidonic metabolism by human polymorphonuclear leukocytes stimulated by *N*-formyl-Met-Leu-Phe or complement component C5a is independent of phospholipase activation, *Proc. Natl. Acad. Sci. U.S.A.*, 80, 7200, 1983.
102. **Palmer, R. M. J. and Salmon, J. A.**, Comparison of the effects of some compounds on human neutrophil degranulation and leukotriene B_4 and thromboxane B_2 synthesis, *Biochem. Pharmacol.*, 34, 1485, 1985.
103. **Meade, C. J., Turner, G. A., and Bateman, P. E.**, The role of polyphosphoinositides and their breakdown products in A23187-induced release of arachidonic acid from rabbit polymorphonuclear leukocytes, *Biochem. J.*, 238, 425, 1986.

104. **Naccache, P. H., Molski, T. F. P., Borgeat, P., White, J. R., and Sha'afi, R. I.,** Phorbol esters inhibit the fMet-Leu-Phe- and leukotriene B_4-stimulated calcium mobilization and enzyme secretion in rabbit neutrophils, *J. Biol. Chem.*, 260, 2125, 1985.
105. **Robinson, J., Badwey, J., and Karnovsky, M.,** Superoxide release by neutrophils: synergistic effects of a phorbol ester and a calcium ionophore, *Biochem. Biophys. Res. Commun.*, 122, 734, 1984.
106. **Volpi, M., Molski, T. F. P., Naccache, P. H., Feinstein, M. B., and Sha'afi, R. I.,** Phorbol 12 myristate, 13 acetate potentiates the action of the calcium ionophore in stimulating arachidonic acid release and production of phosphatidic acid in rabbit neutrophils, *Biochem. Biophys. Res. Commun.*, 128, 594, 1985.
107. **Franson, R. C., Eisen, D., Jesse, R., and Lanni, C.,** Inhibition of highly purified mammalian phospholipase A_2 by non-steroidal anti-inflammatory agents, *Biochem. J.*, 186, 633, 1980.
108. **Victor, M., Weiss, J., Klempner, M., and Elsbach, P.,** Phospholipase A_2 activity in the plasma membrane of human polymorphonuclear leukocytes, *FEBS Lett.*, 136, 298, 1981.
109. **Hirata, F., Schiffman, E., Ventakasubramanian, K., Salomon, D., and Axelrod, J.,** A phospholipase A_2 inhibitory protein in rabbit neutrophils induced by glucocorticoids, *Proc. Nat. Acad. Sci. U.S.A.*, 77, 2533, 1980.
110. **Blackwell, G. J., Carnuccio, R., Di Rosa, M., Flower, G. J., Parente, L., and Persico, P.,** Macrocortin: a polypeptide causing the anti-phospholipase effect of glucocorticoids, *Nature (London)*, 287, 147, 1980.
111. **Carnuccio, R., Di Rosa, M., Flower, R. J., and Pinto, A.,** The inhibition of hydrocortisone of prostaglandin biosynthesis in rat peritoneal leukocytes is correlated with intracellular macrocortin levels, *Br. J. Pharmacol.*, 74, 322, 1981.
112. **Hirata, F.,** The regulation of lipomodulin, a phospholipase inhibitory protein in rabbit neutrophils by phosphorylation, *J. Biol. Chem.*, 256, 7730, 1981.
113. **Touqui, L., Rothut, B., Shaw, A. M., Fradin, A., Vargaftig, B. B., and Russo-Marie, F.,** Platelet activation — a role for a 40k anti-phospholipase A_2 protein indistinguishable from lipocortin, *Nature (London)*, 321, 177, 1986.
114. **Kramer, C., Franson, R., and Rubin, R.,** Regulation of phosphatidylinositol turnover, calcium metabolism and enzyme secretion by phorbol dibutyrate in neutrophils, *Lipids*, 19, 315, 1984.
115. **Hirata, F., Corcoran, B., Venkatasubramanian, K., Schiffman, E., and Axelrod, J.,** Chemoattractants stimulate degradation of methylated phospholipids and release of arachidonic acid in rabbit leukocytes, *Proc. Natl. Acad. Sci. U.S.A.*, 76, 2640, 1979.
116. **Bormann, B., Huang, C.-K., Mackin, W., and Becker, E.,** Receptor-mediated activation of a phospholipase A_2 in rabbit neutrophil plasma membrane, *Proc. Natl. Acad. Sci. U.S.A.*, 81, 767, 1984.
117. **Rubin, R. P., Sink, L. E., Schrey, M. P., Day, A. R., Liao, C. S., and Freer, R. J.,** Secretogogues for lysosomal enzyme release as stimulants of arachidonyl phosphatidylinositol turnover in rabbit neutrophils, *Biochem. Biophys. Res. Commun.*, 90, 1364, 1979.
118. **Rubin, R. P., Sink, L. E., and Freer, R. J.,** On the relationship between formylmethionyl-leucyl-phenylalanine stimulation of arachidonyl phosphatidylinositol turnover and lysosomal enzyme secretion by rabbit neutrophils, *Mol. Pharmacol.*, 19, 31, 1981.
119. **Rubin, R. P., Sink, L. E., and Freer, R. J.,** Activation of (arachidonyl) phosphatidylinositol turnover in rabbit neutrophils by the calcium ionophore A23187, *Biochem. J.*, 194, 497, 1981.
120. **Walsh, C. E., Waite, B. M., Thomas, M. J., and DeChatelet, L. R.,** Release and metabolism of arachidonic acid in human neutrophils, *J. Biol. Chem.*, 256, 7228, 1981.
121. **Swendsen, C. L., Ellis, J. M., Chilton, F. H., O'Flaherty, J. T., and Wykle, R. L.,** 1-*O*-Alkyl-2-acyl-*SN*-glycero-3-phosphocholine: a novel source of arachidonic acid in neutrophils stimulated by the calcium ionophore A23187, *Biochem. Biophys. Res. Commun.*, 113, 72, 1983.
122. **Chilton, F. H., Ellis, J. M., Olson, S. C., and Wykle, R. L.,** 1-*O*-alkyl-2-arachidonyl-*sn*-glycero-3-phosphocholine. A common source of platelet-activating factor and arachidonate in human polymorphonuclear leukocytes, *J. Biol. Chem.*, 259, 12014, 1984.
123. **Cockcroft, S.,** Does phosphatidylinositol breakdown control the Ca^{2+}-gating mechanism?, *Trends Pharmacol. Sci.*, 2, 340, 1981.
124. **Cockcroft, S., Bennett, J. P., and Gomperts, B. D.,** f-MetLeuPhe-induced phosphatidylinositol turnover in rabbit neutrophils is dependent on extracellular calcium, *FEBS Lett.*, 110, 115, 1980.
125. **Tsien, R.,** A non-disruptive technique for introducing calcium buffers and indicators into cells, *Nature (London)*, 290, 527, 1981.
126. **Cockcroft, S.,** The dependence on Ca^{2+} of the guanine nucleotide-activated polyphosphoinositide phosphodiesterase in neutrophil plasma membranes, *Biochem. J.*, 240, 503, 1986.
127. **Billah, M. M., Bryant, R. W., and Siegel, M. I.,** Lipoxygenase products of arachidonic acid modulate biosynthesis of platelet-activating factor (1-*O*-alkyl-2-acetyl-*sn*-glycero-3-phosphocholine) by human neutrophils via phospholipase A_2, *J. Biol. Chem.*, 260, 6899, 1985.
128. **Bokoch, G. M. and Reed, P. W.,** Stimulation of arachidonic acid metabolism in the polymorphonuclear leukocyte by an *N*-formylated peptide, *J. Biol. Chem.*, 255, 10,223, 1980.

129. **Pozzan, T., Lew, D. P., Wollheim, C. B., and Tsien, R. Y.,** Is cytosolic ionized calcium regulating neutrophil activation?, *Science,* 221, 1413, 1983.
130. **Barrowman, M. M., Cockcroft, S., and Gomperts, B. D.,** Differential control of azurophilic and specific granule exocytosis in Sendai virus-permeabilized rabbit neutrophils, *J. Physiol.,* 383, 115, 1987.
131. **Allan, D. and Thomas, P.,** The effects of Ca^{2+} and Sr^{2+} on Ca^{2+}-sensitive biochemical changes in human erythrocytes and their membranes, *Biochem. J.,* 198, 441, 1981.
132. **Scharmann, W.,** Formation and isolation of leukocidin from *Pseudomonas aeruginosa, Infect. Immun.,* 43, 21, 1976.
133. **Scharmann, W.,** Purification and characterisation of leukocidin from *Pseudomonas aeruginosa J. Gen. Microbiol.,* 93, 292, 1976.
134. **Noda, M., Hirayama, T., Kato, I., and Matsuda, F.,** Crystallization and properties of staphylococcal leukocidin, *Biochim. Biophys. Acta,* 633, 33, 1980.
135. **Noda, M., Kato, I., Hirayama, T., and Matsuda, F.,** Fixation and inactivation of staphylococcal leukocidin by phosphatidylcholine and ganglioside G_{M1} in rabbit polymorphonuclear leukocytes, *Infect. Immun.,* 29, 678, 1980.
136. **Woodin, A. M. and Wieneke, A. A.,** The accumulation of calcium by the polymorphonuclear leukocyte treated with staphyloccocal leucocidin and its significance in the extrusion of protein, *Biochem. J.,* 87, 487, 1963.
137. **Hirayama, T. and Kato, I.,** Mode of cytotoxic action of pseudomonal leukocidin on phosphatidylinositol metabolism and activation of lysosomal enzyme in rabbit leukocytes, *Infect. Immun.,* 43, 21, 1984.
138. **Hirayama, T. and Kato, I.,** A rapid stimulation of phosphatidylinositol metabolism in rabbit leukocytes by pseudomonal leukocidin, *FEBS Lett.,* 157, 46, 1983.
139. **Lad, P. M., Olson, C. V., and Grewel, I. S.,** A step sensitive to pertussis toxin and phorbol ester in human neutrophils regulates chemotaxis and capping but not phagocytosis, *FEBS Lett.,* 200, 91, 1986.
140. **Lad, P. M., Olson, C. V., and Grewal, I. S.,** Role of a pertussis toxin substrate in the control of lectin-induced cap formation in human neutrophils, *Biochem. J.,* 238, 29, 1986.
141. **Leeb-Lundberg, L. M. F., Cotecchia, S., Lomasney, J. W., DeBernardis, J. F., Lefkowitz, R. J., and Caron, M. G.,** Phorbol esters promote α_1-adrenergic receptor phosphorylation and receptor uncouping from inositol phospholipid metabolism, *Proc. Natl. Acad. Sci. U.S.A.,* 82, 5651, 1985.
142. **Zavoico, G. B., Halenda, S. P., Sha'afi, R. I., and Feinstein, M. B.,** Phorbol myristate acetate inhibits thrombin-stimulated calcium mobilization and phosphatidylinsitol-4,5-bisphosphate hydrolysis in human platelets, *Proc. Natl. Acad. Sci. U.S.A.,* 82, 3859, 1985.
143. **Takenawa, T., Ishitoya, J., and Nagai, Y.,** Inhibitory effect of prostaglandin E2, forskolin, and dibutyryl cAMP on arachidonic acid release and inositol phospholipid metabolism in guinea pig neutrophils, *J. Biol. Chem.,* 261, 1092, 1986.
144. **Ham, E. A., Soderman, D. D., Zanetti, M. E., Dougherty, H. W., McCauley, E., and Kuehl, J. F. A.,** Inhibition by prostaglandins of leukotriene B_4 release from activated neutrophils, *Proc. Natl. Acad. Sci. U.S.A.,* 80, 4349, 1983.
145. **Dawson, R. M. C.,** Studies on the enzymic hydrolysis of monophosphoinositide by phospholipase preparations from *P. notatum* and ox pancreas, *Biochim. Biophys. Acta,* 33, 68, 1959.
146. **Shukla, S.,** Phosphatidylinositol specific phospholipase C, *Life Sci.,* 30, 1323, 1982.
147. **Irvine, R. and Dawson, R.,** The distribution of calcium-dependent phosphatidylinositol-specific phosphodiesterase in rat brain, *J. Neurochem.,* 31, 1427, 1978.
148. **Dawson, R. M. C., Freinkel, N., Jungalwala, F. B., and Clarke, N.,** The enzymatic formation of myo-inositol 1:2-cyclic phosphate from phosphatidylinositol, *Biochem. J.,* 122, 605, 1971.
149. **Rittenhouse, S. E.,** Human platelets contain phospholipase C that hydrolyses polyphosphoinositides, *Proc. Natl. Acad. Sci. U.S.A.,* 80, 5417, 1983.
150. **Wilson, D. B., Bross, T. E., Hofmann, S. L., and Majerus, P. W.,** Hydrolysis of polyphosphoinositides by purified sheep seminal vesicle phospholipase C enzymes, *J. Biol. Chem.,* 259, 11,817, 1984.
151. **Connolly, T. M., Wilson, D. B., Bross, T. E., and Majerus, P. W.,** Isolation and characterisation of the inositol cyclic phosphate products of phosphoinositide cleavage by phospholipase C. Metabolism in cell-free extracts, *J. Biol. Chem.,* 261, 122, 1986.
152. **Irvine, R. F., Letcher, A., and Dawson, R. M. C.,** Phosphatidylinositol-4,5-bisphosphate phosphodiesterase and phosphomonoesterase activities of rat brain. Some properties and possible control mechanisms, *Biochem. J.,* 218, 177, 1984.
153. **Rodbell, M.,** The role of hormone receptors and GTP-regulatory proteins in membrane transduction, *Nature (London),* 284, 17, 1980.
154. **Allan, D. and Michell, R. H.,** A calcium activated polyphosphoinositide phosphodiesterase in the plasma membrane of human and rabbit erythrocytes, *Biochim. Biophys. Acta,* 508, 277, 1978.
155. **Downes, C. P. and Michell, R. H.,** The control by Ca^{2+} of the phosphoinositide phosphodiesterase and the Ca^{2+}-pump ATPase in human erythrocytes, *Biochem. J.,* 202, 53, 1982.

156. **Allan, D. and Cockcroft, S.**, The fatty acid composition of 1,2-diacylglycerol and polyphosphoinositides from human erythrocyte membranes, *Biochem. J.*, 213, 555, 1983.
157. **Cockcroft, S. and Gomperts, B. D.**, Role of guanine nucleotide binding protein in the activation of polyphosphoinositide phosphodiesterase, *Nature (London)*, 314, 534, 1985.
158. **Cockcroft, S., Taylor, J. A., and Judah, J. D.**, Subcellular localisation of inositol lipid kinases in rat liver, *Biochim. Biophys. Acta*, 845, 163, 1985.
159. **Smith, C. D., Lane, B. C., Kusaka, I., Verghese, M. W., and Snyderman, R.**, Chemoattractant receptor induced hydrolysis of phosphatidylinositol 4,5-bisphosphate in human polymorphonuclear leukocytes membranes: requirement for a guanine nucleotide regulatory protein, *J. Biol. Chem.*, 260, 5875, 1985.
160. **Smith, C. D., Cox, C. C., and Snyderman, R.**, Receptor-coupled activation of phosphoinositide specific phospholipase C by an N protein, *Science*, 232, 97, 1986.
161. **Kikuchi, A., Kozawa, O., Kaibuchi, K., Katada, T., Ui, M., and Takai, Y.**, Direct evidence for involvement of a guanine nucleotide-binding protein in chemotactic peptide-stimulated formation of inositol bisphosphate and trisphosphate in differentiated human leukemic (HL-60) cells, *J. Biol. Chem.*, 261, 11,558, 1986.
162. **Hirayama, T. and Kato, I.**, Two types of Ca^{2+}-dependent phosphorylation in rabbit leukocytes, *FEBS Lett.*, 146, 209, 1982.
163. **Williams, M. B. and Jones, H. P.**, Calmodulin-dependent NAD-kinase of human neutrophils, *Arch. Biochem. Biophys.*, 237, 80, 1985.
164. **Segal, A., Heyworth, P., Cockcroft, S., and Barrowman, M.**, Stimulated neutrophils from patients with autosomal recessive chronic granulomatous disease fail to phosphorylate a Mr-44,000 protein, *Nature (London)*, 316, 547, 1985.
165. **Uhing, R. J., Prpic, V., Jiang, H., and Exton, J. H.**, Hormone-stimulated polyphosphoinositide breakdown in rat liver plasma membranes. Roles of guanine nucleotides and calcium, *J. Biol. Chem.*, 261, 2140, 1986.
166. **Martin, T. F. J., Lucas, D. O., Bajjalieh, S. M., and Kowalchyk, J. A.**, Thyrotropin-releasing hormone activates a Ca^{2+}-dependent polyphosphoinositide phosphodiesterase in permeable GH_3 cells. GTPyS potentiation by a cholera and pertussis toxin-insensitive mechanism, *J. Biol. Chem.*, 261, 2918, 1986.
167. **Merritt, J. E., Taylor, C. W., Rubin, R. P., and Putney, J. W.**, Evidence suggesting that a novel guanine nucleotide regulatory protein couples receptors to phospholipase C in exocrine pancreas, *Biochem. J.*, 236, 337, 1986.
168. **Litosch, I., Wallis, C., and Fain, J. N.**, 5-Hydroxytryptamine stimulates inositol phosphate production in a cell-free system from blowfly salivary glands: evidence for a role of GTP in coupling receptor activation to phosphoinositide breakdown, *J. Biol. Chem.*, 260, 5464, 1985.
169. **Gonzales, R. A. and Crews, F. T.**, Guanine nucleotides stimulate production of inositol trisphosphate in rat cortical membranes, *Biochem. J.*, 232, 799, 1985.
170. **Gilman, A.**, Guanine nucleotide binding regulatory proteins and dual control of adenylate cyclase, *J. Clin. Invest.*, 73, 1, 1984.
171. **Krause, K-H., Schlegel, W., Wollheim, C. B., Anderson, T., Waldvogel, F. A., and Lew, D. P.**, Chemotactic peptide activation of human neutrophils and HL-60 cells. Pertussis toxin reveals correlation between inositol trisphosphate generation, calcium ion transients, and cellular activation, *J. Clin. Invest.*, 76, 1348, 1985.
172. **Lad, P. M., Olson, C. V., and Smiley, P. A.**, Association of the *N*-formyl-Met-Leu-Phe receptor in human neutrophils with a GTP-binding protein sensitive to pertussis toxin, *Proc. Natl. Acad. Sci. U.S.A.*, 82, 869, 1985.
173. **Okajima, F. and Ui, M.**, ADP-ribosylation of the specific membrane protein by islet-activating protein, pertussis toxin, associated with inhibition of a chemotactic peptide-induced arachidonate release in neutrophils: a possible role of the toxin substrate in Ca^{2+}-mobilizing biosignaling, *J. Biol. Chem.*, 259, 13,863, 1984.
174. **Goldman, D. W., Chang, F. H., Gifford, L. A., Goetzl, E. J., and Bourne, H. R.**, Pertussis toxin inhibition of chemotactic factor-induced calcium mobilization and function in human polymorphonuclear leukocytes, *J. Exp. Med.*, 162, 145, 1985.
175. **Backlund, P. S., Jr., Meade, B. D., Manclark, C. R., Cantoni, G. L., and Aksamit, R. R.**, Pertussis toxin inhibition of chemotaxis and the ADP-ribosylation of a membrane protein in a human-mouse hybrid cell line, *Proc. Natl. Acad. Sci. U.S.A.*, 82, 2637, 1985.
176. **Becker, E. L., Kermode, J. C., Naccache, P. H., Yassin, R., Marsh, M. L., Munoz, J. J., and Sha'afi, R. J.**, The inhibition of neutrophil granule secretion and chemotaxis by pertussis toxin, *J. Cell Biol.*, 100, 1641, 1985.
177. **Molski, T. F. P., Naccache, P. H., Marsh, M. L., Kermode, J., Becker, E. L., and Sha'afi, R. I.**, Pertussis toxin inhibits the rise in intracellular concentration of free calcium that is induced by chemotactic factors in rabbit neutrophils, *Biochem. Biophys. Res. Commun*, 124, 644, 1984.

Chapter 8

ROLE OF CALCIUM IN NEUTROPHIL ACTIVATION

Francesco Di Virgilio, Susan Treves, Daniel Lew, and Tullio Pozzan

TABLE OF CONTENTS

I. Introduction 200

II. Methods for Studying Calcium Homeostasis 200
A. Cell Free Systems (Subcellular Fractions) 200
B. Leaky Cells 200
C. Intact Cells 200
1. ^{45}Ca Fluxes 200
2. Chlorotetracycline 201
3. Cell Hybrids and Photoproteins 201
4. Tetracarboxylates 201
a. Quin2 Localization in Neutrophils 201
b. Side Effects of Quin2 Loading 202

III. Ca^{2+} Homeostasis in Resting Neutrophils 202

IV. Ca^{2+} Homeostasis in Activated Neutrophils 204
A. Ca^{2+} Ionophores 204
B. Chemoattractants 205
C. Phorbol Esters 206

V. Biochemistry of the Calcium Signal: Inositol Trisphosphate 207
A. Effect of $InsP_3$ on Permeabilized Neutrophils 207
B. $InsP_3$ Production in Intact Neutrophils: Kinetics and Ca^{2+} Dependence 207

VI. Role of $[Ca^{2+}]_i$ in Activation of Neutrophil Functions 208
A. NADPH-Oxidase 208
B. Exocytosis 210
C. Phagocytosis 211
D. Chemotaxis 212

VII. Concluding Remarks 213

Acknowledgments 214

References 214

I. INTRODUCTION

The importance of Ca^{2+} in cell physiology began to be appreciated in the 19th century, when Ringer noticed that the presence of this ion in the perfusion buffer was fundamental for cardiac contractility.[1] Now, more than 100 years later, the role of Ca^{2+} as a second messenger is universally recognized.[2-4] The present chapter deals with the regulation of Ca^{2+} homeostasis and the role of Ca^{2+} in neutrophil activation. We will mainly discuss these problems on the basis of data obtained during the past 4 years using tetracarboxylate Ca^{2+} indicators, photoproteins, permeabilized cells, and subcellular fractions. We will only briefly discuss more traditional methods such as tracer fluxes and the hydrophobic Ca^{2+} indicator, chlorotetracycline.

II. METHODS FOR STUDYING CALCIUM HOMEOSTASIS

A. Cell-Free System (Subcellular Fractions)

A classical method for studying Ca^{2+} homeostasis in eukaryotic cells has been, and still is, that of isolating subcellular fractions and studying their capacity to bind and/or accumulate Ca^{2+} and to release it upon stimulation. From this point of view, the plasma membrane of neutrophils has been studied in great detail. Inside-out plasma membrane vesicles can be generated by various methods;[7-9] since the polarity of such vesicles is opposite that of intact cells, they accumulate, rather than expel Ca^{2+}, against a concentration gradient. Only very recently has the capacity of other intracellular organelles, such as mitochondria and microsomes, to buffer Ca^{2+} been fully appreciated.[10]

B. Leaky Cells

The use of leaky cells to study Ca^{2+} homeostasis was first described in 1980,[11] but has only recently become popular.[12-15] The basic principle of this technique is straightforward: the plasma membrane is made permeable to low molecular weight solutes by a number of means. The most popular are treatment of cells with detergents, such as saponin or digitonin, which preferentially act on cholesterol-rich membranes, or the use of brief pulses of high voltage discharge, which makes small holes in the plasma membrane. Under these conditions the Ca^{2+} of the medium is assumed to be in rapid equilibrium with the Ca^{2+} in the space enclosed by the plasma membrane. Thus, by measuring the Ca^{2+} in the extracellular medium, either with a Ca^{2+} selective electrode or with an indicator, one can obtain information on the mechanism of Ca^{2+} homeostasis and on the relative contribution of the different intracellular organelles.

C. Intact Cells

1. ^{45}Ca Fluxes

Extensive reviews have been published on the basic principles underlying the use of radioactive tracers and the special problems posed by ^{45}Ca.[16] Neutrophils give rise to an additional problem, as these cells are short-lived, even in vivo, and consequently, isotopic equilibrium can not be reached, complicating quantitative interpretations. The most obvious difficulties encountered in ^{45}Ca flux experiments are that (1) they do not necessarily reflect a net movement of Ca^{2+} ions and (2) they cannot be unequivocally taken to indicate true changes in $[Ca^{2+}]_i$. A classical example of an erroneous interpretation of ^{45}Ca fluxes is that of phorbol myristate acetate, PMA. This compound was shown to promote an increase in ^{45}Ca uptake and release.[17,18] Thus, it was suggested that PMA caused a rise in $[Ca^{2+}]_i$, primarily by releasing Ca^{2+} from intracellular stores. When techniques to directly measure the $[Ca^{2+}]_i$ became available, it was demonstrated that PMA did not increase, but in fact caused a slight decrease in the $[Ca^{2+}]_i$.[19,20]

2. Chlorotetracycline

Chlorotetracycline (CTC) has been largely used in the past to monitor changes in Ca^{2+} homeostasis in intact cells.[21-23] This antibiotic is sufficiently hydrophobic to penetrate into intact cells and changes in CTC fluorescence have been detected after the addition of agonists to neutrophils in suspension.[22,23] However, the use of this compound as a Ca^{2+} indicator suffers from major drawbacks: (1) CTC fluorescence increases upon Ca^{2+} binding, but the dissociation constant for Ca^{2+} is approximately 0.4 m*M*, at physiological pH in an aqueous solution.[21] This concentration is three to four orders of magnitude larger than the $[Ca^{2+}]_i$ of resting cells.[24] (2) CTC has a Ca^{2+}:Mg^{2+} selectivity of less than 1.[21] (3) The intracellular localization of CTC is largely unknown, yet this probe is often erroneously reported to measure ''membrane bound calcium'',[22,25] a rather ambiguous physiochemical definition. Because of its amphipathic nature, CTC probably localizes at the interphase of cell membranes, where it binds Ca^{2+}. It is quite likely that part of the CTC signal comes from Ca^{2+}-rich organelles. Thus, rather than measuring variations in the cytoplasmic Ca^{2+}, one probably measures changes in the Ca^{2+} content of such organelles. Although studies with CTC have offered some insight into the events occurring during neutrophil activation, its poor specificity has prevented any firm conclusion on the role of $[Ca^{2+}]_i$.

3. Cell Hybrids and Photoproteins

The first direct measurement of $[Ca^{2+}]_i$ ever to be published in neutrophils was that of Hallett and Campbell, using the photoprotein obelin.[26] This important contribution was made possible by the application of an elegant microinjection technique, previously used to introduce high molecular weight substances, such as bacterial toxins, into the cytoplasm of small cells.[27] First, erythrocyte ghosts are loaded with the Ca^{2+} sensitive photoprotein obelin; then, the obelin-loaded ghosts are fused with rabbit inflammatory neutrophils using Sendai virus.[26,28] Unfortunately, this technique is time consuming and laborious. In addition, with obelin, as with other calcium-sensitive photoproteins, there are serious calibration difficulties which have yet to be solved. Another drawback arises from the fact that once photons of light have been emitted, the photoprotein is irreversibly inactivated. Thus, strong cell stimulation uses up a significant amount of the probe. Probably due to these inconveniences, this technique has not gained much popularity and none of the authors of this chapter have ever used this method. The interested reader is referred to the original paper and to recent reviews by Hallett and Campbell for technical details.[26,28,29]

4. Tetracarboxylates

After the introduction of quin2,[30,31] $[Ca^{2+}]_i$ measurements have become routine in most laboratories interested in cell biology and tens of papers using this technique have appeared in the literature (for review, see References 32 to 34). The new generation Ca^{2+} indicators, fura-2 and indo-1,[35] have only recently become commercially available and, to our knowledge, very few papers on their application to neutrophils have been published.[36]

The reader is referred to recent reviews published on the use of tetracarboxylate Ca^{2+} indicators for a general discussion on the principles, advantages, and major drawbacks of this technique.[31-34] Here we will only mention the major points which are relevant to the understanding of the next sections.

a. Quin2 Localization in Neutrophils

The selective cytoplasmic trapping of quin2 is one of the most important characteristics of this methodology. It has been suggested that in neutrophils, quin2/AM may be sequestered in intracellular organelles, since these cells are particularly rich in esterase-containing granules.[117] Some experimental evidence indicates, however, that this does not occur:

1. Direct morphological observation by fluorescent microscopy reveals a diffuse, rather than punctuate, fluorescence, suggesting cytoplasmic distribution[37]
2. Detergents such as digitonin which preferentially permeabilize cholesterol-rich membranes, such as the plasma membrane, release practically all the trapped quin2 from the cell. This release of quin2 parallels the release of cytoplasmic markers, but not that of mitochondrial or lysosomal enzymes[38]
3. Determination of extracellular quin2 reveals that no appreciable amount of indicator is released under conditions where up to 90% of the tertiary, 60% of the secondary, and 40% of the primary granules are secreted.[38,104]

b. Side Effects of Quin2 Loading

There are two principal side effects which accompany quin2 loading, those related to the capacity of quin2 to buffer Ca^{2+} transients and those independent of Ca^{2+} buffering. Let us consider the first problem, which is by far the most serious one. Quin2 is not only a fluorescent Ca^{2+} indicator, but also a high-affinity Ca^{2+} chelator. Because of its spectral overlap with endogenous fluorescent chromophores (mostly reduced pyridine nucleotides) relatively high concentrations (millimolar range) need to be trapped in order to swamp autofluorescence.[32-34] Thus, one must keep in mind that after quin2 loading, the intracellular Ca^{2+} buffering is greatly increased. In other words, as far as $[Ca^{2+}]_i$ transients are concerned, quin2-loaded neutrophils behave quantitatively and sometimes qualitatively differently from unloaded cells. This effect can be easily illustrated experimentally. When low quin2 concentrations are used and neutrophils are treated with EGTA plus the Ca^{2+} ionophore ionomycin, one observes a quick and large increase in fluorescence, indicating release of Ca^{2+} from membrane-bound intracellular stores. This increase is followed by a rapid return of $[Ca^{2+}]_i$ to resting levels (Figure 1 A). If this same experiment is performed with a high concentration of intracellular quin2 the increase in fluorescence is much smaller and the return to resting $[Ca^{2+}]_i$ is very slow (Figure 1 B). A completely different pattern is observed when Ca^{2+} comes from the extracellular medium (but see below).

On the other hand, the "buffering drawback" can be used advantageously under certain conditions, for example when the contribution of intracellular Ca^{2+} stores to a particular response has to be "eliminated",[39] or when one is interested in the relationship between amplitude and/or speed of the rise in $[Ca^{2+}]_i$ and cellular functions (see below).

As far as the problems unrelated to Ca^{2+} buffering are concerned, they have not been studied in great detail in neutrophils. The simplest method to test for such side effects is to use a stimulus which does not depend on intracellular Ca^{2+} transients, such as PMA. We have demonstrated that neither the PMA-induced secretion nor superoxide anion generation are affected by quin2 loading.[20] Finally, neutrophils seem to be particularly resistant to the ATP depleting effects of intracellular formaldehyde generation, a by-product of quin2/AM hydrolysis.[39]

III. Ca^{2+} HOMEOSTASIS IN RESTING NEUTROPHILS

In resting neutrophils the $[Ca^{2+}]_i$ is around 100 n*M*, measured with quin2,[19,20,38-43] fura-2,[36] or photoproteins,[26,28,29] i.e., similar to that of most other mammalian cells. The maintenance of such a large electrochemical gradient for Ca^{2+} across the plasma membrane is an essential feature of all eukaryotic cells. As for other cells, the plasma membrane is a key site in the regulation of Ca^{2+} homeostasis. In fact, although organelles are capable of buffering calcium, the final, long-term regulators of Ca^{2+} homeostasis are the Ca^{2+} extrusion mechanism(s) localized on the plasma membrane.[24] Indeed, neutroplasts[44] (that is, neutrophils which have been depleted of nuclei, granules, and most intracellular organelles) maintain a normal $[Ca^{2+}]_i$,[45] even though they do not possess intracellular Ca^{2+} stores.[45] Various

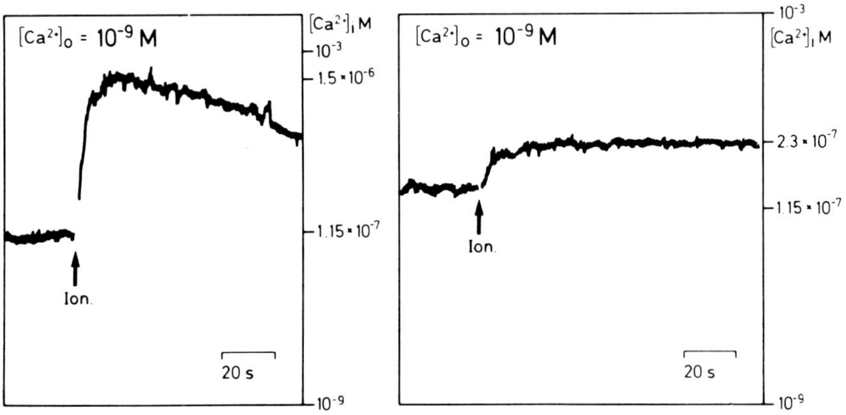

A B

FIGURE 1. Effect of intracellular quin2 on changes in $[Ca^{2+}]_i$ in neutrophils exposed to ionomycin (ion) in calcium-free medium. Intracellular quin2 concentration was 0.22 nmol/10^6 cells (0.6 m*M*) in panel A and 0.9 nmol/10^6 cells (2.6 m*M*) in panel B. (Reproduced from *The Journal of Cell Biology*, 1984, 99, 1212, by copyright permission of the Rockefeller University Press.)

groups have reported the presence, in plasma membrane enriched fractions, of a Ca^{2+}-ATPase showing high affinity for Ca^{2+}, as well as the presence of ATP and Ca^{2+}-dependent phosphorylated intermediates.[7-9] No evidence (see also below) for a significant Na^+/Ca^{2+} exchange mechanism has been provided. The Ca^{2+} ATPase exhibits a high affinity for Ca^{2+}, its K_m ranging between 0.2 and 0.4μ*M*, i.e., two- to fourfold higher than resting $[Ca^{2+}]_i$ levels.[7-9] Ca^{2+} uptake by inside-out plasma membrane vesicles shows an absolute requirement for ATP (K_m = 0.36 + 0.2 m*M*[7]), whereas other trinucleotides are inactive.[7-9] The Ca^{2+} ATPase is extremely sensitive to micromolar concentrations of vanadate,[7-9] an important difference when compared to other intracellular ATPases or the mitochondrial uniporter.

The steady state $[Ca^{2+}]_i$ reached in permeabilized neutrophils, under conditions mimicking resting cytoplasmic conditions, is around 200 n*M*, a concentration which is quite similar to that of intact cells. With the use of specific inhibitors it has been possible to study the contribution of intracellular organelles to the maintenance of Ca^{2+} homeostasis.[10] One such system which has been shown to accumulate Ca^{2+} in an ATP-dependent manner, is unaffected by mitochondrial inhibitors such as oligomycin, uncouplers, antimycin, or Ruthenium Red. Ca^{2+} is clearly taken up in vesicular stores against a concentration gradient, the energy being provided by the hydrolysis of ATP. This Ca^{2+} uptake is inhibited by vanadate, but the concentrations required are at least two orders of magnitude higher than those necessary to inhibit the plasma membrane Ca^{2+} pump.[10] Since neutrophils are known to contain few mitochondria, the finding that permeabilized neutrophils maintain a steady state $[Ca^{2+}]_i$ of 600 n*M*, even in the absence of ATP, was quite surprising.[10] Furthermore, this steady state exhibits certain characteristics of a mitochondrial "set point" since it can be inhibited by mitochondrial poisons; it does not require ATP nor is it affected by oligomycin. These findings suggest that mitochondria may participate, in combination with other intracellular pools, in bringing about the decrease in $[Ca^{2+}]_i$ to submicromolar levels. Nevertheless, it is only the nonmitochondrial pool together with the plasma membrane Ca^{2+} ATPase which lower $[Ca^{2+}]_i$ to resting levels.[10]

As to the nature of these other nonmitochondrial stores, at the moment not much is known. Their presence in neutrophils, as well as in other cell types, is critical as they release Ca^{2+} upon receptor activation. In certain tissues, this intracellular Ca^{2+}-storing organelle has putatively been identified as a subfraction of the endoplasmic reticulum,[12,46] though no definite proof has been provided for this conclusion. In particular, neutrophils are poor in endoplasmic reticulum, but they contain a "normal" amount of intracellular Ca^{2+} stores. We think that neither the endoplasmic reticulum nor any of the known intracellular organelles play a major role in intracellular Ca^{2+} homeostasis. Recent experiments from our laboratories[118] suggest that a new light sedimenting organelle, distinct from Golgi, endoplasmic reticulum, and endocytic vesicles, is responsible for most of the intracellular high affinity Ca^{2+} pumping activity and $InsP_3$ sensitivity. We have proposed for this newly identified organelle the name "calciosome."[118] Finally, three major points ought to be mentioned regarding the regulation of Ca^{2+} homeostasis in resting neutrophils. (1) As opposed to other secretory cells, neutrophils do not contain voltage-gated Ca^{2+} channels, since depolarization has no effect on the $[Ca^{2+}]_i$.[47] In addition, voltage-gated Ca^{2+} channel agonists or antagonists do not induce changes in the $[Ca^{2+}]_i$, either at resting $[Ca^{2+}]_i$ or after stimulation.[47,48] This suggestion has been definitely confirmed recently by direct electrophysiological measurements.[36] (2) As previously mentioned, there is no major Na^+/Ca^{2+} exchange in neutrophils since gramicidin D, which collapses the Na^+ gradient and membrane potential, has no effect on $[Ca^{2+}]_i$.[8,45] (3) As in many other cell types the resting $[Ca^{2+}]_i$ is practically insensitive to variations in the extracellular Ca^{2+} concentration, in the range 100 μM to 2 mM.

IV. Ca^{2+} HOMEOSTASIS IN ACTIVATED NEUTROPHILS

To date, studies on the effects of agonists on $[Ca^{2+}]_i$ in intact cells, have been mainly carried out using (1) Ca^{2+} ionophores, (2) chemotactic substances such as formyl-methionyl-leucyl-phenylalanine (fMLP or f-Met-Leu-Phe), C5a, leukotrienes, and aggregated immunoglobulins and (3) protein kinase C activators. A few reports with particulate stimuli such as opsonized particles, latex beads, etc. have also appeared recently.[37,49-52] With one important exception,[37] all these studies have been performed using neutrophils in suspension.

A. Ca^{2+} Ionophores

Two Ca^{2+} ionophores have been used, A23187 and ionomycin. Because of its endogenous fluorescence at the quin2 wavelengths and its ability to transport Mg^{2+} ions as well, A23187 is rarely used with the tetracarboxylates.[40] Both ionophores, however, have been shown to cause redistribution of Ca^{2+} from intracellular stores and influx of Ca^{2+} from the extracellular medium.[38-42] The main evidence indicating that Ca^{2+} ionophores release Ca^{2+} from intracellular stores in intact cells comes from experiments carried out in Ca^{2+}-free media. Under these conditions, both ionophores induce a rise in $[Ca^{2+}]_i$. The amplitude and duration of this $[Ca^{2+}]_i$ rise is, as expected, a function of the amount of intracellular Ca^{2+} buffering;[39] in addition, Ca^{2+} release from stores is very sensitive to the ionophore concentration; near maximal rises in $[Ca^{2+}]_i$ being obtained at nmolar concentrations.[119] Since the only known function of ionophores is that of transporting Ca^{2+} in exchange with H^+ (ionomycin) and/or Mg^{2+} (A23187) across membranes down their concentration gradients, these results indicate that neutrophils possess membrane enclosed Ca^{2+} stores. The size of the ionophore induced $[Ca^{2+}]_i$ rises can give an estimate of the Ca^{2+} which can be readily mobilized.[39]

In addition to releasing Ca^{2+} from intracellular stores, it is obvious that ionophores also cause Ca^{2+} influx through the plasma membrane. Interestingly, in order to reach steady state levels of $[Ca^{2+}]_i$, higher concentrations of the ionophore, compared to those necessary to release intracellular stores, are required. It ought to be mentioned that, as opposed to the Ca^{2+} released from intracellular stores, the steady state $[Ca^{2+}]_i$ caused by the ionophore is independent of the concentration of intracellular Ca^{2+} buffer.[38] Under these conditions only the time required to reach the steady state is affected by the amount of intracellular Ca^{2+} indicator (Figure 2).

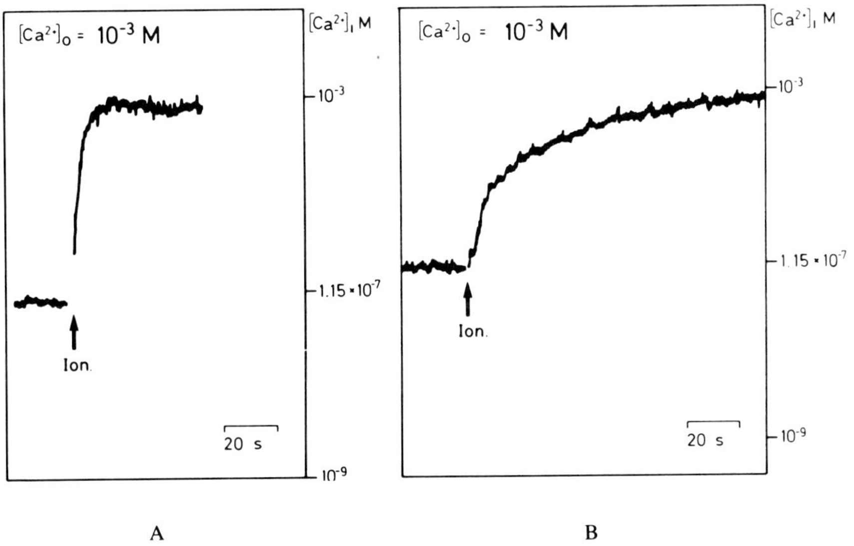

FIGURE 2. Effect of intracellular quin2 on changes in $[Ca^{2+}]_i$ in neutrophils exposed to ionomycin (ion) in the presence of extracellular Ca^{2+}. Intracellular quin2 concentration was 0.22 nmol/10^6 cells (0.6 m*M*) in panel A and 0.9 nmol/10^6 cells (2.6 m*M*) in panel B. (Reproduced from The Journal of Cell Biology, 1984, 99, 1212, by copyright permission of the Rockefeller University Press.)

B. Chemoattractants

The most commonly used chemoattractants are fMLP, C5a, and LTB_4. All three agonists are capable of generating large rises in $[Ca^{2+}]_i$ at exceedingly low concentrations, i.e., $10^{-9}M$ or less.

One way to assess the contribution of intracellular Ca^{2+} stores, to the total $[Ca^{2+}]_i$ increase, is to generate a rise in $[Ca^{2+}]_i$ in normal medium and then compare it with that obtained with the same batch of cells in EGTA-containing medium.[38-42,47] Using this simple procedure, it has been concluded that the rise in $[Ca^{2+}]_i$ induced by chemotactic factors has two components: a very rapid increase, due to release of Ca^{2+} from intracellular stores and a second, slower component, due to Ca^{2+} influx across the plasma membrane.[38-42,47] By using different intracellular quin2 concentrations, thus modulating the increase in $[Ca^{2+}]_i$ caused by the agonist, one can obtain an estimate of the importance of intracellular Ca^{2+} stores in triggering neutrophil responses (see below).

It is worth mentioning that in the absence of extracellular Ca^{2+}, if fMLP and/or LTB_4 are added after the ionophore, they are unable to release Ca^{2+} into the cytosol.[38,39,53] This result suggests that the pool of intracellular Ca^{2+} released following receptor-ligand interaction is the same as that released by ionophores. Furthermore, since ionophores are unable to displace bound Ca^{2+}, these results suggest that the pool of Ca^{2+} released during neutrophil activation is not membrane bound, but membrane enclosed.[38,42]

No quantitative differences in the amount of ionophore or agonist releasable-Ca^{2+} pools have been observed between neutrophils obtained from healthy donors or patients affected by diseases in which an abnormality in the handling of intracellular calcium has been suspected (cystic fibrosis,[54] essential hypertension[55]). A relevant exception are neutrophils of patients affected by Barter's syndrome.[56]

In a variety of neuroendocrine tissues, it is known that the major pathway for Ca^{2+} influx is via voltage-dependent calcium channels.[57] As mentioned above, however, these channels do not seem to be present in neutrophils.[47]

The key question is, thus, if and how the activation of surface receptors leads to an increased Ca^{2+} influx across the plasma membrane. The existence of a receptor-stimulated Ca^{2+} influx is supported by the following experimental results. (1) If neutrophils are incubated in the absence of extracellular Ca^{2+} and then stimulated with fMLP, the introduction of Ca^{2+} back into the medium results in a new, larger and slower rise in $[Ca^{2+}]_i$.[47] (2) The addition of La^{3+} completely inhibits the second phase of the fMLP-induced Ca^{2+} rise to values observed in Ca^{2+}-free medium.[47] (3) The influx of Mn^{2+} across the membrane is increased several fold upon neutrophil stimulation.[47] (4) ^{45}Ca uptake is stimulated by a number of agonists.[5,17]

Until recently, it was not clear whether the increased Ca^{2+} influx is due to the opening of a particular channel or to a carrier-mediated mechanism, nor whether the channel is truly opened as a consequence of ligand-receptor binding or whether its opening is modulated by a product of the phosphoinositide cycle. One of the authors (T. P.) has recently suggested the term SMOC (second messenger operated channel)[58] to distinguish the latter type of channel from the classical ROC (receptor operated channel), such as the acetylcholine receptor, where the channel is part of the receptor itself. The existence of an unspecific cation channel activated by $[Ca^{2+}]_i$ has been recently demonstrated by the patch clamping technique.[36]

C. Phorbol Esters

A few important observations should be mentioned at this stage on the effect of active tumor promoters, such as PMA, and thus on the role of protein kinase C on Ca^{2+} homeostasis. As mentioned in a previous section, it was initially suggested that PMA activates neutrophil functions, such as the respiratory burst, membrane depolarization, and enzyme secretion, by increasing the $[Ca^{2+}]_i$, as indicated by ^{45}Ca flux experiments[17] and net Ca^{2+} movements.[18] However, after the introduction of fluorescent Ca^{2+} indicators, it became clear that this was not the case.[19] In fact, when neutrophils from a variety of different sources, are challenged with phorbol esters, no effect or a small decrease in $[Ca^{2+}]_i$ is observed. This observation has been further extended and it has been subsequently demonstrated that (1) pretreatment of quin2-loaded neutrophils with PMA, decreases the rise in $[Ca^{2+}]_i$ obtained with fMLP and C5a[19] and (2) if neutrophils are stimulated with fMLP and then challenged with PMA, the $[Ca^{2+}]_i$ rapidly falls back to resting levels.[19,59] These results, as well as several others, indicate that protein kinase C is involved in the modulation of Ca^{2+} homeostasis, though it may have multiple modes of action.

In particular, activation of protein kinase C:

1. Inhibits the hydrolysis of polyphosphoinositides, and thus the amount of $InsP_3$ generated by ligand receptor interaction.[59,60]
2. Causes depolarization of the plasma membrane potential, thus reducing the driving force for Ca^{2+} entry.[61] Recently we[62] have suggested that plasma membrane depolarization, a well-known consequence of stimulation with chemotactic peptides or phorbol esters, may play a feedback inhibitory role in Ca^{2+} fluxes. Predepolarization of the plasma membrane potential either by elevating the extracellular $[K^+]$ or with gramicidin D decreases both the release of Ca^{2+} from intracellular stores and the influx through the plasma membrane. Interestingly, the inhibition of $[Ca^{2+}]_i$ changes by predepolarization, is observed only at low fMLP concentrations, i.e., doses which alone have no effect on the plasma membrane potential. It appears that modulation of agonist-induced Ca^{2+} changes by the membrane potential, is a general phenomenon in white blood cells since it has been also observed in B and T lymphocytes.[63]
3. Activates the plasma membrane Ca^{2+} pumping activity by increasing its Vmax. This effect was demonstrated by Lagast et al.[19] in inside-out plasma membrane vesicles isolated from PMA-treated neutrophils and more recently confirmed by Richard and Sheterline.[64] in intact neutrophils.

V. BIOCHEMISTRY OF THE CALCIUM SIGNAL: INOSITOL TRISPHOSPHATE

For a detailed discussion on the biochemistry of the polyphosphoinositide cycle, the reader is referred to Chapter 7 of this book as well as to recent reviews.[65-67] Here we will only briefly discuss the main points relevant to the regulation of Ca^{2+} homeostasis in neutrophils.

For years, the potential role of a phospholipid metabolite generated during receptor stimulation, in the regulation of Ca^{2+} homeostasis, was considered, but never went past a plausible working hypothesis. Possible candidates included phosphatidic acid,[68] leukotriene B_4 (but see Reference 69), or inositol lipid metabolism itself.[70] Work carried out in neutrophils seemed to contradict the hypothesis that polyphosphoinositides could be primarily involved in the regulation of Ca^{2+} homeostasis. In fact, Cockcroft et al.[71] suggested, on the basis of indirect evidence, that a rise in $[Ca^{2+}]_i$ is necessary to cause polyphosphoinositide breakdown. The revolutionary discovery that exogenous inositol 1,4,5 trisphosphate ($InsP_3$) could mobilize Ca^{2+} from permeabilized pancreatic acinar cells,[14] and the extension of these findings to other cell types,[65] shed new light in this field.

A. Effect of $InsP_3$ on Permeabilized Neutrophils

The addition of $InsP_3$ to permeabilized neutrophils causes an almost immediate (<2 sec) release of Ca^{2+} from intracellular stores.[10] At submicromolar concentrations, $InsP_3$ releases part of the Ca^{2+} accumulated in an ATP-dependent manner, but has no effect on the mitochondrial Ca^{2+} pool.[10] This effect is specific as it only occurs with Ins $1,4,5P_3$; in addition, this release of Ca^{2+} is transient and the released Ca^{2+} is rapidly pumped back into the vesicular pool.[10] If the ATP-dependent Ca^{2+} influx mechanism is blocked, the addition of $InsP_3$ results in a larger Ca^{2+} release.[10] Thus, these vesicles appear to have independent influx and efflux Ca^{2+} pathways, a Ca^{2+} ATPase (influx component), and an $InsP_3$-sensitive efflux component.

In order to demonstrate the involvement of $InsP_3$ in the receptor-mediated mobilization of intracellular Ca^{2+} in intact cells, at least three conditions must be satisfied: (1) demonstrate that neutrophils produce $InsP_3$ in response to agonists and that the kinetics of its formation are compatible with the rate of intracellular Ca^{2+} mobilization; (2) demonstrate that a rise in $[Ca^{2+}]_i$ is not necessary to induce $InsP_3$ formation; and (3) demonstrate that $InsP_3$ mobilizes Ca^{2+} from intracellular stores. The experiment described above only satisfies the third requirement. What about the other two criteria?

There are major technical problems for the assessment of $InsP_3$ production in intact neutrophils. These cells are in fact, short-lived and can not be radioactively labeled for long periods of time using the classical technique for labeling the polyphosphoinositide pool. Nevertheless, these problems can be partially overcome by culturing neutrophils for 18 hr[41] or by acutely labeling cells with very high doses of myo-[^{3}H] inositol.[72] Alternatively, one can measure production of $InsP_3$ in the human promyelocytic leukemia cell line, HL-60,[73,74] which acquires neutrophil-like characteristics, including the expression of chemotactic receptors, after differentiation with dimethylsulfoxide.

B. $InsP_3$ Production in Intact Neutrophils: Kinetic and Ca^{2+} Dependence

The accumulation of inositol phosphates in myo-[^{3}H] inositol-labeled neutrophils stimulated with fMLP and LTB_4 has been measured by several groups.[41,53,73] Treatment with these two chemotactic peptides causes the rapid generation of $InsP_3$ (the earliest time which has been measured is 8 sec). Of interest, it has been found that any concentration of fMLP which is capable of increasing $[Ca^{2+}]_i$ is also capable of causing the accumulation of $InsP_3$ even if to a very small extent.[73,120] By HPLC, which allows one to discriminate between the two isomeric forms, Ins $1,4,5P_3$ and Ins 1,3,4 P_3, it has been found that first the Ins 1,4,5 P_3 isomer is produced, followed by a slower, more sustained rise in Ins 1,3,4 P_3.[53,73]

In addition, if the intracellular Ca^{2+} stores are depleted and the rise in $[Ca^{2+}]_i$ inhibited, $InsP_3$ (1,4,5 isomer) production still occurs,[41] while the formation of Ins 1,3,4 P_3 is severely curtailed.[74] It has been recently demonstrated that the combination of an elevated $InsP_3$ and $[Ca^{2+}]_i$, activates the phosphorylation of Ins 1,4,5 P_3 to Ins 1,3,4,5 tetrakisphosphate, $InsP_4$, and subsequent dephosphorylation to Ins 1,3,4 P_3.[75] The function of this pathway is at present unknown, but it has been suggested that it could be involved in calcium influx across the plasma membrane. In conclusion, the dose dependency, rapid production, and Ca^{2+} independence are all compatible with the postulated role of Ins 1,4,5 P_3 as a Ca^{2+}-mobilizing second messenger.

VI. ROLE OF $[Ca^{2+}]_i$ IN THE ACTIVATION OF NEUTROPHIL FUNCTIONS

A. NADPH-Oxidase

The activation of NADPH oxidase, an enzyme complex located on the plasma membrane of phagocytic cells, has a key role in host defense against invading microorganisms.[121] In fact, patients affected by chronic granulomatous disease (CGD) lack either the oxidase or some of its components.[76] These individuals are susceptible to a number of infections, even though their neutrophils can phagocytose and secrete to the same extent as those of healthy controls. A variety of different agents, such as phagocytosable particles,[77,78] deoxycholate,[79] digitonin,[80] saponin,[81] phospholipase C,[82] Concanavalin A,[83] antineutrophil antibodies,[84] fatty acids,[85] cytochalasin E,[86] trypsinized liver microsomes,[87] endotoxin,[88] Ca^{2+} ionophores,[89] phorbol esters,[90] split products of complement,[91] and formylated peptides,[92] have been shown to activate this enzyme complex. For years activation of NADPH oxidase was thought to be triggered by a rise in $[Ca^{2+}]_i$. This hypothesis was mainly based on the demonstration that: (1) a "Ca^{2+} only" signal, triggered by the Ca^{2+} ionophore A23187, can stimulate O_2^- production,[89] and (2) removal of extracellular Ca^{2+} results in inhibition or reduction of O_2^- production, at least with some stimuli. However, a number of observations did not fit into this simple model. For example, among ionophores, the most powerful NADPH-oxidase activator is X537A,[89] a compound which has a low Ca^{2+} specificity and also translocates monovalent ions. Nonetheless, until 1983 this was the most popular hypothesis. Since then, considerable experimental evidence, obtained in different laboratories, has cast serious doubts on the simple hypothesis that a rise of $[Ca^{2+}]_i$ is a necessary and sufficient signal for NADPH oxidase activation.[29,38,50] For example, it has been demonstrated that agonists such as LTB_4 or PAF are potent stimulators of the $[Ca^{2+}]_i$ rise, but poor activators of the respiratory burst.[53,93] There are also clear discrepancies between the doses of chemotactic peptides that are required to maximally elevate $[Ca^{2+}]_i$ and the doses required to stimulate superoxide anion generation.[61] For sure, however, the most definitive evidence against a primary role of Ca^{2+} in the activation of NADPH-oxidase has come from experiments performed with tumor-promoting phorbol esters. As mentioned above, these compounds are believed to exert their biological effects by directly binding to and activating protein kinase C.[94] PMA is, in fact, one of the most potent activators of the respiratory burst and has minimal (it decreases) effects on $[Ca^{2+}]_i$.[19] We have demonstrated that a rise in $[Ca^{2+}]_i$ is not only unnecessary, but Ca^{2+} itself seems superfluous for full activation of superoxide anion production,[20] provided one finds a way to efficiently activate protein kinase C in the complete absence of cytoplasmic Ca^{2+}.

In conclusion, there are few doubts of the possibility of dissociating the respiratory burst from $[Ca^{2+}]_i$ rises, but from a physiological point of view, the main question is whether the rise in $[Ca^{2+}]_i$ which accompanies superoxide anion generation with some agonists, and in particular with chemotactic peptides, is needed for full expression of oxidase activity. In this respect, the most studied agonist is the formylated peptide fMLP. NADPH-oxidase activity triggered with this agonist has been shown to require an increase of the $[Ca^{2+}]_i$.[39] In our opinion the clearest evidence in favor of this conclusion is:

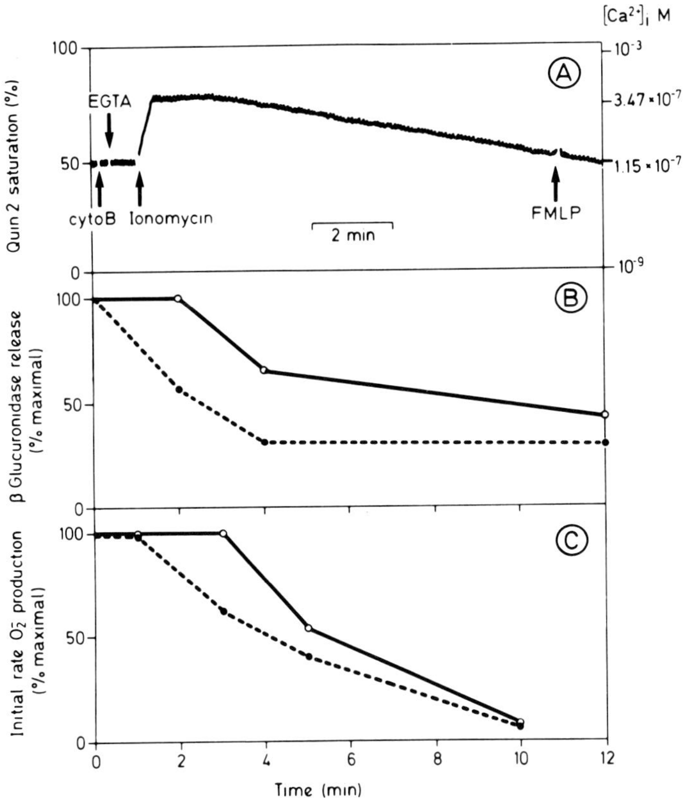

FIGURE 3. Effect of depletion of intracellular calcium on $[Ca^{2+}]_i$ transients, exocytosis and O_2^- production induced by fMLP ($10^{-7}M$. Cells were incubated in calcium-free medium. Cytochalasin B (5)μg/mℓ), and EGTA (1 m*M*) were added, followed by ionomycin (1 μ*M*). fMLP was then added at various times after ionomycin as indicated in B and C, and β-glucuronidase release or O_2^- production were measured. (A) Kinetics of $[Ca^{2+}]_i$ changes; (B) β-glucuronidase release measured 5 min after addition of fMLP and expressed as percent of maximal values (fMLP and ionomycin added simultaneously) after subtraction of basal levels; (C) Initial rate of O_2^- production. All experiments were performed on the same batch of neutrophils. ●, unloaded cells; ○, loaded cells (0.6 nmol/10^6 PMN). (Reproduced, from *The Journal of Cell Biology*, 1984, 99, 1212, by copyright permission of the Rockefeller University Press.)

1. When quin2-loaded neutrophils are incubated in the absence of Ca^{2+} plus EGTA and stimulated with ionomycin, the ionophore fails to trigger oxidase activation. However, addition of fMLP at variable times after ionomycin induces a respiratory burst whose intensity is closely dependent on the time interval between the additions of ionomycin and fMLP (Figure 3).
2. Another example indicating a role for Ca^{2+} in oxidase activation is the dependence of both initial rate and maximal extent of O_2^- generation on the intracellular quin2 concentration, i.e., increasing the $[quin2]_i$ proportionally decreases superoxide anion generation.[39]

Thus, our interpretation of these results is that, if the amount of diacylglycerol (DAG) produced after agonist stimulation is alone insufficient to activate protein kinase C, complete activation of O_2^- production may be obtained with the combination of DAG and $[Ca^{2+}]_i$ rises. That this, in fact, may be the case is suggested by the finding that if subthreshold concentrations of PMA or ionomycin alone are used as stimuli, O_2^- is not generated, while the combination of both agonists, triggers full NADPH-oxidase activity.[20]

Activation of NADPH oxidase has also been studied in vesicles obtained from the plasma membrane of neutrophils or in cell homogenates and the following conclusions can be drawn from those experiments. (1) The oxidase, per se, has no requirement for Ca^{2+}, neither in the triggering phase, nor at any subsequent stage, provided protein kinase C has been preactivated in intact cells.[95] If anything, in the isolated system, Ca^{2+} seems to be inhibitory. (2) In vitro, protein kinase C has been shown to be capable of phosphorylating a component of NADPH oxidase.[96] It has been argued, however, that this scheme for NADPH-oxidase stimulation involving DAG, protein kinase C, and Ca^{2+} may be an oversimplification, and the generation of a third, yet unidentified signal has been postulated by some authors.[97,122,123]

B. Exocytosis

Neutrophils are equipped with three types of secretory granules: azurophilic (or primary), specific (or secondary), and gelatinase-containing (also called tertiary granules or secretory vesicles).[41,42] Under normal conditions during phagocytosis, the content of these granules is in part discharged into phagosomes, but some of their content can also be released into the extracellular medium. This second route is preferentially followed (1) when neutrophils are presented with a substance which is too big to be phagocytosed (frustrated phagocytosis) or (2) when they are pretreated with cytochalasin B, a fungal metabolite which binds to actin, thereby disrupting intracellular microfilaments.[100,101] The precise mode of action of cytochalasins is not fully understood and various side effects are likely to accompany their addition.[101]

Similar to O_2^- generation, a rise in $[Ca^{2+}]_i$ was thought to be the only necessary and sufficient signal for stimulating secretion in neutrophils, as well as in other cell types.[102] When techniques to measure the $[Ca^{2+}]_i$ became available, it was demonstrated that the rise in $[Ca^{2+}]_i$ caused by receptor stimulation was alone insufficient to explain agonist-induced secretion. This conclusion was essentially based on the same experimental protocols described for NADPH oxidase stimulation. As stated above for O_2^- production, however, it appears that both a rise in $[Ca^{2+}]_i$ and an additional signal generated by receptor activation are needed for optimal exocytosis.

The molecular basis for the synergism between Ca^{2+} and the other signal(s) is still not clear. In addition it ought to be stressed that any generalization on the role of Ca^{2+} in secretion, as far as neutrophils are concerned, is going to be quite inaccurate. In fact, as stated above, these cells possess three types of secretory granules, and it has been known for quite some time that specific and azurophilic granules differ greatly as to their sensitivity to stimuli: PMA (and to a lesser extent ConA) is a strong stimulator of specific and tertiary granule secretion,[98,99,103] while it has a marginal effect on azurophilic granule release. Chemotactic peptides, on the other hand, cause secretion of all three types of granules. The basis for this selectivity has become clearer in recent years. While PMA potently stimulates protein kinase C, but does not increase $[Ca^{2+}]_i$, chemotactic peptides stimulate both branches of the activatory pathway. Thus, one can speculate that the three types of granules have different $[Ca^{2+}]_i$ requirements for secretion. This has been recently demonstrated by Lew et al. in quin2-loaded neutrophils stimulated with ionomycin.[104] While the $[Ca^{2+}]_i$ threshold for ionomycin-induced secretion is similar (around 300 n*M*) for all three types of granules and is unaffected by cytochalasin treatment, the $[Ca^{2+}]_i$ giving half maximal release (EC50) is very different for the different granules, i.e., about fivefold higher for azurophilic compared to specific and tertiary granules.[104] Treatment with cytochalasin B significantly decreases the EC50 of $[Ca^{2+}]_i$ rise (by about twofold), but the differences among the granule populations remain.[104] Similar conclusions concerning the Ca^{2+} dependency of secretion have also been reached in permeabilized neutrophils.[104] The picture emerging from these results is that secretion of the different types of granules has a defined and characteristic requirement for $[Ca^{2+}]_i$.

Furthermore, the three types of granules also vary in their sensitivity to other signal(s) generated by receptor activation. For example, both the $[Ca^{2+}]_i$ threshold and EC50 for granule release are decreased by one order of magnitude when fMLP is added in the presence of ionomycin.[104] In addition, a large secretion of secondary (but not primary) granule content is obtained with PMA at vanishingly low (10 n*M*) $[Ca^{2+}]_i$ in intact cells.[20] The main problem, from a mechanistic point of view, is whether a large rise in $[Ca^{2+}]_i$ alone is sufficient to set in motion the secretory apparatus of neutrophils or whether it acts indirectly by generating another excitatory signal(s).[104] The main evidence in favor of the second hypothesis is the observation that, whereas secretion induced by agonists linked to receptor activation is completed in less than 60 sec with no appreciable lag phase, that induced by Ca^{2+} ionophores is always accompanied by a lag of about 15 sec, even though the kinetics of the $[Ca^{2+}]_i$ rises are similar.[104] A likely, though unproven, explanation could be that the rise in $[Ca^{2+}]_i$, no matter how large, does not trigger secretion by itself in intact cells, but it activates another metabolic event which in turn switches on the secretory machinery. Contrary to this interpretation is the observation that secretion can be obtained in permeabilized cells simply by raising the $[Ca^{2+}]_i$.[105] It is generally assumed that in permeabilized cells any other signal (if generated) would diffuse outside the cell and dilute into the infinite volume of the extracellular space. However, this experiment, too, is inconclusive because even in permeabilized cells, ionic gradients can easily form within the cell.[106] In addition, if the other signal generated by Ca^{2+} is hydrophobic (for example DAG), then, for thermodynamic reasons, it would tend to remain in the cell membrane.

Although a reasonably coherent picture is starting to emerge from the mass of data, including in the activatory signal cascade, inositol trisphosphate, Ca^{2+}, DAG, and protein kinase C, there are a number of results which still remain contradictory. To mention the most striking one, in permeabilized neutrophils micromolar concentrations of a nonhydrolyzable analogue of GTP, GTPγS, is a powerful stimulus for β-glucuronidase secretion.[107] This secretion is not only totally Ca^{2+}- independent, but is increased by increasing the concentration of EGTA (or BAPTA) in the medium. Furthermore, the stimulation by GTPγ S cannot be mimicked by PMA or diacylglycerol, suggesting the involvement of a pathway different from that regulated by protein kinase C.[107] Obviously, whether this result has any relevance to secretion in intact cells remains to be demonstrated. For example, it could be speculated that the presence of millimolar EGTA or BAPTA causes an extensive depletion of intracellular heavy metals, and this in turn may remove an inhibitory constraint on secretion. Indeed, it has been shown that Zn^{2+} blocks antigen-induced histamine release from the rat basophilic leukemic cell line 2H3.[108] A thorough examination of the role of heavy metals in secretion has not been carried out yet. It must be pointed out, however, that cells heavily loaded with quin2 (which is also a potent heavy metal chelator) do not secrete unless stimulated with some sort of agonist.[38,39]

C. Phagocytosis

For many years it has been known that the activity of cytoskeletal-associated proteins is regulated by Ca^{2+}; thus, the idea that calcium ion gradients which arise in the peripheral cytoplasm play an active role in the process of engulfment seemed quite likely.[109] Nevertheless, no evaluation of this postulate could be critically conducted. More recently, studies with obelin[29] have demonstrated that (1) ingestion of latex particles occurs without detectable changes in $[Ca^{2+}]_i$ and (2) trapping of EGTA into the cytoplasm does not drastically affect ingestion of latex particles.[29] Nevertheless, one can argue that phagocytosis of latex particles is an unphysiological phenomenon, yet no one can argue against the physiological significance of phagocytizing a yeast or a bacterium coated with C3b/C3bi or IgG! Neutrophils, in fact, carry on their cell membrane Fc and C3b/C3bi receptors which play a physiological role during phagocytosis. As opposed to the results reported above for latex particles, studies conducted on the role of $[Ca^{2+}]_i$ in phagocytosis triggered by these two types of receptors have demonstrated that rise in $[Ca^{2+}]_i$ is a rapid event which follows the interaction of an opsonized particle with quin2-loaded neutrophils.[110] In fact, when neutrophils in suspension are challenged with IgG-, C3b/C3bi-coated yeast particles, phagocytosis is accompanied, after an initial lag of 2 to 3 min, by a rise in $[Ca^{2+}]_i$ which peaks after about 4 min and remains elevated for more than 10 min.[110] This rise in $[Ca^{2+}]_i$ is not observed in the presence of unopsinized yeast. Similar observations have been reported by Young et al. in the macrophage cell line J774 during phagocytosis of opsonized red blood cells.[52] Increasing the intracellular Ca^{2+} buffering capacity, i.e., increasing $[quin2]_i$, reduces the elevation of the $[Ca^{2+}]_i$ triggered by the opsonized yeast.[110] However, while phagocytosis of IgG-coated yeast is inhibited by increasing $[quin2]_i$ (both the percent as well as the total number of ingested particles), C3b/C3bi-induced phagocytosis is unaffected.[110] Other experiments using C3b/C3bi-opsonized oil particles have confirmed the independence of C3b/C3bi-mediated ingestion from $[Ca^{2+}]_i$. In fact, phagocytosis can occur in the complete absence of extracellular Ca^{2+} ($[Ca^{2+}]_o = 10^{-9}$ *M*), i.e., in cells with no residual ionophore accessible Ca^{2+} pools and in which $[Ca^{2+}]_i$ is reduced tenfold below resting, and in cells pretreated with the Ca^{2+} ionophore, i.e., under conditions in which $[Ca^{2+}]_i$ is elevated tenfold above resting and local elevations of $[Ca^{2+}]_i$ are unlikely.[110] When the same conditions are applied to Fc-receptor mediated phagocytosis, particle ingestion is reduced by more than 60%.[110]

These results suggest that in human neutrophils two mechanisms controlling phagocytosis coexist: (1) Ca^{2+}-dependent ingestion, which may require localized intracellular Ca^{2+} gradients and is at the basis of Fc-receptor mediated phagocytosis, and (2) a Ca^{2+}-independent mechanism triggered by C3b/C3bi receptors. The existence of local gradients of $[Ca^{2+}]_i$ during phagocytosis was recently demonstrated directly using computerized image analysis on quin2-loaded neutrophils.[37] Sawyer et al.[37] observed that when a neutrophil ingests opsonized zymosan, there is an elevation of $[Ca^{2+}]_i$ in the pseudopod surrounding the ingested particle. On the contrary, they showed that ingestion of latex beads is not accompanied by an appreciable increase in the $[Ca^{2+}]_i$ in any region of the cell cytoplasm, confirming the observation of Hallett and Campbell[29] in suspensions of obelin-loaded granulocytes. In conclusion, it appears that changes in the $[Ca^{2+}]_i$ might occur during phagocytosis, but they seem to be related more to the mechanism of receptor triggering rather than to the process of engulfment per se. Even more striking is the picture emerging from recent studies conducted on adherent macrophages. When these cells are loaded with aequorin (by the scrape-loading technique), and challenged with IgG-coated red blood cells, no change in $[Ca^{2+}]_i$ takes place.[111]It is still unclear whether this negative result is due to the use of a different methodology to measure $[Ca^{2+}]_i$ or to a true difference between cell types. Recently, Di Virgilio et al.[124] have shown that, although ligation of the F_c-receptor causes a rise in $[Ca^{2+}]_i$ of both J774 and inflammatory mouse macrophages, phagocytosis of IgG-coated erythrocytes was unaffected by extensive $[Ca^{2+}]_i$ depletion in these cells. Thus, the question of whether $[Ca^{2+}]_i$ plays any role in phagocytosis remains largely unexplained, and if the use of single-cell analysis with computerized image processing might solve the experimental discrepancies, the biochemistry underlying phagocytosis is still somewhat of a black box; in particular how cytoskeletal proteins can function in the absence of $[Ca^{2+}]_i$ changes is, for the time being, a big mystery.

D. Chemotaxis

For a general discussion on chemotaxis and the effect of manipulating extracellular Ca^{2+}, on this function in granulocytes, the reader is referred to previous reviews on this subject and on the specific chapter in this book. Here, we will briefly consider the few available data on direct $[Ca^{2+}]_i$ measurements during chemotaxis. As mentioned above, one of the most important physiological characteristics of neutrophils is their capacity to orient themselves and move in a gradient of chemotactic factors. During oriented locomotion, neutrophils show morphological polarity which is characterized by (1) a region of cytoplasmic ruffling and lamellopodia formation at the front of the cell, (2) a mid-region containing the nucleus, and (3) a narrow region often associated with retractive fibers at the rear of the cell. Various authors have demonstrated an asymmetric distribution of chemotactic and phagocytic receptors on locomoting neutrophils, the majority of the receptors being localized in the front and mid-portion of the cell.[112,113] Measurements with quin2 in neutrophil suspensions indicate that at concentrations which cause chemoattraction, the agonists also induce large rises in $[Ca^{2+}]_i$. Nevertheless, these studies are inconclusive since they are performed in suspensions of cells exposed to homogeneous concentrations of stimuli, a situation which is completely different from that under which cells undergo oriented locomotion. A direct study of $[Ca^{2+}]_i$ in neutrophils under true chemotactic conditions has been only recently performed by Sawyer et al.[37] with quin2 using a computerized image analysis system. It was found that when neutrophils migrate towards opsonized zymosan, the $[Ca^{2+}]_i$ is highest in the lamellar region. When the cells started to engulf the particles, the $[Ca^{2+}]_i$ was found highest in the pseudopod region and periphagosomal cytoplasm. Thus, the regional increments in $[Ca^{2+}]_i$ appear to be strategically located to modulate chemotaxis. This observation is, however, phenomenological, and, to date, there have been no in-depth studies aimed at modulating $[Ca^{2+}]_i$ in order to assess the causal role of this cation in cell locomotion.

Other experiments of great importance in the elucidation of the molecular mechanisms leading to chemotaxis have been performed with pertussis toxin (PT). Pretreatment of neutrophils with PT, which ADP-ribosylates a G protein involved in phospholipase C activation (for a general review see Reference 114), inhibits chemoattractant-stimulated locomotion,[115-116] suggesting that some of the intracellular messengers ($[Ca^{2+}]_i$, DAG, or others) regulated by this G protein are involved in the control of cell locomotion.

VII. CONCLUDING REMARKS

Our knowledge of Ca^{2+} homeostasis has greatly increased during the past 3 to 4 years. We now know quantitatively the $[Ca^{2+}]_i$ of resting neutrophils, the different kinetics of the $[Ca^{2+}]_i$ transients obtained with different agonists, and the importance of intracellular Ca^{2+} stores relative to influx via Ca^{2+} channels. Finally, we are beginning to understand the early events triggered by chemotactic receptors and the relative roles played by second messengers. Most of the studies reviewed in the present chapter have been carried out on large numbers of cells; thus, cell heterogeneity may be at the basis of some discrepancies. In particular, some neutrophil responses, such as phagocytosis and chemotaxis, involve only a region of the cell; thus, it is quite likely that sequential regional $[Ca^{2+}]_i$ transients are the relevant parameters to be followed for those process. The mechanisms underlying phagocytosis and how engulfment proceeds remain largely obscure. In other systems, evidence supports the role of protein kinase C in endocytosis, but at the moment it is not clear whether this applies to phagocytosis as well. There are many problems which must be tackled in the near future, for example:

1. Relatively little is known on the nature of intracellular organelles which accumulate and release Ca^{2+}, the site of action, and molecular mechanism by which $InsP_3$ functions.
2. How Ca^{2+} influx across the plasma membrane is regulated. The recent paper by Von Tscharner [36] suggests that the release of Ca^{2+} from intracellular stores must precede and should in turn regulate the influx of Ca^{2+} from the extracellular milieu, via nonspecific cation channels.
3. The molecular structure of chemotactic and phagocytic receptors is still largely unknown; the nature of their transducing mechanism and whether they all use the same

phospholipase C and G protein as a common end-reaction remains to be established (see review by Snyderman et al.[114]).

4. We and others have suggested that protein kinase C plays a fundamental role in neutrophil activation; unfortunately, quantitative information on the physiological activators of this key enzyme are still lacking.
5. The role and existence of other intracellular messengers is still open to question.

AKNOWLEDGMENTS

This work was supported in part by institutional funds from the CNR Center for the Study of the Physiology of Mitochondria, by grants from the Ministero della Pubblica Istruzione of Italy (40% to 60%), in part by a grant from CNR (Special Project Oncology, No. 850074544) and in part by grant No. 3.829.087 from Swiss Nat. Found. S. T. is a recipient of a Banca Popolare di Padova, Rovigo e Treviso fellowship. During part of this work F. Di V. was holder of an EMBO long-term fellowship to work in the Department of Physiology, Columbia University, New York.

REFERENCES

1. **Ringer, S.,** A further contribution regarding the influence of the different constituents of the blood on the contraction of the heart, *J. Physiol.,* 4, 29, 1883.
2. **Campbell, A. K.,** *Intracellular Ca^{2+} Its Universal Role as a Regulator,* John Wiley & Sons, Chichester, 1983.
3. **Rasmussen, H.,** *Calcium and cAMP Synarchic Messengers,* John Wiley & Sons, New York, 1981.
4. **Thomas, M. V.,** *Techniques in Calcium Research,* Academic Press, New York, 1982.
5. **Molski, T. F. P., Naccache, P. H., Sha'afi, R. I.,** Stimulus-dependent changes in calcium metabolism in rabbit neutrophils, *Cell Calcium,* 4, 57, 1983.
6. **Naccache, P. H., Volpi, M., Showell, H. J., Becker, E. L., and Sha'afi R. I.,** Chemotactic factor-induced release of membrane calcium in rabbit neutrophils, *Science,* 203, 461, 1979.
7. **Lagast, H., Lew, D. P., and Waldvogel, F. A.,** Adenosine triphosphate dependent Ca^{2+} pump in the plasma membrane of guinea pig and human neutrophils, *J. Clin. Invest.,* 73, 107, 1984.
8. **Volpi, M, Naccache, P. H., and Sha'afi, R. I.,** Calcium transport in inside out membrane vescicles prepared from rabbit neutrophils, *J. Biol. Chem.,* 258, 4153, 1983.
9. **Schneider, C., Mottola, C., and Romeo, D.,** Calcium dependent adenosine triphosphatase activity and plasma membrane phosphorylation in the human neutrophil, *Biochem. J.,* 182, 655, 1979.
10. **Prentki, M., Wollheim, C. B., and Lew, D. P.,** Ca^{2+} homeostasis in permeabilized human neutrophils: characterization of Ca^{2+} sequestering pools and the action of inositol 1,4,5 trisphosphate, *J. Biol. Chem.,* 259, 13, 777, 1984.
11. **Murphy, E., Coll, K., Terell, L. R., and Williamson, J. R.,** Hormonal effects on Ca^{2+} homeostasis in isolated hepatocytes, *J. Biol. Chem.,* 255, 6600, 1980.
12. **Prentki, M., Biden, T. J., Janijc, D., Irvine, R. F., Beridge, M. J., and Wollheim, C. B.,** Rapid mobilization of Ca^{2+} from rat insulinoma microsomes by inositol 1,4,5 trisphosphate, *Nature (London),* 309, 562, 1984.
13. **Biden, T. J., Prentki, M., Irvine, R. F., Berridge, M. J., and Wollheim, C. B.,** Inositol 1,4,5 trisphosphate mobilizes intracellular Ca^{2+} from permeabilized insulin secreting cells, *Biochem. J.,* 223, 467, 1984.
14. **Streb, H., Irvine, R. F., Berridge, M. J., and Schultz, I.,** Release of Ca^{2+} from a non mitochondrial intracellular store in pancreatic acinar cells by inositol 1,4,5 trisphoshate, *Nature (London),* 306, 67, 1983.
15. **Knight, D. E. and Baker, P. F.** Calcium dependence of catecholamine release from bovine adrenal medullary cells after exposure to high electric fields, *J. Membr. Biol.,* 68, 107, 1982.
16. **Borle, A. B.,** Control, regulation and modulation of cell calcium, *Rev. Physiol. Biochem. Pharmol.,* 90, 13, 1981.
17. **Sha'afi, R. I. and Naccache, P. H.,** Ionic events in neutrophil chemotaxis and secretion, in *Advances in Inflammatory Research,* Vol. 3, Weissman, G., Ed., Raven Press, New York, 1981, 115.

18. **Mottola, C. and Romeo, D.,** Calcium movement and membrane potential changes in the early phase of neutrophil activation by phorbol myristate acetate: a study with ion selective electrode, *J. Cell Biol.*, 93, 129, 1982.
19. **Lagast, H., Pozzan, T., Waldvogel, F. A., and Lew, D. P.,** PMA stimulates the ATP-dependent calcium transport by the plasma membrane of neutrophils, *J. Clin. Invest.*, 73, 878, 1984.
20. **DiVirgilio, F., Lew, D. P., and Pozzan, T.,** Protein kinase C activation of physiological processes in human neutrophils at vanishingly small cytosolic Ca^{2+} levels, *Nature (London)*, 310, 691, 1984.
21. **Caswell, A. H.,** Methods for measuring intracellular calcium, *Int. Rev. Cytol.*, 56, 145, 1979.
22. **Naccache, P. H., Showell, H. J., Becker, E. L., and Sha'afi, R. I.,** Changes in ionic movements across rabbit polymorphonuclear leukocyte membranes during lysosomal enzyme release, *J. Cell Biol.*, 75, 635, 1977.
23. **Smolen, J. E. and Weissman, G.,** The effect of various stimuli and calcium antagonists on the fluorescence response of chloratetracycline-loaded human neutrophils, *Biochim. Biophys. Acta*, 720, 172, 1982.
24. **Tsien, R. Y. and Rink, T. J.,** Measurement of free Ca^{2+} in cytoplasm, in *Electrophysiological Techniques*, Vol. 2, Barber, J. L., Ed., John Wiley & Sons, New York, 1983, 249.
25. **Naccache, P. H., Showell, H. J., Becker, E. L., and Sha'afi, R. I.,** Involvement of membrane calcium in the response of rabbit neutrophils to chemotactic factors as evidenced by the fluorescence of chloratetracycline, *J. Cell Biol.*, 83, 179, 1979.
26. **Hallett, M. B. and Campbell, A. K.,** Measurements of changes in cytoplasmic free Ca^{2+} in fused cell hybrids, *Nature (London)*, 295, 155, 1982.
27. **Loyter, A., Zakai, N., and Kulka, R. G.,** Ultramicroinjection of macromolecules or small particles into animal cells, *J. Cell Biol.*, 66, 292, 1975.
28. **Campbell, A. K., Dormer, R. L., and Hallett, M. B.,** Coelentrate photoproteins as indicators of cytoplasmic free Ca^{2+} in small cells, *Cell Calcium*, 6, 69, 1985.
29. **Campbell, A. K. and Hallett, H. B.,** Measurement of intracellular calcium ions and oxygen radicals in polymorphonuclear leukocyte-erythrocyte "ghost" hybrids, *J. Physiol. (London)*, 338, 537, 1983.
30. **Tsien, R. Y., Pozzan, T., and Rink, T. J.,** T cell mitogens cause early changes in $[Ca^{2+}]_i$ and membrane potential in lymphocytes, *Nature (London)*, 295, 68, 1982.
31. **Tsien, R. Y., Pozzan, T., and Rink, T. J.,** Calcium homeostasis in intact lymphocytes: cytoplasmic free calcium monitored with a new intracellular trapped fluorescent indicator, *J. Cell Biol.*, 94, 325, 1982.
32. **Rink, T. J. and Pozzan, T.,** Using quin2 in cell suspension, *Cell Calcium*, 6, 133, 1985.
33. **Tsien, R. Y. and Pozzan, T.,** Intracellular Ca^{2+} fluorescent indicators, *Meth. Enzymol.*, in press.
34. **Capponi, S., Lew, D. P., Schlegel, W., and Pozzan, T.,** Use of intracellular calcium and membrane potential indicators in neuroendocrine cells, *Meth. Enzymol.*, 124, 116, 1986.
35. **Grinkiewicz, G., Poenie, M., and Tsien, R. Y.,** A new generation of Ca^{2+} indicators with greatly improved fluorescence properties, *J. Biol. Chem.*, 260, 3440, 1985.
36. **Von Tscharner, V., Prod'hom, B., Baggiolini, M., and Reuter, H.,** Ion channels in human neutrophils activated by a rise in free cytosolic calcium concentration, *Nature (London)*, 324, 369, 1986.
37. **Sawyer, D. W., Sullivan, J. A., and Mandell, G. L.,** Intracellular free calcium localization in neutrophils during phagocytosis, *Science*, 230, 663, 1985.
38. **Pozzan, T., Lew, D. P., Wollheim, C. B., and Tsien, R. Y.,** Is cytosolic free calcium regulating neutrophil functions?, *Science*, 221, 1413, 1983.
39. **Lew, D. P., Wollheim, C. B., Waldvogel, F. A., and Pozzan, T.,** Modulation of cytosolic free calcium transients by changes in intracellular calcium buffering capacity: correlation with exocytosis and O_2^- production in human neutrophils, *J. Cell Biol.*, 99, 1212, 1984.
40. **DiVirgilio, F. and Gomperts, B. D.,** Cytosol Mg^{2+} modulates Ca^{2+} ionophore induced secretion from rabbit neutrophils, *FEBS Lett.*, 13, 315, 1983.
41. **DiVirgilio, F., Vicentini, L. M., Treves, S., Riz, G., and Pozzan, T.,** Inositol phosphate formation in fMet-Leu-Phe stimulated human neutrophils does not require an increase in the cytosolic free Ca^{2+} concentration, *Biochem. J.*, 229, 361, 1985.
42. **Treves, S. and Pozzan, T.,** Stimulus-activation coupling in neutrophils: an overview, in *Clinical Immunology: Vol. 1, Human Inflammatory Diseases*, Marone, G., Lichtenstein, L. M., Condrelli, M., and Fauci, A. S., Eds., B. C. Decker, Toronto, 1988, 263.
43. **White, J. R., Naccache, P. H., Molski, J. F. P., Borgeat, P., and Sha'afi, R. I.,** Direct demonstration of increased intracellular concentration of free calcium in rabbit and human neutrophils following stimulation by chemotactic factors, *Biochem. Biophys. Res. Commun.*, 113, 44, 1983.
44. **Roos, D., Voedman, A. A., and Meerhogh, J.,** Functional activity of enucleated human polymorphonuclear leucocytes, *J. Cell Biol.*, 97, 368, 1983.
45. **Gennaro, P., Pozzan, T., and Romeo, D.,** Monitoring $[Ca^{2+}]_i$ in C5a stimulated neutrophils: loss of receptor modulated calcium stores and calcium uptake in granule free neutrophils, *Proc. Natl. Acad. Sci. U.S.A.*, 81, 1416, 1984.

46. **Sreb, H., Bayerdorffer, E., Haase, W., Irvine, R. F., and Schulz, I. J.,** Effect of inositol 1,4,5-trisphosphate on isolated subcellular fractions of rat pancreas, *J. Membr. Biol.*, 81, 241, 1984.
47. **Andersson, T., Dahlgren, C., Pozzan, T., and Lew, D.,** Characterization of fMet-Leu-Phe receptor mediated Ca^{2+} influx across the plasma membrane of human neutrophils, *Mol. Pharmacol.*, 30, 437, 1987.
48. **Della Bianca, V., Grzeskowiak, G., DeTogni, P., Cassatella, M., and Rossi, F.,** Inhibition by verapamil of neutrophil response to formylmethionmylleucylphenylalanine and phorbol myristate acetate. Mechanism involving Ca^{2+} changes, cyclic AMP and protein kinase C, *Biochim. Biophys. Acta*, 845, 223, 1985.
49. **DiVirgilio, F., Lew, D. P., and Pozzan, T.,** Protein kinase C activation of physiological processes in human neutrophils at vanishingly small cytosolic Ca^{2+} levels, *Nature (London)*, 310, 691, 1984.
50. **Hallett, M. B. and Campbell, A. K.,** Two distinct mechanisms for stimulation of oxygen radical production by polymorphonuclear leukocytes, *Biochem. J.*, 216, 459, 1983.
51. **McNeil, P. L., Swanson, J. A., Wright, S. C., Silverstein, S. C., and Taylor, D. L.,** Fc-receptor mediated phagocytosis occurs in macrophages without an increase in average $[Ca^{2+}]_i$, *J. Cell Biol.*, 102, 1586, 1986.
52. **Young, J. D. E., Ko, S. S., and Cohn, Z. A.,** The increase in intracellular free Ca^{2+} associated with IgG 2b/1 Fc receptor-ligand interactions: role in phagocytosis, *Proc. Natl. Acad. Sci. U.S.A.*, 81, 5430, 1984.
53. **Lew, D. P., Monod, A., Waldvogel, F. A., and Pozzan, T.,** Role of cytosolic free calcium and phospholipase C in leukotriene B4 stimulated secretion in human neutrophils: comparison with the chemotactic peptide fMLP, *Eur. J. Biochem.*, 162, 161, 1987.
54. **Suter, S. S., Lew, D. P., Ballaman, J., and Waldvogel, F. A.,** Intracellular calcium handling in cystic fibrosis: normal cytosolic calcium and intracellular calcium stores in neutrophils, *Ped. Res.*, 19, 346, 1985.
55. **Lew, D. P., Favre, L., Waldvogel, F. A., and Vollotton, M.,** Cytosolic free calcium and intracellular calcium stores in neutrophils from hypertensive patients, *Clin. Sci.*, 69, 227, 1985.
56. **DiVirgilio, F., Calo, L., Cantaro, S., Favaro, S., Piccoli, A., and Borsatti, A.,** Resting and stimulated cytosolic free calcium levels in neutrophils from patients with Barter's syndrome, *Clin. Sci.*, 72, 483, 1987.
57. **Miller, R.,** Multiple calcium channels and neuronal function, *Science*, 235, 46, 1987.
58. **Meldolesi, J. and Pozzan, T.,** Pathways of Ca^{2+} influx at the plasma membrane: voltage-, receptor- and second messenger operated channels, *Exp. Cell Res.*, 171, 271, 1987.
59. **Naccache, P. H., Molski, T. F. P., Borgeat, P., White, J. R., and Sha'afi, R. I.,** Phorbol esters inhibit the fMet-Leu-Phe and leukotriene B_4 stimulated calcium mobilization and enzyme secretion in rabbit neutrophils, *J. Cell Biol.*, 260, 2125, 1985.
60. **Vicentini, L. M., DiVirgilio, F., Ambrosini, A., Pozzan, T., and Meldolesi, J.,** Second messenger generation and secretion in PC12. Role of Ca^{2+} and phosphoinositide hydrolysis, *Biochem. Biophys. Res. Commun.*, 127, 310, 1985.
61. **DiVirgilio, F., Lew, D. P., Andersson, T., and Pozzan, T.,** Plasma membrane potential modulates chemotactic peptide-stimulated cytosolic free Ca^{2+} changes in human neutrophils, *J. Biol. Chem.*, 262, 4574, 1987.
62. **DiVirgilio, F., Lew, D. P., Andersson, T., Treves, S., and Pozzan, T.,** Role of plasma membrane potential in granulocyte activation, *Ann. N.Y. Acad. Sci.*, 488, 525, 1986.
63. **Oettgen, H. C., Terhorst, C., Cantley, L. C., and Rosoff, P. M.,** Stimulation of the T3-T cell receptor complex induces a membrane potential-sensitive calcium influx, *Cell*, 40, 583, 1985.
64. **Richard, J. E. and Sheterline, P.,** Evidence that phorbol ester interferes with stimulated Ca^{2+} redistribution by activating Ca^{2+} efflux in neutrophil leukocytes, *Biochem. J.*, 231, 623, 1985.
65. **Berridge, M. J. and Irvine, R. K.,** Inositol trisphosphate, a novel second messenger in cellular signal transduction, *Nature (London)*, 312, 315, 1984.
66. **Berridge, M. J.,** Inositoltrisphosphate and diacyclglyerol as second messengers, *Biochem. J.*, 220, 345, 1984.
67. **Berridge, M. J.,** The molecular basis of communication within the cell, *Sci. Am.*, 253, 124, 1985.
68. **Serhan, C. N., Fridovich, J., Goetzl, E. J., Dunham, P. B., and Weissmann, G.,** Leukotriene B_4 and phosphatidic acid are Ca^{2+} ionophores, *J. Biol.Chem.* 257, 4746, 1982.
69. **Holmes, R. P. and Yoss, N. L.,** Failure of phosphatidic acid to translocate calcium across phosphatidylcholine membranes, *Nature (London)*, 305, 637, 1983.
70. **Michell, R. H., Kirk, C. J., Jones, L. M., Downes, C. D., and Crepe, J. A.,** The stimulation of inositol lipid metabolism that accompanies calcium mobilisation in stimulated cells: defined characteristics and unanswered questions, *Philos. Trans. R. Soc. London B*, 296, 123, 1981.
71. **Cockcroft, S., Bennett, J. P., and Gomperts, B. D.,** The dependence on Ca^{2+} of phosphatidylinositol breakdown and enzyme secretion in rabbit neutrophils stimulated by fMLP or ionomycin, *Biochem. J.*, 200, 502, 1982.
72. **Bradford, R. G. and Rubin, R. P.,** Characterization of formyl methionyl-leucyl-phenylalanine stimulation of inositoltrisphosphate accumulation in rabbit neutrophils, *Mol. Pharmacol.*, 27, 74, 1985.

73. **Andersson, T., Schlegel, W., Monod, A., Krause, K. H., Stendahl, O., and Lew, D., P.,** Leukotriene B_4 stimulation of phagocytes results in the formation of inositol 1,4,5 trisphosphate, *Biochem. J.*, 240, 333, 1986.
74. **Lew, D. P., Monod, A., Krause, K. H., Waldvogel, F. A., Biden, T. J., and Schlegel, W.,** The role of cytosolic free Ca^{2+} in the generation of inositol 1,4,5 trisphosphate and inositol 1,3,4 trisphosphate in HL60 cells, *J. Biol. Chem.*, 261, 13, 121, 1986.
75. **Biden, T. J. and Wollheim, C. B.,** Calcium regulates the inositoltetrakisphosphate pathway in intact and broken preparations of insulin secreting Rim5F cells, *J. Biol. Chem.*, 261, 11, 931, 1986.
76. **Klebanoff, S. J. and Clark, R. A.,** *The Neutrophil: Function and Clinical Disorders,* North-Holland, Amsterdam, 1979.
77. **Patriarca, P., Cramer, R., Moncalvo, S., Rossi, F., and Romeo, D.,** Enzymatic basis of metabolic stimulation in leukocytes during phagocytosis: the role of activated NADPH-oxidase, *Arch. Biochem.*, 145, 255, 1971.
78. **Rossi, F., Romeo, D., and Patriarca, P.,** Mechanism of phagocytosis-associated oxidative metabolism in polymorphonuclear leukocytes and macrophages, *J. Reticuloendothel. Soc.*, 12, 127, 1972.
79. **Graham, R. C., Karnowsky, M. J., Shafer, A. W., Glass, E. A., and Karnowsky, M. L.,** Metabolic and morphological observations on the effect of surface-active agents on leukocytes, *J. Cell Biol.*, 32, 629, 1972.
80. **Zatti, M. and Rossi, F.,** Relationship between glycolysis and respiration in surfactant treated leukocytes, *Biochim. Biophys. Acta,* 148, 553, 1967.
81. **Rossi, F. and Zatti, M.,** Mechanism of the respiratory stimulation in saponine-treated leukocytes. The KCN-insensitive oxidation of NADPH, *Biochim. Biophys. Acta,* 153, 296, 1968.
82. **Patriarca, P., Zatti, M., Cramer, R., and Rossi, F.,** Stimulation of the respiration of polymorphonuclear leukocytes by phospholipase C, *Life Sci.*, 9, 841, 1970.
83. **Romeo, D., Zabucchi, G., and Rossi, F.,** Reversible metabolic stimulation of polymorphonuclear leukocytes and macrophages by Concanavalin A., *Nature (New Biology),* 243, 111, 1973.
84. **Rossi, F., Zatti, M., Patriarca, P., and Cramer, R.,** Stimulation of the respiration of polymorphonuclear leukocytes by anti-leukocyte antibodies, *Experientia,* 26, 491, 1970.
85. **Kakinuma, K.,** Effects of fatty acids on the oxidative metabolism of leukocytes, *Biochim. Biophys. Acta,* 348, 76, 1974.
86. **Nakagawara, A., Takeshige, K., and Minakami, S.,** Induction of phagocytosis-like metabolic pattern in polymorphonuclear leukocytes by cytochalasin E, *Exp. Cell Res.*, 87, 392, 1974.
87. **Takeshige, K., Nakagavara, A., Hatal, T., and Minakami, S.,** Metabolic pattern of polymorphonuclear leukocyte induced by trypsin-digested microsomes, *J. Biochem.*, 77, 559, 1975.
88. **Strauss, B. S. and Stetson, C. A., Jr.,** Studies on the effect of certain macromolecular substances on the respiratory activity of the leukocytes of peripheral blood, *J. Exp. Med.*, 112, 635, 1960.
89. **Romeo, D., Zabucchi, G., Miani, N., and Rossi, F.,** Ion movement across the leukocyte plasma membrane and excitation of their metabolism, *Nature (London),* 253, 542, 1975.
90. **Repine, J. E., White, J. G., Clawson, C. C., and Holmes, B. M.,** The influence of phorbol myristate acetate on oxygen consumption by polymorphonuclear leukocytes, *J. Lab Clin. Med.*, 83, 911, 1974.
91. **Tedesco, F., Trani, S., Soranzo, M. R., and Patriarca, P.,** Stimulation of glucose oxidation in human polymorphonuclear leukocytes by C3-sepharose and soluble C567, *FEBS Lett.*, 51, 232, 1975.
92. **Becker, E. L., Sigman, M., and Oliver, J. M.,** Superoxide production induced in rabbit polymorphonuclear leukocytes by synthetic chemotactic peptides and A23187, *Am. J. Pathol.*, 95, 81, 1979.
93. **Dewald, B. and Baggiolinin, M.,** Platelet activating factor as a stimulus of exocytosis in human neutrophils, *Biochim. Biophys. Acta,* 888, 42, 1986.
94. **Nishizuka, Y.,** The role of protein kinase C in cell surface signal transduction and tumor promotion, *Nature (London),* 308, 693, 1984.
95. **Lew, D. P. and Stossel, T. P.,** Effect of calcium on superoxide production by phagocytic vescicles from rabbit alveolar macrophages, *J. Clin. Invest.*, 75, 1740, 1981.
96. **Papini, E., Grzeskowiak, M., Bellavite, P., and Rossi, F.,** Protein kinase C phosphorylates a component of NADPH oxidase in neutrophils, *FEBS Lett.*, 190, 204, 1985.
97. **Cooke, E. and Hallet, M. B.,** The role of C-kinase in the physiological activation of the neutrophil, *Biochem. J.*, 232, 323, 1985.
98. **Dewald, B., Bretz, U., and Baggiolini, M.,** Release of gelatinase from a novel secretory compartment of human neutrophils, *J. Clin. Invest.*, 70, 518, 1982.
99. **Wright, D. G., Bralove, D. A., and Gallin, J. I.,** The differential mobilization of human neutrophil granules, *Am. J. Pathol.*, 87, 273, 1977.
100. **Zurier, R. B., Hoffstein, S., and Weissmann, G.,** Cytochalasin B: effect on lysosomal enzyme release from human leukocytes, *Proc. Natl. Acad. Sci. U.S.A.*, 70, 844, 1973.
101. **Treves, S., DiVirgilio, F., Vaselli, G. M., and Pozzan, T.,** Effect of cytochalasins on cytosolic free calcium concentration and phosphoinositide metabolism in leukocytes, *Exp. Cell Res.*, 168, 285, 1987.

102. **Rubin, R. P.,** *Calcium and Secretion,* Plenum Press, New York, 1982.
103. **Estensen, R. D., White, J. G., and Holmes,** Specific degranulation of human polymorphonuclear leukocytes, *Nature (London),* 248, 347, 1974.
104. **Lew, D. P., Monod, A., Waldvogel, F. A., Dewald, B., Baggiolini, M., and Pozzan, T.,** Quantitative analysis of the cytosolic free calcium dependency of exocytosis from three subcellular components in intact human neutrophils, *J. Cell Biol.,* 102, 2197, 1986.
105. **Smolen, J. E. and Stoehr, S. J.,** Micromolar concentrations of free calcium provoke secretion of lysozyme from human neutrophils permeabilized with saponin, *J. Immunol.,* 134, 1859, 1985.
106. **Biden, T. J., Wollheim, C. B., and Schlegel, W.,** Inositol 1,4,5 trisphosphate and intracellular Ca^{2+} homeostasis in clonal pituitary cells (GH3), *J. Biol. Chem.,* 261, 7223, 1986.
107. **Borrowman, M. M., Cockcroft, S., and Gomperts, B. D.,** Two roles for guanine nucleotides in the stimulus-secretion sequence of neutrophils, *Nature (London),* 319, 504, 1986.
108. **Beaven, M. A., Rogers, J., Moore, J. P., Hesketh, R. T., Smith, G. A., and Metcalfe, J. C.,** The mechanism of the calcium signal and correlation with histamine release in 2H3 cells, *J. Biol. Chem.,* 259, 7129, 1984.
109. **Southwick, F. S. and Stossel, T. P.,** Contractile proteins in Leukocyte function, *Sem. Hematol.,* 20, 305, 1983.
110. **Lew, D. P., Andersson, T., DiVirgilio, F., Pozzan, T., and Stendahl, O.,** Ca^{2+}-dependent and Ca^{2+}-independent phagocytosis in human neutrophils, *Nature (London),* 315, 509, 1985.
111. **McNeil, P. L., Swanson, J. A., Wright, J. D., Silverstein, S. C., and Taylor, D. L.,** Fc-receptor mediated phagocytosis occurs in macrophages without an increase in average $[Ca^{2+}]_i$, *J. Cell Biol.,* 102, 1586, 1986.
112. **Walter, R. J., Berlin, R. D., and Oliver, J. M.,** Asymmetric Fc receptor distribution on human PMN oriented in a chemotactic gradient, *Nature (London),* 286, 724, 1986.
113. **Sullivan, S., Daukas, G., and Zigmond, S. H.,** Asymmetric distribution of the chemotactic peptide receptor on polymorphonuclear leukocytes, *J. Cell Biol.,* 99, 1461, 1984.
114. **Snyderman, R., Smith, C. D., and Verghese, M. W.,** Model for leukocyte regulation by chemoattractant receptors: roles of a guanine nucleotide regulatory protein and polyphosphoinositide metabolism, *J. Leukocyte Biol.,* 40, 785, 1986.
115. **Lad, P. M., Olson, C. V., and Grewal, I. S.,** A step sensitive to pertussis toxin and phorbol esters in neutrophils regulates chemotaxis and capping but not phagocytosis, *FEBS Lett.,* 200, 91, 1986.
116. **Becker, E. L., Kermode, J. C., Naccache, P. H., Yassin, R., Mash, M. L., Munoz, S., and Sha'afi, R. I.,** The inhibition of neutrophil granule enzyme secretion and chemotaxis by pertussis toxin, *J. Cell Biol.,* 100, 1641, 1985.
117. **Hallett, M. B.,** personal communication, 1985.
118. **Volpe, P., Krause, K. H., Hashimoto, S., Pozzan, T., Meldolesi, J., and Lew, D. P.,** Calciosome cytoplasmic organelle: the inositol 1,4,5-trisphosphate-sensitive Ca^{2+} store of nonmuscle cells?, *Proc. Natl. Acad. Sci. U.S.A.,* in press.
119. **Treves, S.,** unpublished observations, 1986.
120. **Di Virgilio, F. and Pozzan, T.,** unpublished observations, 1984.
121. **Rossi, F.,** The NADPH oxidase, *Biochim. Biophys. Acta,* 853, 65, 1986.
122. **Delle Bianca, V., Grzeskowiek, M., Cossatella, M., Zeni, L., and Rossi, F.,** Phorbol myristate acetate potentiates the respiratory burst while inhibits phosphoinositide hydrolysis and calcium mobilization by fMet-Leu-Phe in human neutrophils, *Biochem. Biophys. Res. Commun.,* 135, 556, 1986.
123. **Wymann, P. M., von Tscherner, V., Derenlan, D. A., and Baggiolini, M.,** The onset of the respiratory burst in human neutrophils, *J. Biol. Chem.,* 262, 12048, 1987.
124. **Di Virgilio, F., Meyer, B. C., Greenberg, S., and Silverstein, S. C.,** F_c receptor-mediated phagocytosis occurs in macrophages at exceedingly low cystolic Ca^{2+} levels, *J. Cell Biol.,* 106, 657, 1988.

Chapter 9

CALCIUM-DEPENDENT AND INDEPENDENT MECHANISMS OF CELLULAR CONTROL WITHIN NEUTROPHILS: THE ROLES OF KINASE C, DIACYGLYCEROL, AND UNIDENTIFIED INTRACELLULAR MESSENGERS

Eryl Cooke, Futwan A. Al-Mohanna, and Maurice B. Hallett

TABLE OF CONTENTS

I. INTRODUCTION 220

II. Role of Intracellular Ca^{2+} in Response Triggering 220
A. Experimental Agents 220
B. Receptor Agonists 221
C. Particulate Stimuli 222

III. Nature of Non-Ca^{2+} Intracellular Messengers 224
A. Interactions with the Ca^{2+} Signal 224
B. Independence from Intracellular Ca^{2+} 224

IV. Role of Kinase C in Ca^{2+} -Dependent and -Independent Response Triggering 224
A. Properties of Kinase C 224
B. Phosphorylation Accompanying Cell Activation 225
C. Experimental Activators of Kinase C 226
D. Experimental Inhibitors of Kinase C 228

V. Possible Identity of Non-Ca^{2+} Intracellular Messengers 230
A. Diacylglycerol 230
B. GTP 231
C. Inositol Phosphates 232
D. Cyclic Nucleotides 233
E. Fatty Acids 233
F. Hydrogen Ions (pHi) 233
G. Membrane Potential Changes 234

VI. Implications of Multiple Intracellular Signaling Mechanisms for the Control of Neutrophil Responses 234
A. Inappropriate Activation 234
B. Inappropriate Cellular Location of Response 234

References 236

I. INTRODUCTION

Probably the first indication that an elevation in the concentration of intracellular calcium triggered a neutrophil response was provided by the demonstration that leucocidin produced both an accumulation of calcium within the cell and secretion of proteins.[1] With the development of calcium ionophores as a more specific method for increasing cell membrane permeability for calcium, it was confirmed that raised intracellular calcium concentrations triggered not only secretion, but also oxidase activation.[2,3] Measurement of $^{45}Ca^{2+}$ fluxes across the plasma membrane of the neutrophil showed that an increase in plasma-membrane permeability for Ca^{2+} was also provoked by chemotactic and secretory stimuli.[4,5] It was therefore widely accepted that, in common with other cells, intracellular calcium acted as the sole intracellular trigger for neutrophil responses. The belief in the uniqueness of calcium as the sole trigger enabled experiments which demonstrated that responses could occur in the absence of extracellular calcium to be interpreted as evidence for an intracellular source of calcium. It was not, however, until techniques that enabled the direct measurement of intracellular Ca^{2+} within living neutrophils were developed that the complexity of the role of this ion was fully realized. Ironically, when methods for the direct measurements of intracellular Ca^{2+} in neutrophils became available, the first two reports demonstrated the unsuspected existence of noncalcium intracellular messengers.[6,7] In this chapter, the role of calcium and noncalcium intracellular messengers and their interactions will be discussed. In particular, the evidence of a role for kinase C in coupling stimulation to response will be explored.

II. ROLE OF INTRACELLULAR Ca^{2+} IN RESPONSE TRIGGERING

There is general agreement from the direct measurement of intracellular Ca^{2+} in neutrophils from a number of species that the resting intracellular Ca^{2+} concentration is approximately 0.1 μM (Table 1). Stimuli can be classified by the effect they have on this intracellular level (Table 2). Some stimuli, such as the experimental agents A23187 and ionomycin, elevate intracellular Ca^{2+} to levels as high as 15 μM, while others, such as the receptor agonists, elevate intracellular Ca^{2+} to a lesser extent, reaching levels less than 1 μM, and some other agents do not elevate intracellular Ca^{2+} at all. This classification provides a framework for discussing the role of intracellular Ca^{2+} in triggering neutrophil responses.

A. Experimental Agents

The experimental Ca^{2+} ionophores, A23187 and ionomycin, provide a means of artificially increasing the permeability of the membrane to Ca^{2+} ions. In the presence of extracellular Ca^{2+} concentrations in the millimolar range, intracellular Ca^{2+} rises rapidly to high levels. The intracellular Ca^{2+} concentration is sufficient to saturate the fluorescent probe, quin2,[7-9] and is estimated by the Ca^{2+} -activated photoprotein, obelin, to be approximately 10 μM.[10,11] When the conditions are manipulated, either by reducing the extracellular concentration of Ca^{2+} or by reducing the concentration of the ionophore, lower levels of intracellular Ca^{2+} can be achieved. In these conditions, the intracellular requirement for Ca^{2+} can be titrated. It was found that oxidase activation in both rat and human neutrophils required intracellular Ca^{2+} in excess of 1 μM for maximum stimulation.[7,10,11] The terminal attack complex of complement elevated intracellular Ca^{2+} to micromolar levels and stimulated the oxidase to an extent predicted from ionophore titration.[11] A similar dependence on intracellular Ca^{2+} exists for the secretion of β-glucuronidase when stimulated by ionomycin, no secretion occurring at below 2 μM intracellular Ca^{2+}.[12]

The other group of widely used experimental agents is the phorbol esters, and, in particular, phorbol myristate acetate (tetradecanoyl phorbol acetate or TPA). This agent, in extreme

Table 1
COMPARISON OF SOME MEASUREMENTS OF INTRACELLULAR FREE CA^{2+} CONCENTRATION IN NEUTROPHILS

Ca^{2+} concentration (n*M*)				
Resting	**Stimulated**	**Method**	**Species**	**Ref.**
100—300	ca.700	Photoprotein (consumption)	Rat	6
100—300	550 ± 30	Photoprotein (luminescence)	Rat	10
120	800	Quin-2	Human	7,8
100—200	300—450	Quin-2	Rabbit	98
170—300	—	Set point of permeabilized cells	Human	63
42.5 ± 6	287	Quin-2	Pig	99
74 ± 4	—	Quin-2	Human (single cells)	100
115 ± 25	—	Quin-2	Human	101
94 ± 13	768 ± 173	Fura-2	Rat	102

Table 2
REQUIREMENT OF THE ACTION OF SOME STIMULI FOR INTRACELLULAR Ca^{2+}

Group	Agent	Ca^{2+} conc required for activation	Ref.
O	Unstimulated	75—150 n*M*	(See Table 1)
I	Latex beads		10,11
	C3bi-opsonized particles	No rise required	23
	PMA		13
II	f-Met-Leu-Phe		10,7
	Con A		10
	C5a	0.1—1 μ*M*	105
	Leukotriene B_4		106
	IgG		23
III	A23187		10,11
	Ionomycin		7,12
	Terminal attack complex of complement	1—10 μ*M*	10,11
	Sendai virus		48
	Saponin		104
	Digitonin		53

contrast to the ionophores, stimulates a number of neutrophil responses without elevation of intracellular Ca^{2+}.[13-15] The significance of these latter agents will be discussed in detail in Section IV.C.

B. Receptor Agonists

The receptor agonist, formyl-methionyl-leucyl-phenylalanine, often abbreviated to f-Met-Leu-Phe or fMLP (not fMLF as would be the more conventional single letter designation), or chemotactic peptide, stimulates chemotaxis, oxidase activation, and (in the presence of cytochalasin B) secretion. However, this agent causes only a small rise in intracellular Ca^{2+} concentration to a maximum of approximately 0.8 μ*M* (Table 1). Furthermore, the rise to this level is transient (although a sustained but lower elevation can persist for several minutes[8,10]). This peak of elevated intracellular Ca^{2+} was far less than that found to be necessary for stimulation by ionophores. However, this small intracellular Ca^{2+} was essential for triggering the response. When intracellular Ca^{2+} was "clamped" by intracellular EGTA or by high intracellular concentrations of quin2, responsiveness to chemotactic peptide was lost.[8bc,10]

Several possibilities existed to account for this apparent contradiction in the magnitudes of the absolute requirement for intracellular Ca^{2+} by chemotactic peptide and ionophores. First, was the intracellular Ca^{2+} induced by chemotactic peptide in reality much higher than that measured by the available techniques? For example, since intracellular Ca^{2+} measurements performed on cell populations reported an average Ca^{2+}-dependent signal, the magnitude of an intracellular Ca^{2+} change restricted to a small part of the cytoplasm or to a small number of cells would be significantly underestimated. This possibility would seem to be unlikely because both Ca^{2+}-activated photoprotein and quin-2 report a similar intracellular Ca^{2+} rise, yet the error associated with underestimation due to restriction of Ca^{2+} would be different for each method. In the submicromolar range, the former technique produces a signal which is proportional to $(Ca^{2+})^n$, where $n = 2$ to 3,[15] whereas for the latter technique the signal would be proportional to Ca^{2+}. If the intracellular Ca^{2+} rise were restricted to a fraction of the cytoplasm (or a fraction of the cells), f, the methods would underestimate the local change by f^{-n} for photoprotein[16] and f^{-1} for the fluorescent probe. Since both methods report approximately equivalent intracellular Ca^{2+} changes, f must be close to unity, i.e., the intracellular Ca^{2+} change is not significantly restricted.

The second possibility was that chemotactic peptide generated a second intracellular messenger which was not generated by ionophores, the effect of which was to increase the affinity of the Ca^{2+}-sensitive machinery within the cell for Ca^{2+}. The existence of this non-Ca^{2+} intracellular messenger was elegantly demonstrated by Pozzan and co-workers[7] in the following experiment. In the absence of extracellular Ca^{2+}, ionomycin elevated intracellular Ca^{2+} to only a small extent (see Chapter 8) which was insufficient to activate the cells. Subsequent stimulation by chemotactic peptide resulted in activation of the cells without any further intracellular Ca^{2+} change. It was therefore concluded that chemotactic peptide had generated a non-Ca^{2+} intracellular messenger or messengers which triggered the cells. Further clues to the nature of the non-Ca^{2+} intracellular messengers have now accumulated (see Section IV).

C. Particulate Stimuli

Phagocytosis is the result of a complex series of cellular and molecular events. Following binding of the particle to the cell surface, actin is locally polymerized to produce pseudopodia which engulf the particle. Subsequently, fusion between the phagosome and granule membranes occurs and the nonmitochondrial oxidase is also activated. Although there had been much speculation about the role of calcium in these processes, it was difficult to interpret some early experiments. For example, the significance of the dependence of phagocytosis on extracellular calcium was not clear, since it was not possible to distinguish between the requirements of Ca^{2+} for binding of the particle to the membrane and as an intracellular messenger. Furthermore, although it was widely held that, as with other cellular functions, phagocytosis was regulated by intracellular Ca^{2+}, it was, however, difficult to construct a molecular model for this proposal. In particular, with the discovery of gelsolin[17] it was found that the Ca^{2+}-bound form of this protein caused depolymerization of actin. This consequently led Stossel to propose that a local *fall* in intracellular Ca^{2+} was the signal for actin polymerization,[18] and was consequently the critical event in the initiation of phagocytosis.

Direct measurement of intracellular Ca^{2+} in neutrophils during phagocytosis, however, has revealed that the relationship between intracellular Ca^{2+} and phagocytosis is complex and that there is no absolute requirement for a change in intracellular Ca^{2+}. Rat neutrophils phagocytose unopsonized latex particles rapidly, even when the intracellular Ca^{2+} concen-

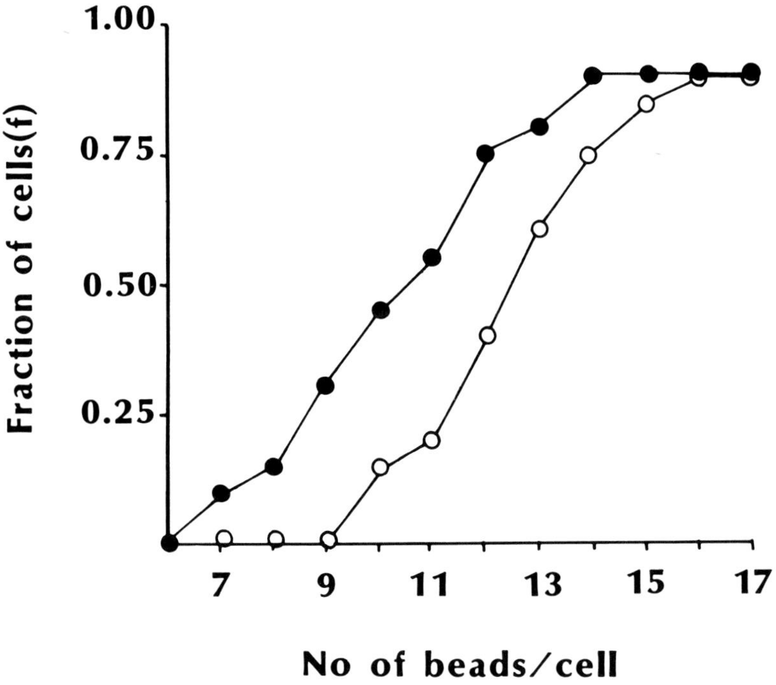

FIGURE 1. Effect of intracellular BAPTA on phagocytosis. Rat neutrophils were loaded with the Ca^{2+} chelator BAPTA by incubation with the acetoxymethyl ester (BAPTA-AM). The final intracellular concentration of chelator was approximately 10 m*M*. The population distribution of phagocytosed fluorescein latex beads (1 m*M*) was determined after 15 min incubation at 37°C and is shown as the fraction of cells with less than the number of beads/cell shown. Closed symbols show the data for BAPTA-loaded cells. The small shift in the distribution may be attributable to a small "toxic" effect of the by-products of BAPTA-AM hydrolysis.

tration is "clamped" at a low level using BAPTA (Figure 1). Accompanying phagocytosis is a large "burst" of oxidase activity.[19] However, in neutrophil-erythrocyte hybrids no accompanying intracellular Ca^{2+} rise was detectable by the Ca^{2+}-activated photoprotein, obelin.[6,10] The possibility that a rise in intracellular Ca^{2+} had actually occurred, but was too localized or small to be detectable, was excluded by the demonstration that intracellular EGTA in sufficient concentration to inhibit the detectable intracellular Ca^{2+} rise induced by chemotactic peptide, f-Met-Leu-Phe, did not prevent the latex bead-induced activation of the cells[10]. Furthermore, no rise in intracellular Ca^{2+} in rat neutrophils was detected by the fluorescent Ca^{2+} indicator fura-2. A similar lack of intracellular Ca^{2+} rise detectable by Ca^{2+}-activated photoprotein has been demonstrated during phagocytosis of unopsonized latex beads by rat macrophages[20] and IgG-coated red blood cells by human macrophages.[21] However, it is interesting to note that intracellular free Ca^{2+} as reported by quin-2 was reported to rise in the macrophage-like cell line, J774, following phagocytosis of opsonized red blood cells.[22] Likewise, in neutrophils, quin-2 reports an intracellular Ca^{2+} rise following phagocytosis of either IgG- or C3bi- coated yeast particles,[23] although phagocytosis of the C3bi particles was not inhibited by suppression of this Ca^{2+} rise. It has therefore been suggested that two mechanisms for phagocytosis coexist in neutrophils, one dependent upon and the other independent of a rise in intracellular Ca^{2+} (see Chapter 8).

III. NATURE OF NON-Ca^{2+} INTRACELLULAR MESSENGERS

Non-Ca^{2+} intracellular messengers generated during stimulation of neutrophils have two possible modes of action. The messenger could either interact with a coexisting Ca^{2+} signal to modify the response to this stimulus or it could act in complete independence from the Ca^{2+} signal. It may be argued that the existence of both types of intracellular messenger may be required for the neutrophil to respond in a coordinated and appropriate manner (see Chapter 1).

A. Interactions with the Ca^{2+} Signal

Since the "Ca^{2+} signal" generated within the cells may cause several intracellular events, there are several possible sites at which other intracellular messengers could interact to modify its effect. At least three important intracellular Ca^{2+} -binding sites have been identified namely, calmodulin, kinase C and gelsolin. Evidence has already been presented to suggest that at least one non-Ca^{2+} intracellular messenger acts to increase the affinity of a key "intracellular Ca^{2+} receptor" for Ca^{2+} (see Section IV.A). The possibility exists that this key site is kinase C (see Section IV). Other intracellular messengers interacting with other key Ca^{2+} -binding proteins cannot, however, be excluded.

Non-Ca^{2+} intracellular messengers may also be generated that modify the Ca^{2+} transient rather than the effect of the Ca^{2+} transient. It has been suggested, for example, that cyclic AMP generated by either the Ca^{2+} -dependent agonist, such as chemotactic peptide, or by a modifying agent such as adenosine (see section V.E), may exert an inhibitory influence on the opening of "Ca^{2+} channels". In contrast, other intracellular messengers, such as IP_4, may be generated, which increase the probability of "Ca^{2+} channel" opening.

B. Independence from Intracellular Ca^{2+}

The possibility also exists that some intracellular messengers act in complete independence of intracellular Ca^{2+}. This would be the case if that part of the intracellular machinery were either fully activated at resting intracellular Ca^{2+} concentration or insensitive to Ca^{2+} in the submicromolar concentration range.

IV. ROLE OF KINASE C IN Ca^{2+} -DEPENDENT AND -INDEPENDENT RESPONSE TRIGGERING

A clue to the role of kinase C as an important intracellular regulatory protein was provided by the discovery that the receptor for the phorbol ester, phorbol myristate acetate, was the enzyme kinase C.[24] Since it had been known for many years that phorbol myristate acetate could stimulate selective release of lysozyme and alkaline phosphatase[25] and was a potent activator of the oxidase system, [26] the possibility existed that kinase C activation mediated this and perhaps other neutrophil functions. The possibility that kinase C played an important part in the activation of neutrophil responses may also be suggested by the existence of non-Ca^{2+} intracellular messengers which increase the affinity of the intracellular machinery for Ca^{2+}. Furthermore, neutrophils seem to be particularly rich in kinase C activity, rat neutrophils containing kinase C activity equivalent to 21,000 pmol of phosphate transferred to histone per minute per gram of cells (1.2 amol/min/cell). There are several properties of kinase C which are relevant to an understanding of the postulated role of this enzyme.

A. Properties of Kinase C

The enzyme, kinase C, also called the Ca^{2+} -activated phospholipid-dependent kinase, was first described by Nishizuka.[27] As its name implies, the enzyme is activated by Ca^{2+} and its activity is also dependent binding to phospholipid (phosphatidyl serine alone producing

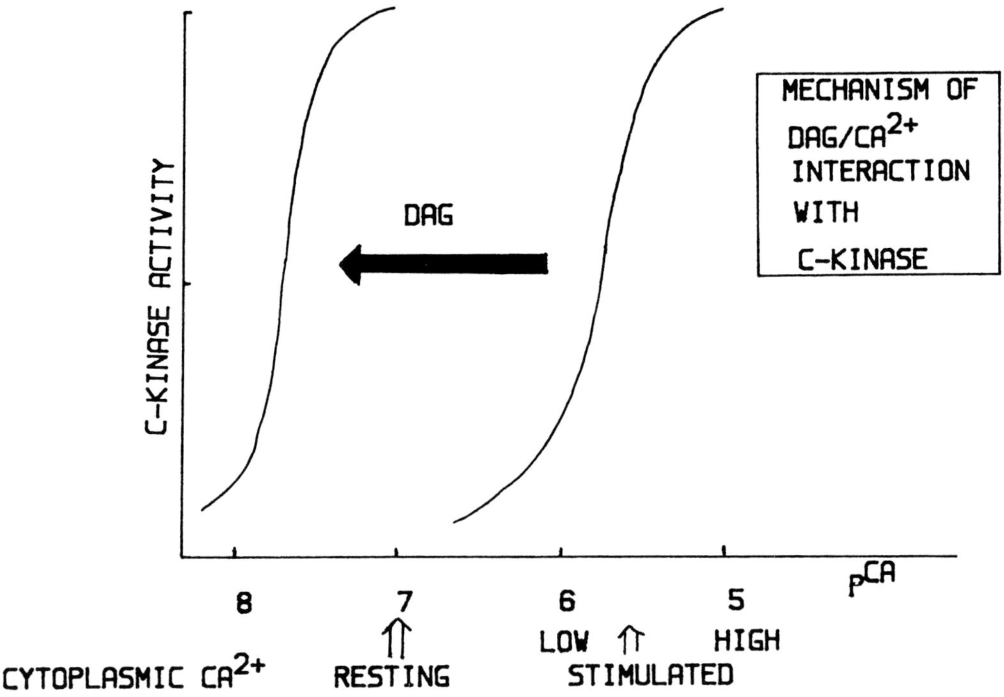

FIGURE 2. Effect of 1,2 diacylglycerol (DAG) on the activation of kinase C by Ca^{2+}.

full activation). Under these conditions, Ca^{2+} in the range 1 to 10 μM is necessary for activation. However, the affinity of the enzyme for Ca^{2+} is greatly increased by certain diacylglycerols,[27] such that the enzyme can be fully activated by submicromolar concentrations of Ca^{2+} (Figure 2). This feature would provide a molecular explanation for the intracellular Ca^{2+} requirement of the three types of stimuli described in Table 1. If diacylglycerol mediates activation of the key regulatory enzyme, kinase C, then it must also be proposed that (1) the ionophores do not produce the diacylglycerol and hence activate by Ca^{2+} alone, requiring high intracellular Ca^{2+} for stimulation; (2) the receptor agonists activate by Ca^{2+} plus diacylglycerol, enabling activation at lower intracellular Ca^{2+} concentrations; and (3) phagocytic stimuli produce sufficient diacylglycerol to enable activation of the cell in the absence of a rise in intracellular Ca^{2+}. Another group of agents which activate kinase C, the fatty acids, may also have physiological importance as intracellular messengers, as discussed later (see Section V.D).

Although attention has focused on the lipid activators of kinase C as possible physiologically important activators of this enzyme, its activation was first described following proteolysis. This route of activation may also have physiological importance in neutrophils.[28] The proteolytically cleaved kinase has no requirement for phospholipid and no sensitivity to Ca^{2+}. Consequently, the kinase is released from the membrane to become free in the cytoplasm when activated by the protease, calpain. This enzyme is activated by micromolar Ca^{2+} concentrations. Thus, at higher intracellular Ca^{2+} concentrations this may represent a mechanism by which the kinase C system is irreversibly activated with access to both membrane - associated and cytoplasmic proteins.

B. Phosphorylation Accompanying Cell Activation

The involvement of protein kinases in cell activation is well documented. It is not surprising that a number of studies have been performed on neutrophils. All such studies agree that

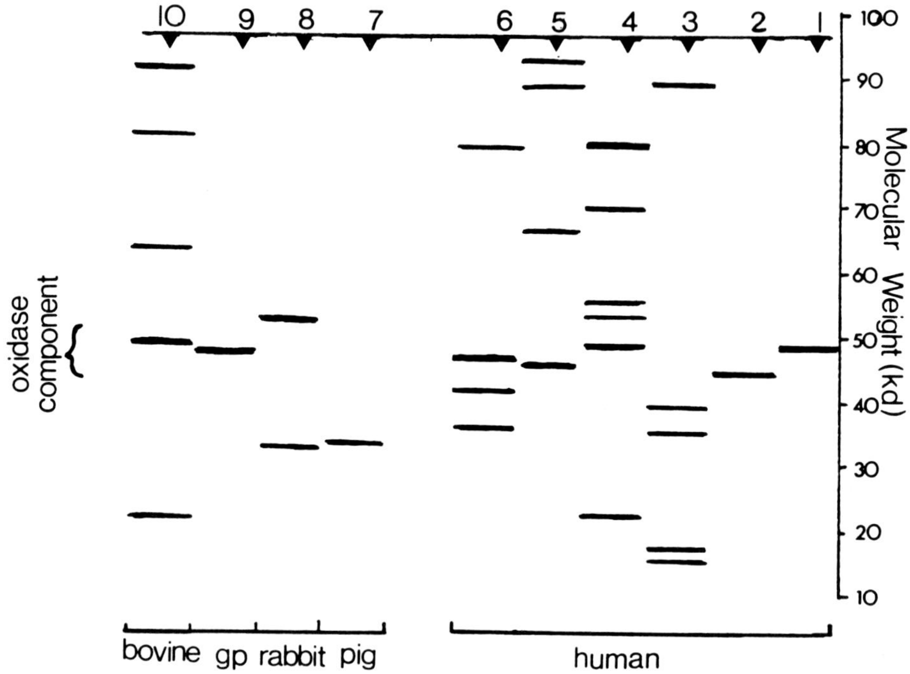

FIGURE 3. Comparison of molecular weights of some key proteins phosphorylated during stimulation of neutrophils. Bars represent the phosphoprotein reported in references (1)120, (2)34, (3)121, (4)122, (5)109, and (6)133 for human, (7)31, (8)123, (9)124 and (10)134 for pig, rabbit, guinea-pig, and bovine, respectively. For comparison, the flavoprotein component of the oxidase is shown by Cross and Jones.[35]

phosphorylation of a number of proteins accompanies stimulation (Figure 3). This phosphorylation is probably mediated by a number of kinases (although the major kinase activity present in the neutrophil is that of kinase C[28]). These studies, however, show little agreement as to the molecular weight of its substrates (Figure 3).

A feature of protein kinase is that phosphorylation occurs at specific sites on the protein, e.g., serine and threonine for kinase C, and, provided these sites are present, the enzyme does exhibit much specificity for the protein. Presumably in the cell, specificity is provided by the location or accessibility of the enzyme or its substrate. In vitro, where these constraints are not present, kinase C has been shown to catalyze the phosphorylation of a number of proteins. While the physiological significance of a number of these observations remains to be established, one substrate is especially interesting, namely, the neutrophil oxidase. It has been demonstrated that phosphorylation of the partially purified inactive oxidase can be catalyzed by kinase C with the result that the oxidase is activated.[30,31] This demonstration may provide the molecular basis for the activation of the oxidase by experimental activators of kinase C, such as phorbol myristate acetate (see section IV.C). Furthermore, a 44 to 47 kdalton protein which becomes phosphorylated during stimulation by PMA, f-Met-Leu-Phe, and latex beads in normal,[32,33] but not autosomal recessive granulomatous disease cells,[34] may be identified as a component of the oxidase.[35] These observations do not, of course, exclude the involvement of other kinases. For example, calmodulin-dependent kinase also modifies the activity of the oxidase in vitro.[36] It remains to be established which is the important route in physiological activation or whether both routes operate.

C. Experimental Activators of Kinase C

A number of experimental agents have been shown to activate kinase C. Although many

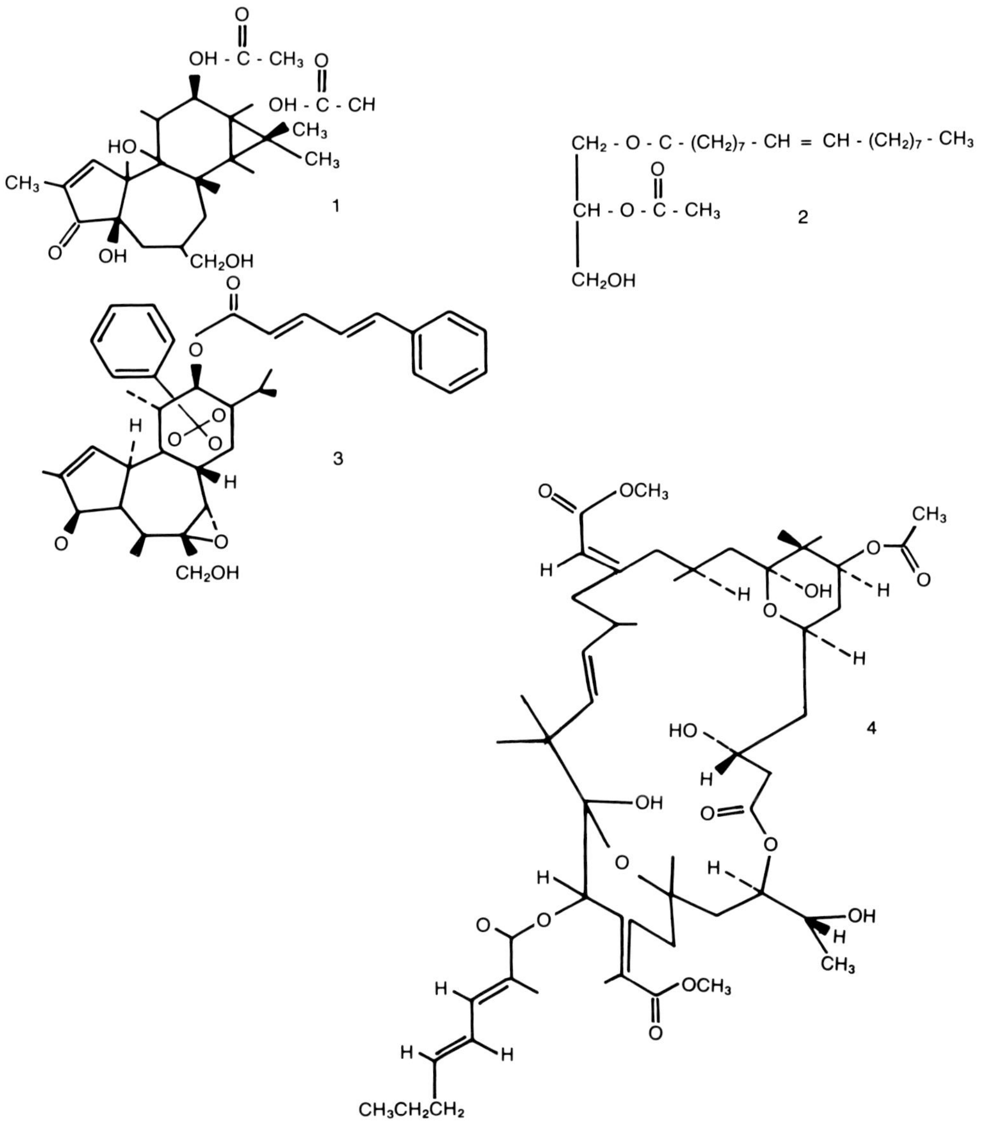

FIGURE 4. Some kinase C activators which stimulated neutrophils: (1) PMA-phorbol myristate acetate;[22,26] (2) oleoyl-acetate glycerol, OAG;[107] (3) mezerein;[108] (4) bryostatin.[109]

of these have either phorbol-like structures or are related to diacylglycerol, there is also a range of structurally different agents which activate kinase C (Figure 4). All of these agents activate the neutrophil oxidase, and there is a marked correlation between their ability to activate the neutrophil oxidase in the intact cell and their ability to activate kinase C in vitro (Figure 5). The most potent activator by far is PMA (phorbol myristate acetate), which is able to activate maximally the oxidase of 2.10^7 cells (at a rate of 2.5 nmol 0_2 consumed per 10^6 cells per minute). This corresponds to a mere $1.46 \cdot 10^5$ molecules of PMA per cell. Assuming a maximum of one kinase C molecule activated per PMA molecule and a maximum rate for the oxidase of 13 mol 0_2 produced per second per mole cytochrome b,[37] it can be calculated that one molecule of kinase C activates at least 14 molecules of the oxidase system. This would suggest a very efficient coupling of oxidase activation to kinase C.

It is not clear, however, whether the mechanism by which PMA and other agents activate

Table 3
INHIBITORS OF KINASE C THAT INHIBIT NEUTROPHIL ACTIVATION INDUCED BY PMA BUT NOT CHEMOTACTIC PEPTIDE (CP,fMLP)

Kinase C inhibitor	Ref.
Retinal	39
Polymixin B	125
C1	115
H7	114
H9	126

azine, are strong inhibitors of secretion and oxidase activation,[40] they are of little use in distinguishing the roles played by different kinases. Recently, sphinganine and sphingosine, which, it may be argued on theoretical grounds, would have a selective effect on phospholipid-dependent enzymes (and thus be a selective inhibitor for kinase C), have also been shown to inhibit activation of neutrophils in response to both PMA and chemotactic peptide.[41]

However, some kinase C inhibitors have the ability to selectively inhibit activation by PMA but not by fMLP (Table 3). This differential inhibition suggests that fMLP utilizes a pathway that is independent of that activated by PMA, and thus probably independent of the activity of kinase C. Retinoids have been used as inhibitors, although some also activate. This property is reflected in the ability of these agents to activate oxidase in the intact neutrophil and inhibit subsequent activation by other kinase C activators.[42] Retinal, when used in sufficient concentration to activate the oxidase (presumably due to binding to kinase C), fails to inhibit subsequent activation by fMLP.[39] In contrast, subsequent activation by PMA or unopsonized latex beads is inhibited.[39] This, therefore, confirms the conclusion reached with kinase C activators that whereas kinase C is involved in the latter mechanisms of activation, it is not involved in the former. It must, however, be recognized that no demonstration that these inhibitors selectively inhibit the physiologically relevant kinase C has been provided, apart from their ability to inhibit PMA. Caution must therefore be exercised in the interpretation of the use of ''kinase C inhibitors'' since the possibility exists that they may also act at other as yet undiscovered (PMA sensitive) sites.

V. POSSIBLE IDENTITY OF NON-Ca^{2+} INTRACELLULAR MESSENGERS

A. Diacylglycerol

It has been suggested that diacylglycerol produced during the breakdown of phosphatidyl inositol would enable activation of kinase C at resting cytoplasmic Ca^{2+} concentrations[27] and thereby act as an important intracellular messenger acting synergistically with Ca^{2+}. However, direct measurements of diacylglycerol production in neutrophils have proved difficult to interpret (see Chapter 7, Section III.C.). There is no detectable increase in diacylglycerol concentration 10 sec after stimulation by fMLP, although an increase was observed later.[43] This latter rise is unlikely to be significant for response coupling since the rise occurred when cellular responses associated with this stimulus, i.e., chemotaxis, secretion, and oxidase activation would already have been triggered. Furthermore, this latter diacylglycerol production did not have the fatty acid composition of phosphatidyl inositol, and therefore may not have arisen from the action of phospholipase C.

The production of phophatidic acid rises sharply after stimulation by fMLP, reaching 5 times its resting concentration within 10 sec.[43] Since this rise in phosphatidic acid probably resulted from the phosphorylation of diacylglycerol, the possibility exists that diacylglycerol was generated with a time course appropriate for it to act as an intracellular messenger.

The enzyme responsible for the generation of phosphatidic acid and hence the reduction in diacylglycerol concentration, diacylglycerol kinase, is associated with the neutrophil plasma membrane[44] and can be inhibited by the agent, R59022. In platelets, R59022 has been shown to elevate diacylglycerol concentrations following stimulation and also to increase phosphorylation of a 40K protein.[45] Since inhibitors of cyclic AMP phosphodiesterase have proved useful in evaluating the role of cyclic AMP, R59022 may also be useful in an analogous way, for examining the second messenger role of diacylglycerol.

Preincubation of rat neutrophils with R59022 (0.1 to 20 μM) had no significant effect on the resting level of oxidase activity, but the response to an optimal concentration of fMLP, 1 μM, was enhanced by up to fourfold.[46] A similar effect has recently also been shown for human neutrophils.[47] This therefore suggests that diacylglycerol was generated by this agonist and that it could enhance the response. Since the enhancement could be prevented by kinase C inhibitors, it was likely that the effect was mediated by this enzyme. Interestingly, the response to the calcium ionophore, A23187, was enhanced to a greater extent, 20 μM, R59022 producing approximately 20-fold enhancement,[46] whereas responses induced by PMA and unopsonized latex beads were not. This may reflect the intracellular Ca^{2+} concentrations existing in the cells under these conditions. Whereas stimulation by fMLP, is totally dependent on a rise in intracellular Ca^{2+},[8-11] activation by unopsonized latex beads.[10,11] does not depend on such a rise.

Since inhibitors of kinase C prevent oxidase activation by latex beads, it was suggested that this response was mediated by kinase C and that diacylglycerol may play a role.[34] The lack of enhancement by an agent expected to elevate the concentration of this putative messenger therefore leads to two possibilities: (1) diacylglycerol played no part in signal transduction by this route, or (2) the concentration of diacylglycerol generated was already sufficiently high to maximally activate kinase C at the existing Ca^{2+} concentration (approximately 0.1 μM). Since a membrane-bound intracellular messenger, such as diacyglyerol, would be restricted to the phagosomal membrane, this latter possibility cannot be excluded. It was also noted that a characteristic of the responses to both fMLP and latex beads following pretreatment with R59022 was a reduction in the rate at which the oxidase activation returned to baseline (Figure 8). This effect may have resulted from maintenance of an intracellular signal in these conditions.

Diacylglycerol may also have a role in modulating the association of actin to the plasma membrane.[103] However, the evidence for this role in neutrophils has not yet been provided. The possibility remains that second messengers other than Ca^{2+} and diacylglycerol also operate in the neutrophil.

B. GTP

It has been proposed that GTP can activate secretion from neutrophils by two distinct mechanisms.[48] First, GTP can act via a novel guanine nucleotide regulatory protein to decrease the dependence of phospholipase C for Ca^{2+}[49-52] (see details in Chapter 7). This may enable diacylglycerol to be generated and activate kinase C and consequently to stimulate secretion at resting intracellular Ca^{2+} concentrations. Second, GTP may be able to activate the secretory mechanism directly and independently of Ca^{2+}.[45] However, the introduction of nonhydrolyzable analogues of GTP into permeabilized neutrophils also evoked secretion, but the mechanism by which this was achieved was by increasing the affinity of the secretory machinery for Ca^{2+} by approximately tenfold.[53] Thus, GTP may be acting in a manner that would explain the shift in Ca^{2+} sensitivity observed when comparing physiological stimuli to experimentally induced elevation of Ca^{2+}.

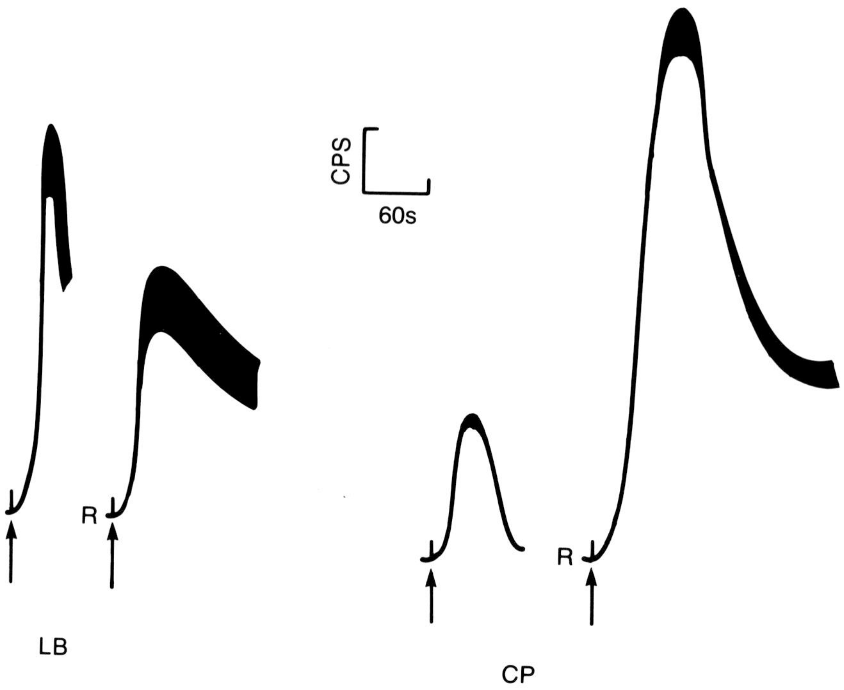

FIGURE 8. Effect of the diacyglycerol kinase inhibitor R59022 on oxidase activation by latex particles and chemotactic peptide. Luminol-dependent chemiluminescence traces are shown for cells stimulated with unopsonized latex beads diameter = 1 μm; 4.10^9) at the arrows marked LB and fMLP (1 μ*M*) at the arrows marked CP. Traces preceded by the label "R" indicate preincubation with R59022 (30 μ*M*).

Two GTP-binding proteins, involved in the transduction of the signal from hormone receptors to adenylate cyclase, have been identified in a number of cell types, Ns activating and Ni inhibiting. In neutrophils, Ns has been identified,[54] and chlorea toxin, which ADP-ribosylates and activates Ns, causes marked elevations of cyclic AMP, which subsequently inhibits responses to chemotactic stimuli.[55,56] Pertussis toxin, which ADP-ribosylates and activates Ni, however, has no effect on cyclic AMP[57]. This agent does, however, inhibit a number of neutrophil responses, including secretion,[57,58] chemotaxis,[57] oxidase activation,[58,59] and arachidonate release.[60] These inhibitory effects probably result from the inhibition of Ca^{2+} mobilization[58] and phosphatidyl inositol breakdown.[61] It is interesting to note that phagocytosis, which occurs without a requirement for a rise in intracellular Ca^{2+}, is unaffected by pertussis toxin.[62] The possibility, therefore, exists that pertussis toxin acts on a crucial "transducin-like" molecule which is involved in the generation of a rise in intracellular Ca^{2+}.

C. Inositol Phosphates

As described in detail in Chapter 7, inositol phosphates are metabolized during activation of neutrophils. Of particular importance is inositol (1,4,5)tris-phosphate, which can liberate Ca^{2+} for as yet unidentified intracellular stores of calcium.[63] Other inositol phosphates may have physiological roles. For example, inositol (1,3,4,5)tetrakis phosphate generated by phosphorylation of inositol tris-phosphate may open Ca^{2+} channels in the plasma membrane.[64] The possibility that cyclic inositol monophosphate can act as an intracellular messenger has also been suggested.[65] Furthermore, since the activity of both profilin[66] and gelsolin[67] are influenced by phosphatidylinositol 4,5-bisphosphate, it has also been suggested that phospholipid has importance in regulating actin polymerization, and so modify neutrophil activation.

D. Cyclic Nucleotides

It has been suggested that the balance between intracellular cyclic AMP and cyclic GMP in neutrophils regulates a number of cellular responses. The two nucleotides have opposing roles, cyclic GMP acting as a stimulatory signal and cyclic AMP acting as a counteracting inhibitor.[68,69] However, this so called "yin-yang theory" does not seem to be so clear cut as originally envisaged. While initial measurements of cyclic GMP suggested that this agent may have had an important role in triggering some neutrophil responses,[70] its exact role has remained uncertain. Similarly, there is uncertainty as to the role of cyclic AMP, although pharmacological intervention has supported the view that cyclic AMP can exert an inhibitory influence on some neutrophil responses.

Stimulation of neutrophils with chemotactic peptide produces a transient elevation of intracellular cyclic AMP. Phosphodiesterase inhibitors, forskolin, and A_2 adenosine agonists all exaggerate this transient, intracellular cyclic AMP rising to extremely high levels, up to 33 pmol cyclic AMP/10^6 cells for rat neutrophils.[71] This represents an intracellular concentration of approximately 0.12 m*M*. There is general agreement that these extremely elevated levels can exert an inhibitory influence on chemotactic peptide-stimulated responses. This route may be the mechanism by which agents such as adenosine modify neutrophil responsiveness.[72,73] Adenosine, presumably acting through cyclic AMP, also inhibits responses induced by another agent which elevates intracellular Ca^{2+}, the Ca^{2+} ionophore, A23187.[72] Since responses triggered by two intracellular Ca^{2+}-independent stimuli, latex beads and PMA, were not inhibited,[72,73] the possibility exists that the inhibitory effect of high intracellular cyclic AMP is specific for the Ca^{2+} route. The relevance of these observations to the normal physiology of the neutrophils, however, is not clear. In contrast, phagocytosis of a number of particle types is unaffected by extracellular cAMP,[74-79] although both enhancement[80] and inhibition[81] have also been reported. No detectable change in the intracellular concentration of cyclic AMP has been reported during phagocytosis of a range of particles,[74-79] although a rise has also been reported.[80-83]

It has been reported that cyclic AMP appeared on the cell surface immediately before phagosome formation.[80] The regulatory subunit of cAMP-dependent kinase also appeared on the surface. No change in cyclic GMP was detected in that report.[80]

The possibility that other cyclic nucleotides play a role in neutrophil responses cannot be excluded. For example, cyclic IMP (cyclic inositol monophosphate) may be generated.[62] This may have relevance to the neutrophil, since cyclic IMP-activated kinase activity has been detected in neutrophils.[84]

E. Fatty Acids

It has been suggested that since fatty acids are activators of kinase C they may play a physiological role in neutrophil activation.[85] Arachidonic acid generated during stimulation of the cells (see Chapter 5) may therefore serve as an intracellular messenger activating kinase C. There is also the possibility that other fatty acids are generated during phagocytosis. For example, phagocytosis of latex particles results in a transient production of fatty acid.[86,87] Although many studies have also shown that extracellularly added fatty acids can activate neutrophils,[88-90] it is likely that this activation resulted from the increased Ca^{2+} influx that accompanies insertion of the fatty acid into the plasma membrane (see Chapter 5).

F. Hydrogen Ions (pHi)

The possibility that intracellular pH changes may regulate some neutrophil responses has also been suggested.[91,92] It has been shown that, following stimulation, intracellular pH changes have been detected. Stimulation of rat neutrophils by PMA or chemotactic peptide results in a decrease in intracellular pH, from aproximately 7.25 to 6.8. The time course of this change, however, is inconsistent with it acting as a trigger for neutrophil responses. For example, the intracellular pH of resting rat neutrophils, as determined by fluorescein, is 7.04. Following stimulation by chemotactic peptide plus cytochalsin B, the intracellular pH decreased after a 40 sec lag. However, intracellular Ca^{2+} and actin polymerization both peaked at approximately 6 sec, with the onset of oxidase activation occurring at a similar time. These key events, therefore, preceded the intracellular pH change. Lowering the intracellular pH artificially with nigericin does not activate the oxidase or inhibit subsequent stimulation by chemotactic peptide (Figure 9). Under these conditions, chemotactic peptide does not reduce pHi further. It is probable that the pH change normally accompanying stimulation of the cells results from the production of hydrogen ions accompanying activation of the hexose monophosphate shunt.

G. Membrane Potential Changes

As discussed in Chapter 10, the membrane potential of neutrophils changes during stimulation. A resting potential of -60 mV represents a voltage difference across the membrane of $1.2 \cdot 10^5$ volts/cm. This voltage field may have a conformational effect on proteins, such as the oxidase or phospholipases, within the membrane. Changes in the membrane potential may therefore influence the conformation and activity of these enzymes. The possibility that such effects play a role in signal transduction in neutrophils has as yet remained unexplored.

VI. IMPLICATIONS OF MULTIPLE INTRACELLULAR SIGNALING MECHANISMS FOR THE CONTROL OF NEUTROPHIL RESPONSES

The existence of both Ca^{2+}-dependent and possibly multiple Ca^{2+}-independent mechanisms in the neutrophil may reflect the complexity of cellular events that must occur for these cells to successfully complete their task. Such a combination of intracellular messengers may enable control of a number of different cellular events and also control of the same event at different cellular locations.

A. Inappropriate Activation

Unless the elements involved are vectorally arranged, the neatly written ''biochemical pathway'' gives a misleading impression of the actual biochemical events occurring within the cell. As discussed in Chapter 1, in processes controlled by diffusion, such as transmission of the intracellular messenger from its site of generation to its intracellular site of action, the result depends upon statistical processes. If a low number of molecules act as the intracellular message, then probability will play an important role. In this situation, the coupling between a stimulus and its response will be subject to the laws of probability. The interaction of a number of different intracellular messengers, both calcium and other non-calcium, will therefore increase the likelihood of an appropriate stimulus evoking an inappropriate response. (See Chapter 1 for more detailed discussion.) This may provide a basis for the inappropriate responses of neutrophils during inflammatory disease.

B. Inappropriate Cellular Location of Response

The normal functioning of the neutrophil depends upon the cell controlling not only the magnitude of a response, but also the cellular location at which the event occurs. For example, intracellular killing of phagocytosed microorganisms without damage to extracellular environment depends upon (1) fusion of the granule membrane specifically with the phagosome membrane in the absence of fusion with plasma membrane, in order to discharge granular enzymes specifically into the phagosome, and (2) activation of the oxidase only in the

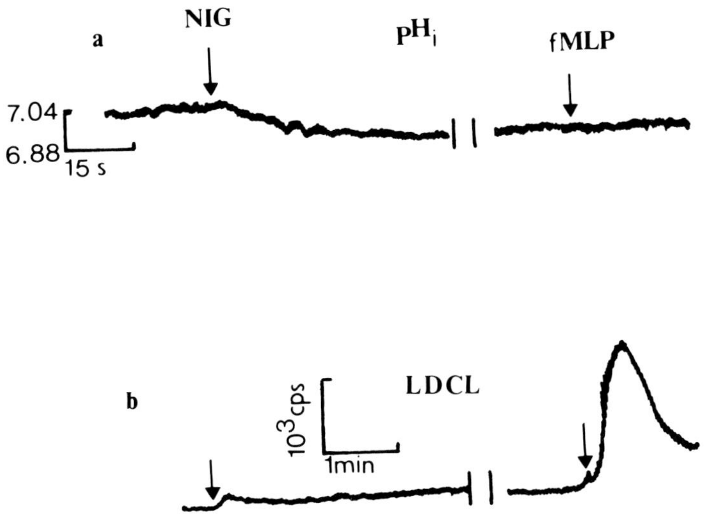

FIGURE 9. Effect of change in intracellular pH on oxidase stimulation. (a) Intracellular pH was determined using fluorescein-loaded rat neutrophils and (b) oxidase activity by luminol-dependent chemiluminescence. Nigericin was added at the first arrow and chemotactic peptide (1 μ*M*) at the second.

phagosomal membrane, without activation in the plasma membrane. Since any part of the plasma membrane can apparently form a phagosome, the cell must employ a mechanism for restricting the effect of any intracellular messenger generated. The possibility exists that hydrophobic messengers such as diacylglycerol or fatty acids would be restricted to the membrane of the forming phagosome. Such messengers would, therefore, provide a mechanism for activating the oxidase only in the phagosomes or of restricting fusion events to granule-phagosome. It is interesting to note that the generation of diacylglycerol or fatty acid on one side of a lipid bilayer would lead to a change in curvature of the membrane that may be a necessary prelude to fusion.[93]

Hydrophilic messengers such as Ca^{2+} would be released into the cytoplasm and thus gain access to more intracellular sites. Cellular events requiring whole cell involvement, such as chemotaxis, may thus require such messengers. It is interesting to note that when chemotaxis is inhibited by the actin-polymerizing inhibitors, such as cytochalsin B or botulinum C2 toxin,[94] that chemotactic peptide produces whole-cell responses such as secretion (i.e., unrestricted fusion of granule membranes with plasma membrane) and activation of the oxidase at the plasma membrane (i.e., not restricted to phagosomes).[95] The possibility also exists that activation of kinase C by proteolysis occurs at high intracellular Ca^{2+}.[28] This would result in activated kinase gaining access to the cytoplasm and, consequently, may lead to a more general cell response.

If, under some pathological conditions, the control over cellular location of reponse were lost, secretion of degradative enzymes and activation of plasma membrane-located oxidase, leading to the extracellular generation of reactive oxygen species, would result. It is possible to suggest two molecular scenarios that would result in loss of such control: (1) inhibition of actin polymerization by fungal metabolites or bacterial toxins: and (2) pathologically high intracellular Ca^{2+} concentrations, produced, for example, by the terminal attack complex of complement[19,96] or by envelope viruses.[48,97] Obviously, further study of the nature of noncalcium intracellular messengers and their interaction with Ca^{2+} will be required before a clear understanding of the physiological and pathological control of neutrophil responses is achieved.

NOTE ADDED IN PROOF

Since completion of this chapter, papers have appeared which throw further light on two controversial areas discussed here. The use of sphingosine and sphinganine for providing evidence for kinase C has been seriously questioned, the former being unlikely to be specific for the enzyme (Bazzi and Nelsestuen, *Biochem. Biophys. Res. Commun.*, 146, 203, 1987) and the latter inhibiting neutrophil activity by nonspecific toxicity (Pittet et al., *J. Biol. Chem.*, 262, 10072, 1987). Further evidence has also been provided that phagocytosis can occur in the absence of a rise in cytoplasmic free Ca^{2+} (Levitz et al., *Infect. Immun.*, 55, 2783, 1987; Di Virgilio et al., *J. Cell Biol.*, 106, 657, 1988).

REFERENCES

1. **Woodin, A. M. and Wienke, A. A.,** The accumulation of calcium by polymorphonuclear leucocytes treated with staphylococcal leucocidin, *Biochem. J.*, 87, 487, 1963.
2. **Schell-Frederick, E.,** Stimulation of the oxidative metabolism of polymorphonuclear leukocytes by the calcium ionophore A23187, *FEBS Lett.*, 48, 37, 1974.
3. **Romeo, D., Zabucchi, G., Miani, N. and Rossi, F.,** Ion movement across leukocyte plasma membrane and excitation of their metabolism, *Nature (London)*, 253, 542, 1975.
4. **Boucek, M. M. and Snyderman, R.,** Calcium influx requirement for human neutrophil chemotaxis: inhibition by lathanum ions, *Science*, 193, 905, 1976.
5. **Nacchache, P. H., Showell, H. J., Becker, E. L., and Sha'afi, R. I.,** Transport of sodium, potassium and calcium across rabbit polymorphonuclear leukocyte membranes: effect of chemotactic factors, *J. Cell Biol.*, 73, 428, 1977.
6. **Hallett, M. B. and Campbell, A. K.,** Measurement of changes in cytoplasmic free calcium in fused cell hybrids, *Nature (London)*, 295, 155, 1982.
7. **Pozzan, T., Lew, D. P., Wollheim, C. B., and Tsien, R. Y.,** Is cytosolic calcium the trigger for neutrophil activation?, *Science*, 221, 1413, 1983.
8. **Lew, D. P., Wollheim, C. B., Waldrogel, F. A., and Pozzan, T.,** Modulation of cytosolic free calcium transients by changes in intracellular calcium buffering capacity: correlation with exocytosis and O_2^- production in human neutrophils, *J. Cell Biol.*, 99, 1212, 1984.
9. **White, J. R., Naccache, P. H., Molski, T. F. P., Borgeat, P., and Sha'afi, R. I.,** Direct demonstration of increased intracellular concentration of free calcium in rabbit and human neutrophils following stimulation by chemotactic factor, *Biochem. Biophys. Res. Commun.*, 113, 44, 1983.
10. **Campbell, A. K. and Hallett, M. B.,** Direct measurement of intracellular free calcium and oxygen radicals in polymorphonuclear leucocyte-erythrocyte "ghost" hybrids, *J. Physiol. (London)*, 338, 537, 1983.
11. **Hallett, M. B. and Campbell, A. K.,** Is intracellular Ca^{2+} the trigger for oxygen radical production by polymorphonuclear leucocytes?, *Cell Calcium*, 5, 1, 1984.
12. **Lew, D. P., Monod, A., Waldvogel, F. A., Dewald, B., Bagglioni, M., and Pozzan, T.,** Quantitative analysis of the cytosolic free calcium dependency of exocytosis from three subcellular components in intact human neutrophils, *J. Cell Biol.*, 102, 2197, 1986.
13. **Sha'afi, R. I., White, J. R., Molski, T. F. P., Schefcyk, J., Volpi, M., Naccache, P. H., and Feinstein, M. B.,** Phorbol-12-myristate 13-acetate activates rabbit neutrophils without an apparent rise in the level of intracellular free calcium, *Biophys. Biochem. Res. Commun.*, 114, 638, 1983.
14. **DiVirgilio, F., Lew, D. P., and Pozzan, T.,** Protein kinase activation of physiological processes in human neutrophils at vanishingly small cytosolic Ca^{2+} levels, *Nature (London)*, 310, 691, 1984.
15. **Campbell, A. K., Dormer, R. L., and Hallett, M. B.,** Coelenterate photoproteins as indicators of cytoplasmic free Ca^{2+} in small cells, *Cell Calcium*, 6, 69, 1985.
16. **Hallett, M. B. and Campbell, A. K.,** Applications of coelenterate luminescent proteins, in *Clinical and Biochemical Luminescence* , Vol. 12, Kricka, L. J. and Carter, T. J. N., Eds., Marcel Dekker, New York, 1982, 89.
17. **Yin, H. L. and Stossel, T. P.,** Control of cytoplasmic actin gel-sol transformation by gelsolin, a calcium-dependent regulatory protein, *Nature (London)*, 281, 583, 1979.
18. **Yin, H. L. and Stossel, T. P.,** The mechanism of phagocytosis, in *Phagocytosis — Past and Future,* Karnovsky, M. L. and Bolis, L., Eds., Academic Press, New York, 1982, 13.

19. **Hallett, M. B., Luzio, J. P., and Campbell, A. K.,** Stimulation of Ca^{2+}-dependent chemiluminescence in rat polymorphonuclear leucocytes by polystyrene beads and the non-lytic action of complement, *Immunology,* 44, 569, 1981.
20. **Hallett, M. B. and Campbell, A. K.,** Direct measurement of intracellular free Ca^{2+} in rat peritoneal macrophages; correlation with oxygen radical production, *Immunology,* 50, 487, 1984.
21. **McNeil, P. L., Swanson, J. A., Wright, J. D., Silverstein, S. C., and Taylor, D. L.,** Fc-mediated phagocytosis occurs in macrophages without an increase in average $(Ca^{2+})_i$, *J. Cell Biol.,* 102, 1586, 1986.
22. **Young, T. D. E., Ko, S. S., and Cohn, Z. A.,** The increase in intracellular free calcium associated with IgG_{2b}/FC receptor-ligand interactions: role in phagocytosis, *Proc. Natl. Acad. Sci. U.S.A.,* 81, 5430, 1984.
23. **Lew, D. P., Andersson, T., DiVirgilio, F., Pozzan, T., and Stendahl, O.,** Ca^{2+}-dependent and Ca^{2+}-independent phagocytosis in human neutrophils, *Nature (London),* 315, 509, 1985.
24. **Neidel, J. E., Kuhn, L. J. T., and Vanderbark, G. R.,** Phorbol diester receptor copurifies with protein kinase C, *Proc. Natl. Acad. Sci. U.S.A.,* 80, 34, 1983.
25. **Estensen, R. D., White, J. G., and Holmes, B.,** Specific degranulation of human polymorphonuclear leukocytes, *Nature (London),* 248, 347, 1974.
26. **DeChatelet, L. R., Shirley, P. S., and Johnson, R. B., Jr.,** Effect of phorbol myristate acetate on the oxidative metabolism of human polymorphonuclear leukocytes, *Blood,* 47, 545, 1976.
27. **Nishizuka, Y.,** The role of protein kinase C in cell surface signal transduction and tumor promotion, *Nature (London),* 308, 693, 1984.
28. **Melloni, E., Pontremoli, S., Michetti, M., Sacco, O., Sparatore, B., Salamino, F., and Horecker, B. L.,** Binding of protein kinase C to neutrophil membranes in the presence of Ca^{2+} and its activation by a Ca^{2+}-requiring proteinase, *Proc. Natl. Acad. Sci. U.S.A.,* 82, 6435, 1985.
29. **Helfman, D. M., Barnes, K. C., Kinade, J. M., Vogler, W. R., Shoji, M., and Kuo, F. F.,** Phospholipid-sensitive and Ca^{2+}-dependent protein phosphorylating system in various types of leukemic cells from patients and from the leukemic cell-lines HL60 and K562, *Cancer Res.,* 43, 2955, 1983.
30. **Cox, J. A., Jeng, A. J., Sharkey, N. A., Blumberg, J. M., and Tauber, A. I.,** Activation of human neutrophil NADPH-oxidase by protein kinase C, *J. Clin. Invest.,* 76, 1932, 1986.
31. **Papini, E., Grzeskowiak, M., Bellavite, P., and Rossi, F.,** Protein kinase C phosphorylates a component of NADPH oxidase of neutrophils, *FEBS Lett.,* 190, 204, 1985.
32. **Heyworth, P. and Segal, A. W.,** Further evidence for the involvement of a phophoprotein in the respiratory burst oxidase of human neutrophils, *Biochem. J.,* 239, 723, 1986.
33. **Gennaro, R., Florio, C., and Romeo, D.,** Activation of protein kinase C in neutrophil cytoplasts. Localization of protein substrates and possible relationship with stimulus-response coupling, *FEBS Lett.,* 180, 185, 1985.
34. **Segal, A. W., Heyworth, P. G., Cockcroft, S., and Barrowman, M. M.,** Stimulated neutrophils from patients with autosomal recessive chronic granulomatous disease fail to phosphorylate a Mr 44,000 protein, *Nature (London),* 316, 547, 1985.
35. **Cross, A. R. and Jones, O. T. G.,** The effect of the inhibitor diphenylene iodonium on the superoxide generating system of neutrophils, *Biochem. J.,* 237, 111, 1986.
36. **Jones, H. P., Ghai, G., Petrone, W. F., and McCord, J.,** Calmodulin-dependent stimulation of the NADPH oxidase of human neutrophils, *Biochim. Biophys. Acta,* 714, 152, 1982.
37. **Cross, A. R., Parkinson, J. F., and Jones, O. T. G.,** Mechanism of the superoxide-producing oxidase of neutrophils; oxygen is necessary for fast reduction of cytochrome b-245, *Biochem. J.,* 226, 881, 1985.
38. **Dale, M. M. and Penfield, A.,** Synergism between phorbol ester and A23187 in superoxide production by neutrophils, *FEBS Lett.,* 175, 170, 1984.
39. **Cooke, E. and Hallett, M. B.,** The role of c-kinase in the physiological activation of the neutrophil oxidase; evidence from pharmacological manipulation of c-kinase activity in intact cells, *Biochem. J.,* 232, 323, 1985.
40. **Naccache, P. H., Molski, T. F. P., Becker, E. L., Showell, H. J., and Sha'afi, R. I.,** Calmodulin inhibitors block neutrophil degranulation at a step distal from the mobilization of calcium, *Biochem. Biophys. Res. Commun.,* 97, 62, 1980.
41. **Wilson, E., Olcott, M. C., Bell, R. M., Merrill, A. H., and Lambeth, J. D.,** Inhibition of oxidative burst in human neutrophils by shingoid long-chain bases; role of protein kinase C in activation of the burst, *J. Biol. Chem.,* 261, 12,616, 1986.
42. **Cooke, E. and Hallett, M. B.,** Retinoids interact with the mechanism of neutrophil oxidase activation, *Biochem. Soc. Trans.,* 14, 954, 1986.
43. **Cockcroft, S. and Allen, D.,** The fatty acid composition of phosphatidylinositol, phosphatidate and 1,2, diacylglycerol in stimulated human neutrophils, *Biochem. J.,* 222, 557, 1984.
44. **Cockcroft, S., Baldwin, J. M., and Allan, D.,** The Ca^{2+}-activated polyphosphoinositide phosphodiesterase of human and rabbit neutrophil membranes, *Biochem. J.,* 221, 477, 1984.

45. **de Chaffoy de Courcelles, D., Roevens, P., and Van Belle, H.,** R59022, a diacylglycerol kinase inhibitor; its effect on diacylglycerol and thrombin-induced C-kinase activation in intact platelets, *J. Biol. Chem.*, 260, 15,762, 1985.
46. **Cook, E., Al-Mohanna, F. A., and Hallett, M. B.,** Diacylglycerol inhibitor, R59022, potentiates neutrophil oxidase activation by Ca^{2+}-dependent stimuli, *Biochem. Pharmacol.*, 36, 3459, 1987.
47. **Muid, R. E., Penfield, A., and Dale, M. M.,** The diacylglycerol kinase inhibitor R59022 enhances the superoxide generation from human neutrophils induced by stimulation of the fmlp, IgG and C3b receptors, *Biochem. Biophys. Res. Commun.*, 143, 630, 1987.
48. **Barrowman, M. M., Cockcroft, S., and Gomperts, B. D.,** Two roles for guanine nucleotides in the stimulus-secretion coupling of neutrophils, *Nature (London)*, 319, 504, 1986.
49. **Cockcroft, S. and Gomperts, B. D.,** Role of guanine nucleotide binding protein in the activation of polyphosphoinositide phosphodiesterase, *Nature (London)*, 314, 534, 1985.
50. **Smith, C. D., Cox, C. C., and Snyderman, R.,** Receptor-coupled activation of phosphoinositide-specific phospholipase C by N protein, *Science*, 232, 97, 1986.
51. **Cockcroft, S.,** The dependence on Ca^{2+} of the guanine nucleotide-activated polyphosphoinositide phosphodiesterase in neutrophil plasma membranes, *Biochem. J.*, 240, 503, 1986.
52. **Bradford, P. G. and Rubin, R. P.,** Guanine nucleotide regulated phospholipase C activity in permeabilized rabbit neutrophils, *Biochem. J.*, 239, 97, 1986.
53. **Smolen, J. E. and Stoehr, S. J.,** Guanine nucleotides reduce the free calcium requirement for granule of granule constituents from permeabilized human neutrophils, *Biochim. Biophys. Acta*, 889, 171, 1986.
54. **Lad, P. M., Glovsky, M. M., Smiley, P. A., Klempner, M., Reissinger, D. M., and Richards, J. M.,** The p-adrenergic receptor in the human neutrophil plasma membrane; receptor cyclase coupling is associated with amplified GTP activation, *J. Immunol.*, 132, 1466, 1984.
55. **Rivkin, I., Rosenblatt, J., and Becker, E. L.,** The role of cAMP in the chemotactic responsiveness and spontaneous motility of rabbit peritoneal neutrophils. The inhibition of neutrophil movement and the elevation of cAMP levels by catecholamines, prostaglandins, theophylline and cholera toxin, *J. Immunol.*, 115, 1114, 1975.
56. **Hill, H. R., Estenson, R. D., Quie, P. G., Hogan, N. A., and Goldberg, N. D.,** Modulation of human neutrophil chemotactic responses by cAMP, *Metabolism*, 24, 447, 1975.
57. **Becker, E. L., Kermode, J. C., Naccache, P. H., Yassin, R., Marsh, M. L., Munoz, J. J., and Sha'afi, R. I.,** The inhibition of neutrophil granule enzyme secretion and chemotaxis by pertussis toxin, *J. Cell Biol.*, 100, 1641, 1985.
58. **Goldman, D. W., Chang, F. H., Gifford, L. A., Goetzl, E. J., and Bourne, H. R.,** Pertussis toxin inhibition of chemotactic factor-induced calcium mobilization and function in human polymorphonuclear leukocytes, *J. Exp. Med.*, 162, 145, 1985.
59. **Verghese, M. W., Smith, C. D., and Snyderman, R.,** Potential role for a guanine nucleotide regulatory protein in chemoattractant receptor mediated polyphosphoinositide metabolism, Ca^{2+} mobilization and cellular responses by leukocytes, *Biochem. Biophys. Res. Commun.*, 127, 450, 1985.
60. **Okajima, F. and Ui, M.,** ADP-ribosylation of the specific membrane by islet-activating factor, pertussis toxin, associated with inhibition of chemotactic peptide-induced arachidonate release in neutrophils, *J. Biol. Chem.*, 259, 13,863, 1984.
61. **Smith, C. D., Lane, B. C., Kusaka, I., Verghese, M. W., and Snyderman, R.,** Chemoattractant receptor-induced hydrolysis of phosphoinositol 4,5-bisphophate in human polymorphonuclear leukocyte membranes, *J. Biol. Chem.*, 260, 5875, 1985.
62. **Lad, P. M., Olson, C. V., and Grewal, I. S.,** A step sensitive to pertussis toxin and phorbol esters in neutrophils regulates chemotaxis and capping but not phagocytosis, *FEBS Lett.*, 200, 91, 1986.
63. **Prentki, M., Wollheim, C. B., and Lew, D. P.,** Ca^{2+} homeostasis in permeabilized human neutrophils: characterization of Ca^{2+} sequestering pools and the action of inositol 1,4,5 trisphosphate, *J. Biol. Chem.*, 259, 13,777, 1984.
64. **Irvine, R. F. and Moor, R. M.,** Micro-injection of 1,3,4,5 tetrakisphosphate activates sea urchin eggs by a mechanism that depends on external Ca^{2+}, *Biochem. J.*, 240, 917, 1986.
65. **Majerus, P. W., Connolly, T. M., Deckmyn, H., Ross, T. S., Bross, T. E., Ishii, H., Bansal, V. S., and Wilson, D. B.,** The metabolism of phosphoinositide-derived messenger molecules, *Science*, 234, 1519, 1986.
66. **Lassing, I. and Lindberg, U.,** Specific interaction between phosphotidylinositol 4,5 bisphosphate and profilactin, *Nature (London)*, 314, 472, 1985.
67. **Janmey, P. A. and Stossel, T. P.,** Modulation of gelsolin function by phosphatidylinositol 4,5-bisphosphate, *Nature (London)*, 325, 362, 1987.
68. **Goldberg, N. D., Haddox, M. K., Nicol, S. E., Glass, R. D., Sandford, C. H., Kuehe, F. A., Jr., and Estensen, R.,** The yin-yang hypothesis of biological control; opposing influences of cyclic AMP and cyclic GMP in the regulation of cell proliferation and other biological processes, *Adv. Cyclic Nucleotide Res.*, 5, 307, 1975.

69. **Weissman, G., Goldstein, I., Hoffstein, S., Chauvet, G., and Robinson R.**, Yin-yang modulation of lysosomal enzyme release from polymorphonuclear leukocytes by cyclic nucleotides, *Ann. N.Y. Acad. Sci.*, 256, 222, 1975.
70. **Smith, R. J. and Ignarro, L. J.**, Bioregulation of lysosomal enzyme secretion from human neutrophils; roles of guanosine 3′5′monophosphate and calcium in stimulus-secretion coupling, *Proc. Natl. Acad. Sci. U.S.A.*, 72, 108, 1975.
71. **Roberts, P. A., Patel, A., Barrow, S. C. L., Hallett, M. B., and Campbell, A. K.**, Cyclic AMP in PMN measured by chemiluminescence energy transfer, in *Analytical Applications of Bioluminescence and Chemiluminescence*, Kricka, L. J., Stanley, P. E., Thorpe, G. H. G., and Whitehead, T. P., Eds., Academic Press, New York, 1984, 175.
72. **Cronstein, B. N., Kramer, S. B., Weissman, G., and Hirschhorn, R.**, Adenosine: a physiological modulator of superoxide generation by human neutrophils, *J. Exp. Med.*, 158, 1160, 1983.
73. **Roberts, P. A., Newby, A. C., Hallett, M. B., and Campbell, A. K.**, Inhibition by adenosine of reactive oxygen metabolite production by human polymorphonuclear leucocytes, *Biochem. J.*, 227, 669, 1985.
74. **Stossel, T. P., Murad, F., Mason , R, J., and Vaughan, M.**, Regulation of glycogen metabolism in polymorphonuclear leukocytes, *J. Biol. Chem.*, 245, 6628, 1970.
75. **Stolc, V.**, Regulation of iodine metabolism in human leukocytes by adenosine 3′5′ monophosphate, *Biochim. Biophys. Acta*, 264, 285, 1972.
76. **Seybeth, H. W., Schmidt-Gayk, H., Jakobs, K. H., and Hackenthal, E.**, Cyclic AMP in phagocytosing granulocytes and alveolar macrophages, *J. Cell Biol.*, 57, 567, 1973.
77. **Schell-Frederick, E. and Van Sande, J.**, Evidence that cyclic AMP is not the chemical mediator of the metabolic changes accompanying phagocytosis, *J. Reticuloendothelial. Soc.*, 15, 139, 1974.
78. **Manganiello, V., Evans, W. H., Stossel, T. P., Mason, R. J., and Vaughan, M.**, The effect of polystyrene beads on cyclic AMP concentration in leukocytes, *J. Clin. Invest.*, 50, 2741, 1971.
79. **Stossel, T. P., Mason, R. J., Hartwig, J., and Vaughan, M.**, Quantitative studies of phagocytosis by polymorphonuclear leukocytes, *J. Clin. Invest.*, 51, 615, 1972.
80. **Park, B. H., Good, R. A., Beck, N. P., and Davis, B. B.**, Concentration of cyclic AMP in human leukocytes during phagocytosis, *Nature (London), New Biol.*, 229, 27, 1979.
81. **Cox, J. P. and Karnovsky, M. L.**, The depression of phagocytosis by exogenous cyclic nucleotide, prostaglandin and theophylline, *J. Cell Biol.*, 59, 480, 1973.
82. **Herlin, T., Petersen, C. S., and Esmann, V.**, The role of calcium and adenosine 3′5′ monophosphate in the regulation of glycogen metabolism in phagocytosing neutrophils, *Biochim. Biophys. Acta*, 542, 63, 1978.
83. **Pryzwansky, K. B., Steiner, A. L., Spitznagel, J. K., and Kapoor, C. L.**, Compartmentalization of cyclic AMP during phagocytosis by human neutrophilic granulocytes, *Science*, 211, 407, 1981.
84. **Tsung, P-K., Sakamoto, T., and Weissmann, G.**, Protein kinase and phosphatases in human polymorphonuclear leucocytes, *Biochem. J.*, 145, 437, 1975.
85. **McPhail, L. C., Clayton, C., and Snyderman, R.**, A potential second messenger role for unsaturated fatty acids: activation of Ca^{2+}-dependent protein kinase, *Science*, 224, 622, 1984.
86. **Elsbach, P. and Farrow, S.**, Cellular triglyceride as a source of fatty acid for lecithin synthesis duing phagocytosis, *Biochim. Biophys. Acta* , 176, 438, 1969.
87. **Elsbach, P., Patriarca, P., Pettis, P., Stossel, T. P., Mason, R. J., and Vaughan, M.**, The appearance of lecithin- ^{32}P, synthesized from lysolecithin- ^{32}P, in phagosomes of polymorphonuclear leukocytes, *J. Clin. Invest.*, 51, 1910, 1972.
88. **Kakinuma, K.**, Effect of fatty acids on the oxidative metabolism of leukocytes, *Biochim. Biophys. Acta.*, 348, 76, 1974.
89. **Badwey, J. A., Curnutte, J. T., and Karnovsky, M. L.**, cis-Polyunsaturated fatty acids induce high levels of superoxide production by human neutrophils, *J. Biol. Chem.*, 256, 12,640, 1981.
90. **Bromberg, Y. and Pick, E.**, Unsaturated fatty acids as second messengers of superoxide generation in macrophages, *Cell Immunol.*, 79, 240, 1983.
91. **Simochowitz, L. and Cragoe, E. J., Jr.**, Regulation of human neutrophil chemotaxis by intracellular pH, *J. Biol. Chem.*, 261, 6492, 1986.
92. **Simochowitz, L.**, Intracellular pH modulates the generation of superoxide radicals by human neutrophils, *J. Clin. Invest.*, 72, 1793, 1985.
93. **Allan, D., Thomas, P., and Michell, R. H.**, Rapid transbilayer diffusion of 1,2-diacylglycerol and its relevance to control of membrane curvature, *Nature (London)*, 276, 289, 1978.
94. **Al-Mohanna, F. A., Ohishi, I., and Hallett, M. B.**, Botulinum C_2 toxin potentiates activation of the neutrophil oxidase; further evidence for a role for actin polymerization, *FEBS Lett.*, 219, 40, 1987.
95. **Hallett, M. B. and Campbell, A. K.**, Two distinct mechanisms for stimulation of oxygen-radical production by polymorphonuclear leucocytes, *Biochem. J.*, 216, 459, 1983.

96. **Campbell, A. K., Daw, R. A., Hallett, M. B., and Luzio, J. P.,** Direct measurement of the increase in intracellar free calcium ion concentration in response to the action of complement, *Biochem. J.*, 194, 551, 1981.
97. **Hallett, M. B., Fuchs, P., and Campbell, A. K.,** Sendai virus causes a rise in intracelluar free Ca^{2+} before cell fusion, *Biochem. J.*, 206, 671, 1982.
98. **White, J. R., Naccache, P. H., Molski, T. F. P., Borgeat, P., and Sha'afi, R. I.,** Direct demonstration of increased intracellular concentration of free calcium in rabbit and human neutrophils following stimulation by chemotactic factors, *Biochem. Biophys. Res. Commun.*, 113, 44, 1983.
99. **Rickards, J. E. and Sheterline, P. E.,** Evidence that phorbol ester interferes with stimulated Ca^{2+} redistribution by activating Ca^{2+} efflux in neutrophil leucocytes, *Biochem. J.*, 231, 632, 1985.
100. **Sawyer, D. W., Sullivan, J. A., and Mandell, G. L.,** Intracellular free calcium localised in neutrophils during phagocytosis, *Science*, 230, 663, 1985.
101. **Schell-Frederick, E.,** A comparison of the effects of soluble stimuli on free cytoplasmic and membrane bound calcium in human neutrophils, *Cell Calcium*, 5, 237, 1984.
102. **Al-Mohanna, F. A. and Hallett, M. B.,** The use of fura 2 as an indicator of intracellular free Ca^{2+} in rat neutrophils, *Cell Calcium*, in press.
103. **Burn, P., Rotman, A., Meyer, R. K., and Burger, M. M.,** Diacylglycerol in large α-actinin/actin complexes and in the cytoskeleton of activated platelets, *Nature (London)*, 314, 469, 1985.
104. **Smolen, J. E. and Stoehr, S. J.,** Micromolar concentrations of free calcium provoke secretion of lysosomal enzymes from human neutrophils permeabilized with saponin, *J. Immunol.*, 134, 1859, 1985.
105. **Gennaro, P., Pozzan, T., and Romeo, D.,** Monitoring Ca^{2+} in C5a stimulated neutrophils; loss of receptor modulated calcium stores and calcium uptake in granule free neutrophils, *Proc. Natl. Acad. Sci. U.S.A.*, 81, 1416, 1984.
106. **Lew, D. P., Monod, A., Waldvogel, F. A., and Pozzan, T.,** Role of cytosolic free calcium and phospholipase C in leukotriene B4 stimulated secretion in human neutrophils: comparison with the chemotactic peptide f-mlp, *Eur. J. Biochem.*, 162, 161, 1987.
107. **Fujita, I., Irita, K., Takeshige, K., and Minakami, S.,** Diacylglycerol, 1-oleoyl-2-acetyl- glycerol stimulates superoxide generation from human neutrophils, *Biochem. Biophys. Res. Commun.*, 120, 318, 1984.
108. **Miyake, R., Tanaka, Y., Tsuda, T., Kaibuchi, K., Kikkawa, U., and Nishisuka, Y.,** Activation of protein kinase C by a non-phorbol tumor promoter mezerein, *Biochem. Biophys. Res. Commun.*, 121, 648, 1984.
109. **Berkow, R. L. and Kraft, A. S.,** Bryostatin a non-phorbol macrocyclic lactone activates intact human polymorphonuclear leukocytes and binds to the phorbol ester receptor, *Biochem. Biophys. Res. Commun.*, 131, 1109, and 1985.
110. **Taffet, S., Greenfield, A. R. L., and Haddox, M. K.,** Retinal inhibits TPA activated calcium-activated phospholipid-dependent protein kinase ("C kinase"), *Biochem. Biophys. Res. Commun.*, 114, 1194, 1983.
111. **O'Brian, C. A., Liskamp, R. M, Solomon, D. H., and Weinstein, I. B.,** Inhibition of protein kinase C by tamoxifen, *Cancer Res.*, 45, 2469, 1985.
112. **Horgan, K., Cooke, E., Hallett, M. B., and Mansel, R. E.,** Inhibition of protein kinase C mediated signal transduction by tamoxifen; importance for antitumour activity, *Biochem. Pharmacol.*, 35, 4463, 1986.
113. **Katoh, N., Wise, B. C., Wrenn, R. W., and Kuo, J. F.,** Inhibition by adriamycin of calmodulin-sensitive and phospholipid-sensitive calcium-dependent phosphorylation of endogenous proteins from heart, *Biochem. J.*, 198, 199, 1981.
114. **Sha'afi, R. I., Molski, T. F. P., Huang, C-K., and Naccache, P. H.,** The inhibition of neutrophil responsiveness caused by phorbol esters is blocked by the protein kinase inhibitor H7, *Biochem. Biophys. Res. Commun.*, 137, 50, 1986.
115. **Gerard, C., McPhail, L. C., Marjal, A., Stimler-Gerard, N. P., Bas, D., and McCall, C. E.,** Role of protein kinase in stimulation of human polymorphonuclear leukocytes by various agents, *J. Clin. Invest.*, 77, 61, 1986.
116. **Kimura, K., Saberada, K., and Katoh, N.,** Inhibition by gossypol of phospholipid-sensitive calcium-dependent protein kinase from pig testis, *Biochim. Biophys. Acta*, 839, 276, 1985.
117. **Su, H-D., Mazzei, G. J., Vogler, W. R., and Kuo, J. F.,** Effect of tamoxifen, a nonsteroidal antiestrogen, on phospholipid/calcium dependent protein kinase and phosphorylation of its endogenous substrate proteins from the rat brain and ovary, *Biochem. Pharmacol.*, 34, 3649, 1985.
118. **Kawamoto, S. and Hidaka, H.,** 1-(5-isoquinolinesulfonyl) 2-methylpiperazine (H7) is a selective inhibitor of protein kinase C in rabbit platelets, *Biochem. Biophys. Res. Commun.*, 125, 258, 1984.
119. **Mazzei, G. J., Katoh, N., and Kuo, J. F.,** Polymixin B is a more selective inhibitor for phospholipid-sensitive calcium-dependent protein kinase than for calmodulin-dependent kinase, *Biochem. Biophys. Res. Commun.*, 109, 1129, 1982.

120. **Hayakawa, T., Suzuki, K., Suzuki, S., Andrews, P. C., and Babior, B. M.,** A possible role for protein phosphorylation in the activation of the respiratory burst in human neutrophils. Evidence from studies with cells from patients with chronic granulomatous disease, *J. Biol. Chem.*, 261, 9109, 1986.
121. **Helfman, D. M., Appelbaum, B. D., Vogler, N. R., and Kuo, J. F.,** Phospholipid-sensitive calcium-dependent protein kinase and its substrates in human neutrophils, *Biochem. Biophys. Res. Commun.*, 111, 847, 1983.
122. **Andrews, P. C. and Babior, B. M.,** Endogenous protein phosphorylation by resting and activated human neutrophils, *Blood,* 61, 333, 1983.
123. **White, J. R., Huang, C. K., Hill, J. M., Naccache, P. H., Becker, E. L., and Sha'afi, R. I.,** Effect of phorbol 12-myristate 13-acetate and its analogue 4-phorbol 12, 13-didecanoate on protein phosphorylation and lysosomal enzyme release in rabbit neutrophils, *J. Biol. Chem.*, 259, 8605, 1984.
124. **Okamura, N., Ohashi, S., Nagahisa, N., and Ishibashi, S.,** Changes in protein phosphorylation in guinea-pig polymorphonuclear leukocytes by treatment with membrane perturbing agents which stimulate superoxide anion production, *Arch. Biochem. Biophys.*, 228, 270, 1984.
125. **Naccache, P. H., Molski, T. F. P., and Sha'afi, R. I.,** Polymixin B inhibits PMA but not chemotactic peptide induced effects in rabbit neutrophils, *FEBS Lett.*, 193, 227, 1985.
126. **Wright, C. D. and Hoffman, M. D.,** The protein kinase inhibitors H7 and H9 fail to inhibit human neutrophil activation, *Biochem. Biophys. Res. Commun.*, 135, 749, 1986.
127. **Kase, H., Iwahashi, K., Nakanishi, S., Matsuda, Y., Yamada, K., Takahasha, M., Murakata, C., Sato, A., and Kanedo, M.,** K-252 compounds, novel and potent inhibitors of protein kinase C and cyclic nucleotide-dependent protein kinases, *Biochem. Biophys. Res. Commun.*, 142, 437, 1987.
128. **Mori, T., Takai, Y., Yu, B., Takahishi, J., Nishizuka, Y., and Fujikaura, T.,** Specificity of the fatty acid moieties of diacylglycerol for the activation of Ca^{2+}-activated phospholipid-dependent protein kinase, *J. Biochem. (Tokyo),* 91, 427, 1982.
129. **Castanga, M., Takai, Y., Kaibuchi, K., Sano, K., Kikkawa, U., and Nishizuka, Y.,** Direct activation of calcium-activated phospholipid-dependent protein kinase by tumor-promoting phorbol esters, *J. Biol. Chem.*, 257, 7847, 1982.
130. **Mufson, R. A., Fischer, S. M., Verma, A. K., Gleason, G. L., Slaga, T. J., and Bontwell, R. K.,** Effects of 12-*O*-tetradeconoyl phorbol-13-acetate and mezerein on epidermal ornithine decarboxylase activity, isoproterenol-stimulated levels of cyclic adenosine 3′5′-monophosphate, and induction of mouse skin tumors *in vivo, Cancer Res.*, 39, 4791, 1979.
131. **Levine, L. and Ohuchi, K.,** Retinoids as well as tumour promoters enhance deacylation of cellular lipids and prostaglandins produced in MDCK cells, *Nature (London),* 276, 274, 1978.
132. **Driedge, P. E. and Blumberg, P. M.,** The effects of phorbol diesters on chicken embryo fibroblasts, *Cancer Res.*, 37, 3257, 1977.
133. **Pontremoli, S., Melloni, E., Salamino, F., Sparatore, B., Michetti, M., Sacco, O., and Horecker, B. L.,** Phosphorylation of proteins in human neutrophils activated with phorbol myristate actetate or with chemotactic factor, *Arch. Biochem. Biophys.*, 250, 23, 1986.
134. **Gennaro, R., Florio, C., and Romeo, D.,** Co-activation of protein kinase C and NADPH oxidase in the plasma membrane of neutrophil cytoplasts, *Biochem. Biophys. Res. Commun.*, 134, 305, 1986.

Chapter 10

ION TRANSPORT IN PHAGOCYTES

Elaine K. Gallin and Leslie C. McKinney

TABLE OF CONTENTS

I. Introduction 244

II. Methods 244
  A. Electrical Methods: the Patch-Clamp Technique 244
  B. Isotopic Methods 246
  C. Optical Methods 246

III. Ionic Basis of Resting Membrane Potential 247
  A. Macrophages 247
  B. Neutrophils 247

IV. Role of Membrane Potential in Phagocyte Function 248

V. Ionic Channels 248
  A. Ionic Channels in Macrophages 249
    1. Inwardly Rectifying K^+ Channel 249
    2. Outwardly Rectifying K^+ Channel 250
    3. Calcium- and Voltage-Activated K^+ Channel 250
    4. Chloride Channel 251
    5. Fc-Receptor Channel 251
  B. Ionic Channels in Neutrophils 252

VI. Pumps and Carriers 253
  A. Transport of Na^+ and K^+ 253
  B. Transport of Cl^- 253
  C. Regulation of Intracellular pH 253
  D. Regulation of Cell Function by Intracellular pH 254
    1. Chemotaxis 255
    2. Oxidative Burst 255

VII. Conclusion 255

Acknowledgments 256

References 256

I. INTRODUCTION

Over the past decade, much information has accumulated about ionic transport processes in phagocytic leukocytes. As in other cells, the permeability of the cell membrane to Na^+, K^+, Ca^{2+}, Cl^-, H^+, and other physiologically important ions is tightly regulated. Specialized membrane proteins (channels, pumps, and carriers) control the movement of ions into and out of the cell, either by allowing ions to passively diffuse down their respective electrochemical gradients or by actively transporting them in an energy-dependent manner. In this way, the cell regulates its intracellular ionic milieu, which in turn influences many other cell functions.

This chapter describes the various types of ion channels, pumps, and carriers found in macrophage and neutrophil membranes (see Table 1 for a listing of the transporters described to date) and discusses the role of ion transport in cell functions. Both neutrophils and macrophages express specific responses to extracellular signals, and it is evident that ionic events at the level of the plasma membrane play a role in some of these processes. Since both neutrophils and macrophages originate from the same progenitor cells in the bone marrow and are similar in many ways, it is likely that they utilize some of the same signal-transduction mechanisms.

II. METHODS

Ion transport can be measured electrically as a net flow of charge across a cell membrane or by measuring the movement of radiolabeled ions into or out of a cell. Additionally, optical techniques can be used to measure intracellular ion concentrations using fluorescent probes. We describe below several of the most widely used techniques in leukocyte physiology.

A. Electrical Methods: the Patch-Clamp Technique

Voltage clamp techniques have been used to monitor current flow, or voltage changes induced by current flow, across whole cell membranes, patches of membrane, and artificial lipid bilayers. The earliest studies on phagocytic leukocytes were done using high-resistance (>30 MΩ) intracellular microelectrodes.[1,2] However, because leukocytes are so small (10 to 15 μm), impaling them with microelectrodes frequently resulted in damage to the cells and production of a significant leak current.[3] This technique has therefore been superseded by the "patch clamp" technique,[4] which utilizes low-resistance (5 MΩ) patch electrodes in several different recording configurations (see Figure 1). A recording is initiated by placing the electrode against the surface of the cell; when gentle suction is applied, the electrode forms a seal with the cell membrane (a so-called "cell-attached patch"), isolating the underlying patch of membrane electrically and mechanically. The seal has a very high resistance ($>10^9 \Omega$), which minimizes leak current and reduces current noise. In this way, small current fluctuations (on the order of 10^{-12} A) can be recorded, which represent the opening and closing of single-ion channels in the patch of membrane underlying the electrode. If further suction is applied in the cell-attached mode, the patch of membrane under the electrode is ruptured, the inside of the cell becomes contiguous with the recording electrode, and currents can be measured across the whole cell membrane. In this "whole-cell" recording mode, the measured current represents the sum of the currents flowing through all open channels, and single-channel events are not usually evident unless the cell has a very low conductance (for definition of conductance, see Section V on Ionic Channels). The whole-cell configuration has the disadvantage (or advantage, depending on the purpose of the study) of perfusing the inside of the cell with the solution contained in the patch electrode. It takes approximately 10 min for a whole-cell recording to stabilize, and conductances can disappear

Table 1
IONIC CHANNELS, PUMPS, AND CARRIERS IN MACROPHAGES AND NEUTROPHILS

Channel/Pump/ Carrier	Cell type	Gating[a]	Blockers	Possible physiological role	Ref.
Outward rectifying K^+	Macrophages	V	D-600, TEA, 4-AP, Cs^+, Ba^{2+}		15, 16
Inward rectifying K^+	Macrophages	V	Ba^{2+}, Cs^+, Rb^+	Setting membrane potential to E_K	16, 39
Ca^{2+}-activated K^+	Macrophages	V, Ca^{2+}	Charybdotoxin	—	19
Cation channel	Neutrophils	Ca^{2+}	—	Chemotaxis/secretion	61
Cl^- channel	Macrophages	V		—	32
Fc-receptor cation channel	Macrophages	L(IgG)	—	Phagocytosis	34
Na^+/K^+ pump	Macrophages Neutrophils		Ouabain	Maintenance of Na^+/K^+ gradients	21 23
Na^+/H^+ exchanger	Neutrophils		Amiloride	Recovery from acidification; chemotaxis	11, 65, 72
Cl^-/HCO_3^- exchanger	Neutrophils		CHC	Recovery from alkalinization	11
Ca^{2+} pump	Neutrophils		—	Maintenance of $[Ca^{2+}]_i$	96

[a] For ion channels: V = voltage gated; L = ligand gated; Ca = gated by $[Ca^{2+}]_i$.

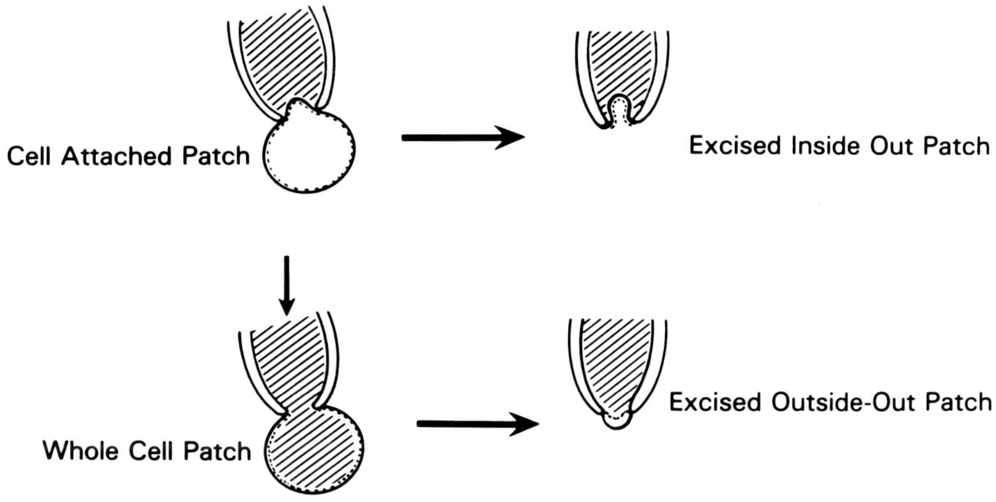

FIGURE 1. Schematic of patch-clamp techniques.

during this period, presumably due to the dialysis of critical intracellular constituents.[5] In some cases, as in the secretory response of the mast cell,[6] intracellular perfusion also blocks cellular responses.

Methods have recently been developed to partially permeabilize the patch of membrane under the electrode while in the cell-attached patch configuration and obtain a "slow patch clamp". Using this technique, ionic exchange between the electrode and the cell takes place, but larger molecules are maintained within the cell. Thus, currents across whole cells can be measured while some of the intracellular constituents necessary for cell functions are preserved.[6]

Because the seal between the electrode and cell membrane is mechanically stable, it is possible to pull the electrode away from the cell along with the underlying membrane, creating an "excised patch". If the electrode is pulled away from the cell in the cell-attached configuration, an "inside-out" patch is created, with the inside of the membrane facing the bathing solution. If it is pulled away while in the whole-cell configuration, an "outside-out" patch is created, with the outside of the membrane facing the bath. Single channels similar to those in the cell-attached patch configuration can be recorded while solutions on either side of the patch membrane are controlled.

B. Isotopic Methods

Electrical measurements of ion flux, while very rapid, measure only net charge transfer and do not yield information about transport systems which are electroneutral. A second way to measure ion flux is to use isotopically labeled ions. Flux techniques for small cells are well established.[7] For influx measurements, suspended cells are layered over oil in a solution containing radioisotope for a given amount of time. For efflux measurements, cells are preexposed to radioactivity and layered over oil in a nonradiolabeled solution. Subsequent centrifugation through the oil layer rapidly separates the cells from the supernatant so that both fractions can be counted. Influx is expressed as the amount of radioactivity taken up per volume cell water or per milligram protein per unit time; efflux is expressed as amount of radioactivity lost by the cell to nonradiolabeled solution. From these measurements, both permeability and conductance values for a specific ion can be calculated if the cell membrane area is known.

Radiolabeled ions can also be used as probes to measure membrane potential and intracellular pH. Triphenylmethylphosphonium ($TPMP^+$) is a cationic, lipophilic molecule that partitions across the membrane according to the Nernst equation. By knowing its intra- and extracellular concentrations, it can be used to estimate resting membrane potential (Vm).[8,9] An accurate estimate of resting Vm depends on the probe reaching equilibrium. This sometimes takes several hours and can be deleterious to the cells. In those cases, tetraphenyl borate (TPB^-), a permeant anion, can be used to increase the rate at which equilibrium of $TPMP^+$ is reached.[9] Control measurements on depolarized cells also must be made so that background radioactivity can be subtracted.

Intracellular pH can be calculated from the steady-state distribution of radiolabeled weak acids or bases, provided that (1) only the uncharged forms of these electrolytes are permeant so that, at steady state, the intracellular and extracellular concentrations of the uncharged forms are the same, and (2) their apparent pKa values in the external and internal fluids are known.[10] Studies performed on phagocytes have used ^{14}C-labeled 5,5-dimethyloxazolidine-2,4-dione (DMO, pKa = 6.1) as a weak acid and trimethylamine (TMA, pKa = 9.6) as a weak base.[10,11] Since equilibration time is related to the concentration of the neutral form of the weak acid or base and can be rather slow, only steady-state values of pH_i or transients, whose time course is slow relative to that of dye equilibration, can be measured. In neutrophils, DMO equilibrates in less than 15 sec and TMA in approximately 2 min.[11] By measuring the change in pH per unit time inside the cell, hydrogen ion flux can be calculated if the buffering power of the cell is known.

C. Optical Methods

A number of different membrane-permeant dyes have been used to optically detect changes in membrane potential or pH in leukocytes. Membrane potential has been assessed using both positively charged (cyanine) and negatively charged (oxonol) dyes.[9] These agents can be calibrated by the addition of valinomycin to cells bathed in solutions of different $[K^+]$, thereby establishing known K^+ diffusion potentials. The dyes partition into intracellular compartments and reflect the average potential throughout the cell. They can be used to measure the average response of a population of cells,[9] or can be used in conjunction with flow microfluorometry to differentiate between subpopulations of cells.[12]

Optical methods for measuring intracellular pH use dyes whose fluorescence is a function of pH. Two fluorescein derivatives commonly used as pH indicators are 6-carboxyfluorescein (CF),[13] and 2′,7′-bis-(2-carboxyethyl)-5(and-6)carboxyfluorescein (BCECF).[14] Both compounds readily permeate as uncharged acetate esters. Once inside the cell, esterases remove the acetate groups: the increase in net negative charge renders the compounds much less permeant. They can be used to measure the average response of a population of cells, or with proper imaging techniques, can be used to measure pH changes in single cells on a time scale of milliseconds.

III. IONIC BASIS OF RESTING MEMBRANE POTENTIAL

A. Macrophages

Values of resting membrane potential of cultured (>24 hr) adherent mouse peritoneal macrophages are near −80 mV,[15] as are those for the mouse-derived macrophage-like cell line J774.1.[16] This value is near the equilibrium potential for potassium (E_k = −85 mV; values for extracellular and intracellular ion concentrations and estimates of the equilibrium potentials for the major ions are given in Table 2). Ion substitution experiments have verified that the resting Vm of these cells is primarily determined by potassium permeability (P_K).[16] In contrast, the resting Vm of freshly isolated human monocytes is approximately −15 mV, and seems to become more negative with time in culture, although this observation has not been systematically verified.[17] Values of resting Vm in mature (5 to 30 days old) macrophages range from −30 to −90 mV.[17-19] During the 1st week in culture, human monocytes are changing morphologically and functionally into mature macrophages, and the shift in resting Vm may be associated with some of these changes. It is not known whether the more negative resting Vm reflects a decrease in leak conductance or the development of a new conductance (for definition of conductance see following section on ionic channels), as has been described in human monocytes.[20]

An additional contribution to the resting potential is made by the electrogenic Na^+/K^+ pump, which adds −7 mV to the resting Vm of mouse spleen macrophages in the steady state.[21] Under nonsteady-state conditions, activation of the Na^+/K^+ pump can produce very large hyperpolarizations. Macrophages that are sodium loaded by placement in cold, K^+-free medium exhibit a −29 mV ouabain-sensitive hyperpolarization after warming and reexposure to K^+.

B. Neutrophils

Unlike the macrophage, no measurements of the resting Vm of neutrophils using microelectrodes have been reported. However, measurements have been made with the radiolabeled probe $TPMP^+$,[9] with ion-sensitive electrodes using tetraphenyl phosphonium (TPP^+),[22] and with the fluorescent dye 3,3′-dipropylthiadicarbocyanine [diS-C_3(5)],[23,24] yielding values of resting Vm for nonadherent neutrophils of −54 mV, −67 mV, and −53 mV, respectively. The neutrophil membrane, like the macrophage, is primarily permeable to K^+, and resting Vm follows $[K^+]_o$ in a manner described by the Nernst equation, except when $[K^+]_o$ is below 10 m*M*.[9,23] At normal $[K^+]_o$ (4.5 m*M*), a small but significant permeability to Na^+ (P_{Na^+}) causes resting Vm to be less negative than that predicted by the Nernst equation for K^+ (−85 mV). Chloride has a low permeability and contributes little to the resting Vm.[23] The relative permeability of the membrane to K^+, Na^+, and Cl^- is 10:1:1.[24] A ouabain-sensitive Na^+/K^+ pump is also present in the neutrophil and is estimated to contribute approximately −9 mV to the steady-state resting potential.[23] In sodium-loaded cells under nonsteady-state conditions, it can contribute as much as −30 mV.

Table 2
IONIC CONCENTRATION GRADIENTS IN NEUTROPHILS AND MACROPHAGES

Ion	Extracellular concentration[a] (meq/liter)	Intracellular concentration		Equilibrium Potential (mV)	
		Neutrophil	Macrophage[b]	Neutrophil	Macrophage[b]
Na^+	136—145	25 m*M* (23)	10 m*M* (29)	46	70
K^+	4.5	120 m*M* (23)	158 m*M* (97, 29)	−95	−85
Cl^-	100—106	80 m*M* (24)	44m*M* (97)	−7	−32
Ca^{2+}	4.3−5.3	~ 100 n*M* (67, 99)	40 n*M* (98)[c] 87 n*M* (59)	150	150
H^+	pH = 7.35—7.45	pH = 7.25 (11, 65, 72)			

[a] Human serum levels
[b] Values from J774 cells.
[c] Value from human alveolar macrophage

IV. ROLE OF MEMBRANE POTENTIAL IN PHAGOCYTE FUNCTION

While the resting Vm of the phagocytic leukocyte, as in other cells, must be important for the transport of various substances into the cell,[25] it is not yet clear whether membrane potential has a role in the higher-order processes of the cell, such as phagocytosis, chemotaxis, and signal transduction. For instance, in the macrophage-like cell line J774.1, Fc-receptor-mediated ingestion of the protozoan parasite *T. gondii* occurs in 120 m*M* K^+ medium, conditions under which the resting Vm is approximately zero.[26] In the neutrophil, cells depolarized by high K^+ can migrate in response to the chemotactic peptide f-Met-Leu-Phe (fMLP),[27] and they also exhibit the same increase in the oxidative burst following stimulation by phorbol ester as do cells in normal (Na^+-containing) medium.[22] In the alveolar macrophage, superoxide anion production induced by the dipeptide f-Met-Phe is only slightly decreased by substituting K^+ for external Na^+.[28] Thus, it seems that a negative resting Vm is not necessary for these processes to occur.

On the other hand, there is some evidence that resting membrane potential is modulated by culture conditions. J774.1 cells grown in suspension have a resting Vm (measured with $TPMP^+$) of −15 mV, while cells that have been adherent for 6 to 8 hr have a resting Vm of −80 mV.[29] This lower V_m is due to the expression of an inward K^+ conductance in these cells following adherence (see section on inwardly rectifying K^+ conductance).

V. IONIC CHANNELS

Ionic channels are integral membrane proteins through which ions passively diffuse at rates exceeding 10^6 ions/sec. Current amplitude is determined by the electrochemical gradient and the conductance of the channel to that particular ion. Conductance is a measure of the ease with which current flows through the channel, and from Ohm's law, is equal to the current divided by the potential across the channel. It is expressed in siemens (S), and its value is the inverse of resistance.

Ion channels are characterized according to the following properties: ionic selectivity (differential permeability to various ions), pharmacology (action of specific agents in blocking or changing the flow of ions), conductance (defined above), gating properties (factors controlling channel opening and closing), and kinetics (rates at which channels open and close). Channel gating can be controlled by voltage, specific chemical ligands, or other factors such as calcium.[30] In addition, recent findings that channel gating can be modulated by phosphorylation provide an important link between biochemical events inside the cell and channel activity.[31]

An important property of some ionic currents is that they flow in one direction more easily than in the other, a property called rectification. Rectification of whole-cell currents can result from rectification at the single-channel level or it can result simply from the voltage dependence of channel gating and the ionic gradient across the membrane. For example, if a K^+ channel is activated (opened) only when the membrane potential of the cell is -40 mV or less negative (more depolarized) and the equilibrium potential for potassium (E_k) is normally -80 mV, then K^+ will only flow out of the cell through this channel. The K^+ current that flows through this type of voltage-dependent channel is referred to as an outwardly rectifying current.

A. Ionic Channels in Macrophages

Recent studies have demonstrated that macrophages exhibit voltage-gated, calcium-gated, and ligand-gated ionic channels. The channels described to date include three different potassium-selective channels[15,16,19] and a large conductance anion-selective channel.[32] One of the K^+ channels is gated by both intracellular ionized calcium $[Ca^{2+}]_i$ and voltage.[19] $[Ca^{2+}]_i$ also modulates the activity of a second type of channel, a nonselective cation channel activated following exposure of macrophages to IgG.[33] This channel differs from the cation-selective channel that results from the binding and cross-linking of ligands to the isolated Fc-receptor.[34]

1. Inwardly Rectifying K^+ Channel

An inwardly rectifying K^+ current that activates at voltages negative to -60 mV was first described in mouse spleen and thioglycolate-induced macrophages that had been cultured for several weeks.[35,36] This current is also present in cultured human macrophages[37] and in the mouse-derived macrophage-like cell line J774.1.[16] It is very similar to the inwardly rectifying K^+ current extensively characterized in both muscle and egg cells,[38] which activates at potentials negative to -50 mV, and has a steep voltage dependence. A unique property of this channel is that its gating depends not only on membrane voltage but also on $[K^+]_o$.[16,38] That is, increasing $[K^+]_o$ shifts the voltage range over which the current is activated to more depolarized potentials. In addition, the whole-cell conductance is proportional to the square root of $[K^+]_o$. Addition of the cations barium, cesium, or rubidium to the bathing medium blocks the current and results in a depolarization of 20 mV or more.[36] Single-channel current fluctuations, whose properties correspond to the macroscopic K^+ currents measured in whole cells, have been described recently in both J774.1 cells[39] and cultured human macrophages.[37] Single-channel conductance is 29 pS in 145 m*M* K, and is proportional to the square root of $[K^+]_o$. Similar inwardly rectifying K^+ channels are present in heart muscle[40] and rat basophilic leukemic cells.[41]

The probable function of the inwardly rectifying channel is to set the resting Vm of the macrophage near the K^+ equilibrium potential and to make the cell sensitive to depolarizing stimuli. Cells exhibiting this conductance can become depolarized by small amounts of inward current, and can exhibit two stable states of resting membrane potential.[35,41] In addition, the dependence of channel gating and conductance on $[K^+]_o$ may be significant since the macrophage is present in sites of dead and dying tissue, and may well be exposed to elevated $[K^+]_o$. A rise in $[K^+]_o$ will increase the conductance, yet preserve a steep current-voltage relationship so that the cells remain sensitive to small current fluctuations.

The time course of expression of this conductance has been followed in J774.1 cells, where it has been shown that, after adherence, several hours are required for the channel to be fully expressed.[16] This correlates well with the finding that the resting Vm of J774.1 cells increases from -14 to -70 mV after plating,[29] and further supports the view that this channel helps to set the resting Vm of the macrophage. In mouse peritoneal macrophages the channel does not appear during the first 4 days in culture[15] and in human peripheral blood monocytes, it may not appear for several weeks after plating.[100] Why this channel is so variably expressed remains unknown.

2. *Outwardly Rectifying K^+ Channel*

Macrophages exhibit an outwardly rectifying K^+ conductance[15] similar to that described in detail in T-lymphocytes.[42] This current activates at potentials positive to −50 mV, inactivates over a time course of milliseconds, and is blocked by 4-aminopyridine. As is the case with the inwardly rectifying channel, it appears to be variably expressed with time in culture. Ypey and Clapham,[15] using resident mouse peritoneal macrophages, reported that the channel was absent during the 1st day following isolation but was present in 96% of the cells cultured for 1 to 4 days. The channel is not found in long-term (2 to 6 weeks) cultured mouse peritoneal[35] and spleen macrophages,[36] perhaps because it is only transiently expressed. In J744.1 cells, the channel is present 1 to 8 hr after adherence, and is not present in long-term adherent cultures where the inwardly rectifying K^+ channel is the predominant channel.[16] The channel is continuously present in the mouse-derived cell line P388D1.[43]

The functional relevance of the outwardly rectifying K^+ channel in the macrophage has not yet been determined. In T-lymphocytes and natural killer cells, agents which block this channel inhibit mitogenesis[44] and cytotoxicity,[45] respectively, implying that this channel plays at least a permissive role in these processes.

3. *Calcium- and Voltage-Activated K^+ Channel*

In 1975, a study using high-resistance intracellular microelectrodes described spontaneous rhythmic hyperpolarizations in human macrophages grown in tissue culture for 2 to 5 weeks, which were accompanied by an increase in permeability to K^+, were blocked by addition of a calcium chelator, and could be induced by the calcium ionophore A23187.[1] It was concluded that the hyperpolarizations were due to the activation of a Ca^{2+}-activated K^+ conductance. Persechini et al.[46] provided further support for the existence of a Ca^{2+}-activated K^+ channel by demonstrating that injection of Ca^{2+} into mouse macrophages could trigger hyperpolarization. Recently, excised patch-clamp experiments in human macrophages have demonstrated a large-conductance K^+ channel (240 pS in symmetrical K^+) activated by both voltage and $[Ca^{2+}]_i$.[19] The channel is blocked by charybdotoxin, a proteinaceous component of scorpion toxin known to block Ca^{2+}-activated K^+ channels in skeletal muscle.[47]

Similar large conductance Ca^{2+}- and voltage-gated channels have been described in a variety of cell types.[48,49] The sensitivity of the channel to $[Ca^{2+}]_i$ varies in different tissues. In pancreatic acinar cells, a change in $[Ca^{2+}]_i$ from 10^{-8} to 10^{-7} *M* greatly increases the probability of channel opening, even at negative voltages.[49] In muscle[48] and the macrophage,[19] 10^{-6} *M* $[Ca^{2+}]_i$ is needed for significant activation. Therefore, it is likely that only a very small fraction (if any) of these channels are open during stimulation.

The physiological relevance of the Ca^{2+}-activated K^+ channel in the macrophage is uncertain. Obviously, it provides a link between membrane potential and internal Ca^{2+} concentration, repolarizing the cell to the K^+ equilibrium potential during instances when $[Ca^{2+}]_i$ rises sufficiently to open the channels. These channels may also play a role in volume regulation or modulating $[K^+]_i$. One model for exocytosis proposes that the activation of Ca^{2+}-activated K^+ channels, present in secretory granule membranes, leads to granule swelling, thereby inducing exocytosis.[50] There is evidence that changes in intracellular K^+ can influence synthetic processes[51] as well as receptor-mediated endocytosis.[52] The contractile machinery of the macrophage may be particularly influenced by changes in K^+, since it contains an actin-modulating protein, acumentin, whose activity is modified by changes in K^+ concentration in the physiological range (100 to 200 m*M*).[53] It has been shown that chemotactic factors can induce membrane hyperpolarization in human macrophages that probably is produced by the activation of Ca^{2+}-activated K^+ channels. However, freshly isolated human peripheral blood monocytes (which lack Ca^{2+}- and voltage-activated K^+ channels) are both chemotactic and phagocytic, which implies that these channels are not essential for those processes.

As is the case for other K^+ conductances present in macrophages, the expression of the Ca^{2+}-activated K^+ channel is time dependent. Freshly isolated human peripheral blood monocytes do not exhibit Ca^{2+}-activated K^+ channels, but macrophages cultured for 2 to 4 days do.[20] After 7 days in culture, 90% of patches from human macrophages contain the channel. During this time, the monocyte matures into a macrophage and undergoes morphological and functional changes that may be related in some way to the appearance of this conductance.

4. *Chloride Channel*

Schwarze and Kolb[32] have described a channel present in excised patches from mouse peritoneal macrophages which has a very large conductance (340 pS), several subconductance states, and is poorly selective for chloride (the selectivity ratio, P_{Cl}:P_{Na}, is 5:1). It was seen in cell-attached patches only after treatment with the Ca^{2+} ionophore A23187; in quiescent patches, excision of the membrane activated the channel. The channel is activated by both depolarizing and hyperpolarizing voltage steps, exhibits voltage dependent inactivation, and behaves as if it were controlled by two independent voltage-sensitive gates. Similar models have been used to describe the behavior of voltage-dependent gap junctions, and it was suggested that the chloride channel may play a role in intracellular communication.[32] Recently, it has been reported that these channels can be elicited by perfusing macrophages with zymosan during cell-attached patch recordings.[54] An anion channel with a similar large conductance and a poor Cl selectivity also has been described in B-lymphocytes.[55]

5. *Fc-Receptor Channel*

Ligand binding to the macrophage Fc-receptor (FcR) and its subsequent cross-linking initiates the ingestion of particles and the secretion of a number of mediators.[56] One of the first indications that ligand binding to the FcR leads to activation of ion channels was obtained by Young et al.[8] who showed that a depolarization of 7 mV occurred in J774.1 cells following the binding and cross-linking of the γ2b/γ1 FcR by IgG or immune complexes. These studies, done by monitoring membrane potential indirectly with $TPMP^+$, showed that depolarization required a multivalent ligand and was inhibited by substituting choline for sodium. Subsequently, purified γ2b/γ1 FcRs were inserted into lipid vesicles and lipid bilayers. It was shown that ligand binding to proteoliposomes containing purified FcR increased their cation permeability[57] and later, that the addition of ligand to bilayers containing the FcR could induce cation-selective ion channels.[34] The channels had a conductance of 60 pS in symmetrical 1 *M* KCl and decreased in activity within several minutes of adding ligand.

Two studies have been done which directly monitored channel activity in intact macrophages or macrophage membranes before and after the addition of IgG. Nelson et al.[17] recorded whole-cell currents as well as single channels in human alveolar macrophages exposed to heat-aggregated IgG (aIgG). The application of aIgG to cells during whole-cell recording produced an inward current that diminished with successive applications of aIgG, indicating that the response desensitized. In cell-attached patches, channel activity was noted only when the electrode contained aIgG and not when aIgG was applied to the bath. The channels had a unitary conductance of 350 pS in symmetrical 140 m*M* NaCl Hanks solution. Changing the permeant cation from Na^+ to K^+ did not affect the reversal potential (that potential at which there is no net current flow), indicating that if the channel is a cation channel, it is nonselective. Unfortunately, anion permeability was not examined. The difference between the value of conductance obtained in this study (350 pS in physiological saline) and that obtained by Young et al.[34] on the isolated FcR (60 pS in symmetrical 1 *M* KCl) is puzzling, and needs to be investigated further. It may reflect the difference in the ionic conditions, and /or the ligands in the two sets of experiments, or a change in receptor behavior following isolation.

There is evidence that the ligand-FcR complex, in addition to acting as an ionic channel itself, may activate a second messenger that indirectly activates ionic channels.[33] In one study, IgG2b was added to the bath during a cell-attached recording from P388D1 cells. Following the addition of IgG2b, multiple current amplitudes were evident, representing either several different types of channels or a single-channel type with different subconductance states; the smallest channels had conductances of 35 to 45 pS and were cation selective. Channel activity could be maintained following excision of the patch, and activity was modulated by changes in $[Ca^{2+}]_i$. Since the ligand was added to the bath, not to the patch electrode, it must have been acting indirectly through a second messenger. Thus, it is unlikely that these channels are the same as those described by Nelson et al. which were activated when aIgG was added to the patch electrode.[17]

Calcium-sensitive (and voltage-insensitive) cation channels have been reported in a number of other cells.[58] Lipton[33] suggested that these channels are activated by increases in intracellular Ca^{2+} levels that may occur following binding and cross-linking of the FcR. However, conflicting data exist regarding the level of $[Ca^{2+}]_i$ during Fc-mediated phagocytosis. Young et al.,[59] using the Ca^{2+} indicator quin-2, demonstrated a transient rise in $[Ca^{2+}]_i$ (from 87 to 400 n*M*) following the addition of 2.4G2 IgG to J774.1 cells. In contrast, a recent study using aequorin in mouse peritoneal macrophages reports no measurable changes in $[Ca^{2+}]_i$ during phagocytosis.[60] It was concluded that either calcium is not involved as a cytoplasmic signal in phagocytosis or changes in calcium concentration occur locally over small areas of the cell and are difficult to detect. It is not known how ion fluxes through ligand-FcR channels or other channels actually mediate phagocytosis and its related events. The ion fluxes and/or the potential changes resulting from FcR-ligand interactions presumably are linked to one or more second messengers which connect the membrane events to ensuing intracellular events.

B. Ionic Channels in Neutrophils

With the exception of a recent report by von Tscharner et al.[61] on the action of the chemotactic factor fMLP on channel activity in neutrophils, there have been no electrophysiological studies of the neutrophil. However, measurements of membrane potential (assessed using indirect probes) and ion fluxes have been made in resting and stimulated neutrophils and indicate that ions may be involved in signal transduction.[62] Following the addition of fMLP, neutrophils transiently depolarize and then repolarize.[63] Unfortunately, the ionic basis of the transient depolarization has not been adequately resolved. Although both Ca^{2+}[64] and Na^{+}[7] influx occur after stimulation with fMLP, fMLP-induced depolarization occurs in the absence of either.[63] The fMLP-induced Na^{+} influx is blocked by amiloride[65,66] and can be accounted for almost entirely by activation of Na^{+}/H^{+} exchange. Ca^{2+} influx is unaffected by amiloride indicating that different permeation pathways are involved for these two cations.[65] Along with the fMLP-induced Ca^{2+} influx, there is a rise in $[Ca^{2+}]_i$ as assessed with quin 2 and fura 2[67] (also see Chapter 8).

The study by von Tscharner et al.[61] has provided some new insight into the ionic events occurring during fMLP stimulation. In these studies, the addition of fMLP to the bath during cell-attached patch recordings increased channel activity in 13 of 24 patches. However, the presence of fMLP in the patch pipette did not increase the probability of channel openings, indicating that the channels were not directly coupled to the fMLP receptor. The channels appeared to be calcium activated since depleting $[Ca^{2+}]_i$ by loading cells with fura-2 prevented fMLP-induced channel activation. Treating cells with saponin, to increase $[Ca^{2+}]_i$ activated channels in the absence of fMLP. Two types of single-channel currents were identified which had conductances of 18 to 25 and 4 to 6 pS, respectively. Ion substitution experiments indicated that the channels were equally permeable to K^+, Na^+, and Ca^{2+}, but were impermeable to Cl^-. Activation of the channels could account for the influx of $[Ca^{2+}]_i$ that occurs following the addition of fMLP.[64] The transient depolarization that has been measured in neutrophils in suspension can be explained by the influx of Na^+ and Ca^{2+} through these channels, but it is difficult to reconcile this with data indicating that the Na^+ flux is amiloride sensitive and that depolarization occurs in the absence of Ca^{2+} and Na^+, as mentioned above. To help resolve these questions, future studies are needed to determine if potential measurements made with indirect probes have been accurate, and what role these and other yet unknown ion channels may play in signal transduction.

VI. PUMPS AND CARRIERS

Besides ion channels, the other important family of proteins involved in ion transport are energy-dependent pumps and carriers. Comparatively little is known about the mechanisms of active ion transport in phagocytes; since more information exists for neutrophils than macrophages, our discussion will necessarily focus on this cell type. Calcium transport is the subject of an entire chapter elsewhere in this book, and is not discussed here.

A. Transport of Na^+ and K^+

As has been previously discussed, both the neutrophil and the macrophage have a ouabain-sensitive Na^+/K^+ pump that is electrogenic and contributes approximately -9 and -7 mV to their respective resting potentials.[21,23] While the stoichiometry of exchange has not been measured directly, these measurements are consistent with calculations showing that, assuming a coupling ratio of 3 Na^+ :2 K^+, the maximum value that can be contributed by the Na^+/K^+ pump under steady-state conditions is 10 mV.[68,69] In the neutrophil, 95% of Na^+ efflux and 63% of K^+ influx are mediated by the pump. The balance of K^+ influx is passive, and follows Ussing's independence principle.[23] The K_m for potassium activation of the pump is 1.5 m*M*. None of the Na^+ influx is not ouabain sensitive, which indicates that the pump is not mediating Na^+/Na^+ exchange when the cells are at rest.

B. Transport of Cl^-

Unlike some cells, in which intracellular chloride passively follows membrane potential, the neutrophil actively maintains $[Cl]_i$ at a value fourfold higher than equilibrium. There are three main pathways for chloride transport in the neutrophil: active uptake, passive diffusion, and Cl^-/Cl^- self-exchange.[24] Active uptake of chloride accounts for 20% of total chloride influx and is mediated by a carrier that can be inhibited by α-cyano-4-hydroxycinnamate (CHC, K_i = 1.7 m*M*), ethacrynate (K_i approximately = 50 μ*M*), and furosemide (K_i approximately = 50 μ*M*). The K_m for chloride is 5 m*M*. About 8% of chloride influx and 30% of chloride efflux is via passive diffusion, through a (putative) nonvoltage-sensitive anion channel. The bulk (70%) of chloride influx and efflux is due to Cl^-/Cl^- self-exchange, via a carrier that may mediate Cl^-/HCO_3^- exchange under other conditions.[11,70] Michaelis constants of the anion exchanger for extracellular Cl^- and the inhibitor CHC are 5 and 0.3 m*M*, respectively; the carrier has no apparent affinity for glucuronate.

C. Regulation of Intracellular pH

The neutrophil, like almost all other animal cells, maintains an intracellular pH considerably more alkaline than one would expect if H^+ were distributed at equilibrium. For the neutrophil, since resting Vm is approximately equal to -60 mV and $pH_O = 7.4$, then from the Nernst equation, at equilibrium, pH_i should be equal to about 6.4 (calculated for 37°C). Measured values of pH_i are near 7.2 (in nominally CO_2-free medium),[11,71,72] which means that the driving force for hydrogen is inward. Thus, to maintain the cell in a steady state with respect to pH, there must be some mechanism for the active extrusion of acid equivalents. Work on a variety of cell types[73-79] has shown that there are two kinds of secondary active transport that regulate pH_i by extruding hydrogen ion equivalents from the cell: Na^+/H^+ exchange and Na-dependent Cl/HCO_3^- exchange. (For a complete description of these and other mechanisms of acid extrusion, see References 10,80,81).

The mechanism of pH_i regulation is studied by imposing intracellular acid or alkaline loads and determining the ion dependence of recovery. There are many ways to manipulate intracellular pH,[10] but most commonly, acidification is produced by exposing the cell to CO_2 or by removal of an NH_4Cl prepulse; conversely, alkalinization is produced by exposing the cell to NH_4Cl or by removal of a CO_2 prepulse. It is important to note that isolation procedures that involve transient exposure of phagocytes to NH_4Cl produce intracellular acidification that can affect subsequent assays. For example, superoxide generation, lysosomal secretion, and phagocytosis have all been shown to artifactually depend on external sodium when phagocytes are pretreated with NH_4Cl. The apparent dependence on Na is due to the fact that cells become acidified during isolation and cannot recover their intracellular pH in the absence of Na.

In the neutrophil, Na^+/H^+ exchange is the primary mechanism used to extrude acid in response to an intracellular acid load.[82] The exchanger transports external sodium for intracellular hydrogen ions with a 1:1 stoichiometry. It has a K_m for external Na^+ of 26 to 35 m*M*, and can transport up to 28 meq of Na^+ per liter of cell water per minute. It is inhibited by amiloride with an apparent K_i of 75 μ*M* in 140 m*M* Na^+-containing medium. In resting neutrophils, nearly all Na^+ influx is passive, and the Na^+/H^+ exchanger appears to be quiescent.[65,66,72] Its activity is increased by an intracellular acid load, and it extrudes acid at a rate that is proportional to intracellular pH.[11,72] The rate of recovery from acidification is reduced 90% by amiloride, which indicates that Na^+/H^+ exchange is primarily responsible for recovery from an acid load in this cell type.[11]

In addition, the neutrophil has a specific mechanism for responding to an intracellular alkaline load, namely, a Cl^-/HCO_3^- exchanger which removes internal HCO_3^- in exchange for external Cl^-. This exchanger is insensitive to the disulfonic stilbene inhibitor, SITS, but is inhibited by α-cyano-4-hydroxycinnamic acid (CHC). The rate of recovery from an alkaline load is proportional to pH_i and is reduced 80%-90% by CHC or by removal of external chloride, indicating that Cl^-/HCO_3^- exchange is the primary mechanism for recovery from alkalinization in the neutrophil.[11] The neutrophil is rather unique[83] in having a specific mechanism for recovering from an alkaline load. Most other tissues appear to passively recover from alkalinization via passive redistribution of acid/base equivalents, or the naturally occurring acid-producing processes in the cell.[10]

It should be noted that while the energy source for pH_i regulation appears to be the sodium and chloride gradients, they do not necessarily determine the final steady-state level of intracellular pH. Calculations show that, if H^+ were in equilibrium with respect to the Na^+ and Cl^- gradients, then intracellular pH would be 7.6, a more alkaline value than is actually observed.[11] To explain this finding, it has been proposed that there is an intracellular regulatory site that is sensitive to pH_i, which sets the steady-state value of intracellular pH.[84-86]

D. Regulation of Cell Function by Intracellular pH

In phagocytes, processes that are known to be modulated by intracellular pH are chemotaxis[65,87] and superoxide production.[88] Degranulation has been shown to be sensitive to intralysosomal pH in neutrophils.[89] The function of many different cell proteins are sensitive to pH,[90-93] so it is not surprising that changes in intracellular pH can affect higher-order cell functions in a nonspecific way. However, it remains an open question as to whether intracellular pH functions as a second messenger, directly responsible for transducing an exogenous signal into an intracellular one. For the discussion below, it is useful to keep in mind three criteria for determining whether a particular process is pH dependent:[94,95] (1) natural stimuli for activation should result in a change in pH_i; (2) artificially induced pH_i changes should elicit the physiological response; and (3) inhibition of the physiological pH_i change must inhibit the response it is thought to regulate.

1. Chemotaxis

In the resting human neutrophil, pH_i is approximately 7.25 and the Na^+/H^+ exchanger is inactive. When stimulated with a 100 n*M* concentration of the chemotactic factor fMLP, Na^+/H^+ exchange becomes prominent, and there occurs a pronounced intracellular alkalinization (pH_i increases from 7.20 to 7.75 in 10 min). There is a 30-fold increase in Na^+ influx which leads to an increase in internal $[Na^+]$ from a resting value of 30 meq/ℓ of cell water to 60 meq/ℓ[66] Amiloride and its analogues which block Na^+/H^+ exchange inhibit chemotaxis and chemokinesis with similar efficacy. There is a positive correlation between the amount of intracellular alkalinization produced by fMLP and the distance moved by a neutrophil in response to a chemotactic gradient; cells that have a more alkaline pH_i migrate farther in response to fMLP.[87] By the criteria listed above, these experiments imply that intracellular pH regulates chemotaxis; however, since it has not been shown that non-fMLP-mediated changes in intracellular pH can induce chemotaxis, it cannot be concluded that a change in pH_i is the sole trigger for chemotaxis.

2. Oxidative Burst

Work by Simchowitz has contributed much to the understanding of the role of intracellular pH in regulating the oxidative burst of the neutrophil.[88] When neutrophils are stimulated with 100 n*M* fMLP, alkalinization and release of superoxide (O_2^-) are stimulated within 5 min. The amount of O_2^- release correlates directly with pH_i, and is enhanced by intracellular alkalinization. At low (4 n*M*) concentrations of fMLP, where there is little or no superoxide release, alkalinization can have a booster effect. However, when alkalinization is blocked by amiloride, O_2^- release is still approximately half of normal, indicating that a change in pH_i is not obligatory for superoxide release to proceed. In the absence of the fMLP stimulus, if one induces alkalinization by exposure to NH_4Cl, CO_2, or monensin, there is no O_2^- release. Thus, as is the case for chemotaxis, intracellular alkalinization can clearly be shown to modulate superoxide production, but by itself is not a sufficient trigger for O_2^- release.

Intracellular pH changes that are related to the activation of the neutrophil are also produced by the phorbol ester 12-*O*-tetra-decanoylphorbol-13-acetate (TPA). Grinstein and Furuya[72] have shown that, upon application of TPA, there is an initial cytoplasmic acidification followed by a compensatory alkalinization due to Na^+/H^+ countertransport. The production of H^+ results from stimulation of the metabolic pathways involved in the production of superoxide.

VII. CONCLUSION

The purpose of this review was to present an overview of the major ion transport processes described to date in the macrophage and neutrophil. In the macrophage, electrophysiological techniques have been used successfully to identify a number of voltage and ligand-gated ion channels. Such techniques are only now being applied to the neutrophil, so corresponding information about ion channels in the neutrophil is limited. In contrast, ion flux studies on

47. **Miller, C., Moczydlowski, E., Latorre, R., and Phillips, M.,** Charybdotoxin, a protein inhibitor of single Ca^{++}- activated K^+ channels from mammalian skeletal muscle, *Nature (London),* 313, 316, 1985.
48. **Barrett, J. N., Magleby, K. L., and Pallota, B. S.,** Properties of single calcium-activated potassium channels in cultured rat muscle, *J. Physiol.,* 331, 211, 1982.
49. **Petersen, O. H. and Maruyama, Y.,** Calcium-activated potassium channels and their role in secretion, *Nature (London),* 307, 693, 1984.
50. **Stanley, E. F. and Ehrenstein, G.,** A model for exocytosis based on the opening of calcium-activated potassium channels in vesicles, *Life Sci.,* 37, 1985.
51. **Shinohara, T. and Piatigorsky, J.,** Regulation of protein synthesis, intracellular electrolytes and cataract formation *in vitro, Nature (London),* 270, 406, 1977.
52. **Larkin, J., Brown, M., Goldstein, J., and Anderson, R.,** Depletion of intracellular potassium arrests coated pit formation and receptor-mediated endocytosis in fibroblasts, *Cell,* 33, 273, 1983.
53. **Southwick, F., Tatsumi, N., and Stossel, T.,** Acumentin, an actin-modulating protein of rabbit pulmonary macrophages, *Biochemistry,* 21, 6321, 1982.
54. **Somogyi, R., Ubl, J., and Kolb, H. A.,** Zymosan and potassium induced activation of single voltage-dependent ion channels in membranes of peritoneal macrophages, in *Advances in Biosciences,* Vol. 66, Riceviti, G., Ed., Pergamon Press, 1987, 275.
55. **Bosma, M.,** Chloride channels in neoplastic B lymphocytes, *Biophys. J.,* 49, 413a, 1986.
56. **Unkeless, J. C., Fleit, H., and Mellman, I. S.,** Structural aspects and heterogeneity of immunoglobulin Fc receptors, *Adv. Immunol.,* 31, 247, 1981.
57. **Young, D. J., Unkeless, J. C., Kaback, H. R., and Cohn, Z. A.,** Mouse macrophage Fc receptor for IgG γ2b/γ1 in artificial and plasma membrane vesicles functions as a ligand-dependent ionophore, *Proc. Natl. Acad. Sci. U.S.A.,* 80, 1636, 1983.
58. **Yellen, G.,** Single Ca^+-activated nonselective cation channels in neuroblastoma, *Nature (London),* 296, 357, 1982.
59. **Young, D. J., Ko, S., and Cohn, Z.,** The increase in intracellular free calcium associated with IgG γ2b/γ Fc receptor-ligand interactions: role of phagocytosis, *Proc. Natl. Acad. Sci. U.S.A.,* 81, 5430, 1984.
60. **McNeil, P. L., Swanson, J. A., Wright, S. D., Silverstein, S. C., and Taylor, D. L.,** Fc-receptor-mediated phagocytosis occurs in macrophages without an increase in average [Ca^{++}], *J. Cell Biol.,* 102, 1586, 1986.
61. **von Tscharner, V., Prod'hom, B., Baggiolini, M., and Reuter, H.,** Ion channels in human neutrophils are activated by a rise in the free cytosolic calcium concentration, *Nature (London),* 324, 369, 1986.
62. **Zigmond, S.,** Chemotaxis by polymorphonuclear leukocytes, *J. Cell Biol.,* 77, 269, 1978.
63. **Seligmann, B. E. and Gallin, J. I.,** Use of lipophilic probes of membrane potential to assess human neutrophil activation, *J. Clin. Invest.,* 66, 493, 1980.
64. **Boucek, M. and Snyderman, R.,** Calcium influx requirement for human neutrophil chemotaxis: inhibition by lanthanum chloride, *Science,* 193, 905, 1976.
65. **Sha'afi, R. I., Molski, T. F. P., and Naccache, P. H.,** Chemotactic factors activate differentiable permeation pathways for sodium and calcium in rabbit neutrophils. Effect of amiloride, *Biochem. Biophys. Res. Commun.,* 99, 1271, 1981.
66. **Simchowitz, L.,** Chemotactic factor-induced activation of Na^+/H^+ exchange in human neutrophils. I. Sodium fluxes, *J. Biol. Chem.,* 260, 13,237, 1985.
67. **Lew, P. D., Wollheim, C. B., Waldvogel, F. A., and Pozzan, T.,** Modulation of cytosolic-free calcium transients by changes in intracellular calcium-buffering capacity: correlation with exocytosis and O_2^- production in human neutrophils, *J. Cell Biol.,* 99, 1212, 1984.
68. **Mullins, L. J. and Noda, K.,** The influence of sodium-free solutions on the membrane potential of frog muscle fibers, *J. Gen. Physiol.,* 47, 117, 1963.
69. **De Weer, P.,** Aspects of the recovery process in nerve, in *Medical and Technical Publishing Co., Ltd., International Review of Science,* Vol. 3, Hunt, C. C., Ed., University Park Press, Baltimore, Md., 1975, 231.
70. **Simchowitz, L., Ratzlaff, R., and De Weer, P.,** Anion/anion exchange in human neutrophils, *J. Gen. Physiol.,* 88, 195, 1986.
71. **Molski, T. F. P., Naccache, D. H., Volpi, M., Wolpert, L. M., and Sha'afi, R. I.,** Specific modulation of the intracellular pH of rabbit neutrophils by chemotactic factors, *Biochem. Biophys. Res. Commun.,* 94, 508, 1980.
72. **Grinstein, S. and Furuya, W.,** Cytoplasmic pH regulation in phorbol ester-activated human neutrophils, *Am. J. Physiol.,* 251, C55, 1986.
73. **Abercrombie, R. F., Putnam, B., and Roos, A.,** The intracellular pH of frog skeletal muscle: its regulation in isotonic solution, *J. Physiol.,* 345, 175, 1983.
74. **Grinstein, S., Cohen, S., and Rothstein, A.,** Cytoplasmic pH regulation in thymic lymphocytes by an amiloride-sensitive Na^+/H^+ antiport, *J. Gen. Physiol.,* 83, 341, 1984.

75. **Schlue, W. R. and Thomas, R. C.,** A dual mechanism for intracellular pH regulation by leech neurones, *J. Physiol.*, 364, 327, 1985.
76. **Galler, S. and Moser, H.,** The ionic mechanism of intracellular pH regulation in crayfish muscle fibres, *J. Physiol.*, 374, 137, 1986.
77. **Moody, W. J.,** The ionic mechanism of intracellular pH regulation in crayfish neurones, *J. Physiol.*, 316, 293, 1981.
78. **Boron, W. F.,** Intracellular pH-regulating mechanism of the squid axon, Relation between the external Na^+ and HCO_3^- dependences, *J. Gen. Physiol.*, 85, 325, 1985.
79. **Boron, W. F., McCormick, W. C., and Roos, A.,** pH regulation in barnacle muscle fibers: dependence on extracellular sodium and bicarbonate, *Am. J. Physiol.*, 240, C80, 1981.
80. **Aronson, P. S.,** Kinetic properties of the plasma membrane Na^+-H^+ exchanger, *Annu. Rev. Physiol.*, 47, 545, 1985.
81. **Mellman, I., Fuchs, R., and Helenius, A.,** Acidification of the endocytic and exocytic pathways, *Annu. Rev. Biochem.*, 55, 663, 1986.
82. **Simchowitz, L.,** Chemotactic factor-induced activation of Na^+/H^+ exchange in human neutrophils. II. Intracellular pH changes, *J. Biol. Chem.*, 260, 13248, 1985.
83. **Vaughn-Jones, R. D.,** Chloride-bicarbonate exchange in the sheep cardiac purkinje fibre, in *Intracellular pH: Its Measurement, Regulation and Utilization in Cellular Functions,* Alan R. Liss, New York, 1982, 239.
84. **Aronson, P. S., Suhn, M. A., and Nee, J.,** Modifier role of internal H^+ in activating the Na^+/H^+ exchanger in renal microvillus membrane vesicles, *Nature (London),* 299, 161, 1982.
85. **Grinstein, S., Cohen, S., Goetz, J. D., and Rothstein, A.,** Na^+/H^+ exchange in volume regulation and cytoplasmic pH homeostasis in lymphocytes, *Fed. Proc.*, 44, 2508, 1985.
86. **Grinstein, S. and Rothstein, A.,** Mechanism of regulation of the Na^+/H^+ exchanger, *J. Membr. Biol.*, 90, 1, 1986.
87. **Simchowitz, L. and Cragoe, E. J., Jr.,** Regulation of human neutrophil chemotaxis by intracellular pH, *J. Biol. Chem.*, 261, 6492, 1986.
88. **Simchowitz, L.,** Intracellular pH modulates the generation of superoxide radicals by human neutrophils, *J. Clin. Invest.*, 76, 1079, 1985.
89. **Klempner, M. S. and Styrt, B.,** Alkalinizing the intralysosomal pH inhibits degranulation of human neutrophils, *J. Clin. Invest.*, 72, 1793, 1983.
90. **Bormann, B. J., Huang, C.-K., Mackin, W. M., and Becker, E. L.,** Receptor-mediated activation of a phospholipase A2 in rabbit neutrophil plasma membrane, *Proc. Natl. Acad. Sci. U.S.A.*, 81, 767, 1984.
91. **Breitweiser, G. E. and Russell, J. M.,** Internal sites are responsible for pH sensitivity of sodium pump fluxes, *Biophys. J.*, 45, 76a, 1984.
92. **Brodwick, M. S. and Eaton, D. C.,** Sodium channel inactivation in squid axon is removed by high internal pH or tyrosine specific reagents, *Science,* 200, 1494, 1978.
93. **Tauber, A. I. and Goetzl, E. J.,** Structural and catalytic properties of the solubilized superoxide generating activity of human polymorphonuclear lymphocytes. Solubilization, stabilization in solution, and partial characterization, *Biochemistry,* 18, 5576, 1979.
94. **Steinhardt, R. A.,** Ionic logic in activation of the cell cycle, in *Ions, Proliferation and Cancer,* Academic Press, New York, 1982.
95. **Busa, W. B. and Nuccitelli, R.,** Metabolic regulation via intracellular pH, *Am. J. Physiol.*, 246, R409, 1984.
96. **Mottola, C., Polzani, L., and Romeo, D.,** The peripheral Ca^{++} pump activity of macrophages and neutrophils, *Cell Calcium,* 1, 371, 1980.
97. **Melmed, R., Karanian, P., and Berlin, R.,** Control of cell volume in the J774 macrophage by microtubule disassembly and cyclic AMP, *J. Cell Biol.*, 90, 761, 1981.
98. **Stickle, D. F., Daniele, R. P., and Holian, A.,** Cytosolic calcium, calcium fluxes, and regulation of alveolar macrophage superoxide anion production, *J. Cell. Physiol.*, 121, 458, 1984.
99. **von Tscharner, V., Deranleau, D. A., and Baggiolini, M.,** Calcium fluxes and calcium buffering in human neutrophils, *J. Biol. Chem.*, 261, 10163, 1986.

INDEX

A

A23187, 118, 121, 124, 179, 220
Actin, 143—145, 148
Actin filament
  behavior of, 150, 151
    in cellular movement, 157—159
    in chemotaxis, 157—159
    in phagocytosis, 152—156
  composition of, 150
  cortical, 143—147
  gels of, 150
  meshwork of, 143—147
  phagocytosis and, 152—156
Actin filament-binding proteins, 146, see also specific types
γ-Actinin, 148
Activated neutrophils, 204—206
Activation of neutrophils, 5—11
Adenosine monophosphate, 130, 180, 181, 234
Adenosine triphosphate (ATP), 182
Adenylate cyclase, 181
Aggregation, 125—126
Agonists, 221—222, see also specific types
AMP, see Adenosine monophosphate
Analog-digital converter, 89, 92
Angular distribution function, 78
Anions, 33, see also specific types
Anisotropic random walk, 68—72
Anisotropy, 64, see also specific types
Antigens, 126, see also specific types
Arachidonic acid
  double dioxygenation of, 120
  functional responses to, 124—125
  metabolism of, 35—36, 116—124
    5-lipoxygenase pathway of, 116—119
    12-lipoxygenase pathway of, 119
    15-lipoxygenase pathway of, 119
    lipoxygenase product profiles and, 119—120
  stimulated release of, 114—116
Artificial surfaces, 125, see also specific types
Assays, see also specific types
  continuous, 28
  degranulation, 27, 28
  DNAase inhibition, 144, 145
  endpoint, 27
  of granule contents, 26
  necrotactic, 74
ATP, see Adenosine triphosphate

B

Bacterial toxins, 17, see also specific types
Band III-like proteins, 146
Biochemistry
  of calcium signal, 207—208
  of neutrophils, 6—10
  nineteenth century, 3—4
Biological effects, 124—130, see also specific types
Biophysics
  of cellular movement, 75—90
    internal programs of cells and, 84—90
    phenomenology and, 76—84
  of disturbed tactic cellular activity, 90—93

C

Calcium, 39, 199—214
  in activation of neutrophil functions, 208—213
  activators of, 168
  biochemistry of signal of, 207—208
  chemoattractants and, 205—206
  chemotaxis and, 212—213
  exocytosis and, 210—211
  fluxes of, 200
  gradients of, 151
  homeostasis of, 200—202
    in activated neutrophils, 204—206
    in resting neutrophils, 202—204
  importance of, 200
  in inositide metabolism, 178—180
  intracellular, 159, 220—224
  ions of, 30—32
  mobilization of, 128—129, 168
  NADPH-oxidase and, 208—210
  phagocytosis and, 211—212
  phorbol esters and, 206
  plasma membrane permeability to, 180
  potassium channel and, 250—251
  PPI-pole and, 183
  redistribution of, 156
  signal of, 224
Calcium ionophores, 179, 204, 220, see also specific types
Carriers, 253—255, see also specific types
Cell area density, 72
Cell-free system, 200
Cells, see also specific types
  activation of, 225—226
  communication of, 89
  directionary sensing activity of, 73
  disturbed tactic activity of, 90—93
  hybrids of, 201
  identification of, 2
  intact, 200—202
  internal programs of, 84—90
  intracellular pH and, 254—255
  intrinsic properties of, 84—90
  leaky, 200
  nonmuscle, 148, 152
  regulation of function of, 254—255
  response of, 16
  size of, 14—15
  tactic activity of, 90—93

viral interactions with, 91
Cell shape anisotropy, 64
Cellular movement, 142
actin filament behavior in, 157—159
biophysics of, see under Biophysics
change in direction of, 84—85
characteristic time of, 82
characterization of, see Characterization of cellular movement
chemotactic factor-stimulated, 145
directed, 72—76, 78
direction-sensing mechanism in, 87—90, 92
functional responses and, 125
intrinsic properties of cells and, 84—90
phagocytosis and, 147
phenomenology of, 76—84
stationary, 64
tactic, 73, 77
three-dimensional, 147—152
Cellular responders, 12, see also specific types
Cellular unpredictability, 17—18
Cellular unreliability, 12—14
CGD, see Chronic granulomatous disease
Characterization of cellular movement, 64—75, see also specific types
anisotropic random walk and, 68—72
chemokinesis and, 64—66
isotropic random walk and, 66—68
microscopical observation and, 64
stationary motion, 64
Chediak-Higashi syndrome, 40
Chemoattractants, 205—206, see also specific types
Chemokinesis, 64—66, 76
Chemostasis, 255
Chemotactic constant, 80
Chemotactic factors, 24, 145, see also specific types
Chemotactic peptide, 118, 121
Chemotaxis, 73, 76—82, 84
actin filament behavior in, 157—159
calcium and, 212—213
functional responses and, 125
soluble agonists for, 172—173
viruses and, 90—91
Chemotropism index, 73
Chloride channel, 251
Chloride transport, 253
Chlorotetracycline (CTC), 201
Chronic granulomatous disease (CGD), 98
Circulating neutrophils, 10
Concentration gradient, 73, 74
Contact angle, 82
Contact guidance, 69—72, 76—78, 82, 83, see also specific types
Continuous assays, 28
Contractile activity, 152
Contractile proteins, 148, 153, see also specific types
Contraction, 148
Cooperativity, 78—82
Correlated walker, 84
Cortex function and structure, 147
Cortical actin filament meshwork, 143—147
Coupling, see also specific types
Coxsackie B virus, 93
CTC, see Chlorotetracycline
Cyclic AMP, 130, 180, 181, 234
Cyclic GMP, 234
Cyclic nucleotides, 37—38, 234
Cyclooxygenase, 35, 116, 120, 121
Cyctochrome b, 99—100
Cytochalasin B, 172
Cytochalasin D, 154
Cytochalasins, 17, 153, 154, see also specific types
Cytochromes, 100, see also specific types
Cytoplasm, 154
Cytoplasmic protrusions, 153
Cytoskeleton, 36—37, 127—129

D

Decision activity, 73
Decision device, 77, 87
Degranulation, 24, 26—28, 126
DG, see Diacylglycerol
Diacylglycerol (DG), 159, 168, 173—176
formation of, 170
intracellular messengers and, 231—235
production of, 174
Diahorase, 101
Dietary fatty acids, 126
Diffusion coefficient, 68, 70, 71, 73, 76, 82
Diffusion law, 68, 71, 76
Diffusion mode, 73
Diffusion process, 82
Dipeptidases, 123
Diphenylene iodonium, 102
Directed locomotion, 72—76, 78
Directional memory, 9, 78, 87
Directionary sensing activity of cells, 73
Direction-sensing mechanisms, 87—90, 92
Displacements, 68, 69, 73
Disturbed tactic cellular activity, 90—93
DNAase, 144
DNAase I inhibition assay, 144, 145
Double dioxygenation, 120
Drift mode, 73
Drift velocity, 76
Dye reductase activities, 101—102

E

Echo virus, 90, 93
Eicosapentaenoic acid, 121, 124
5,8,11-Eicosatrienoic acid, 121, 124
Elastase, 25
Electrical methods of ion transport, 244—246
Electric field, 75
Electron transfer reactions, 102—104
Endothelium, 125
Endpoint assays, 27
Enterovirus, 90

Enzymes, 80, 123—124, see also specific types
Eosinophils, 121
Epoxide, 116
Epoxy acid, 119
Epoxy-eicosatrienoic acid, 119
Esters, 159, 180, 206, see also specific types
Exocytosis, 210—211
Experimental activators, 226—228, see also specific types
Experimental agents, 220—221, see also specific types
Experimental inhibitors, 228—231, see also specific types

F

FACS, see Fluorescence-activated cell sorter
Fatty acid hydroperoxide, 123
Fatty acids, 126, 234, see also specific types
Fibrillar organelle-free cytoplasm, 154
Filamin, 153
Flavoprotein, 100
Fluoride activation, 187
Fluorescence-activated cell sorter (FACS), 145
Force, 82—84, 86, see also specific types
Formyl-methionyl-leucyl-phenylalanine, 221
Functional responses, 124—126, see also specific types
Fusion, 38—39

G

Galvanotactic constant, 80
Galvanotaxis, 73, 75—77, 79, 84
Gamma-glutamyl transpeptidase, 123
Generating functions, 83
Glutathione, 118, 119
Glutathione transferase, 118, 124
GMP, 234
G-proteins, 29—30, 128
Granules, see also specific types
  assays of contents of, 26
  descriptions of, 25—26
  primary, 25
  secondary, 25, 40—41
  secretory, 24
  tertiary, 25—26
GTP, 232—233
GTPase activity, 187
Guanine nucleotide-binding proteins, 29—30, 128
Guanine nucleotide regulation of secretion, 187
Guanine nucleotide regulatory protein, 181—186, see also specific types

H

Hepatitis virus, 93
Herpes simplex virus, 93
History, 2—5
HL-60 cells, 120
Homeostasis, see also specific types
  of calcium, 200—202
    in activated neutrophils, 204—206
    in resting neutrophils, 202—204
Hybrids, 201, see also specific types
Hydrogen, 129—130
  intracellular concentration of, 159
  ions of, 234—235
Hydrolysis
  of leukotriene A, 117—118
  of polyphosphoinositides, 206
  of PPI, 173
Hydroperoxy, 119

I

Immunocytochemistry, 148, 153
Inappropriate activation of neutrophils, 5—11
Inappropriate response, 17
Inflammation, 2, 5, 25, 81, 90, 143, see also specific types
Inflammatory diseases, 98, see also specific types
Influenza, 93
Inhibitors, see also specific types
  of inositide metabolism, 180
  of kinase C, 228—231
  of superoxide generating oxidase, 105—106
Inositide, 178—180
Inositol lipid metabolism, 168—173
Inositol phosphates, 171, 233—234, see also specific types
Inositol trisphosphate, 173—174, 207—208
Instability, 77, 89
Intact cells, 200—202
Intact neutrophils, 178—179
Internal clock, 87, 89
Internal programs of cells, 84—90
Intracellular calcium, 220—224
Intracellular messengers, 224, 231—235, see also specific types
Intracellular pH, 253—255
Ionic channels, 248—253, see also specific types
  in macrophages, 249—252
  in neutrophils, 252—253
Ionic properties of neutrophils, 6—7
Ionomycin, 220
Ions, 30—33, see also specific types
Ion transport, 243—256, see also specific types
  carriers and, 253—255
  chemostasis and, 255
  electrical methods of, 244—246
  intracellular pH and, 254—255
  isotopic methods of, 246
  measurement of, 244
  methods of, 244—247, see also specific types
  optical methods of, 246—247
  oxidative burst and, 255
  pumps and, 253—255
  resting membrane potential and, 247
Iron-receptor channel, 251—252

Isoelectric point, 75
Isotopic methods of ion transport, 246
Isotropic random walk, 66—68
Isozymes, 25, see also specific types

K

Kinase C, 224—231
  experimental activators of, 226—228
  experimental inhibitors, 228—231
  fatty acids and, 234
  properties of, 224—225
Kinetics, 102, 148, see also specific types
  of degranulation, 27—28
  enzyme, 80

L

Langevin equation, 68, 86
Langevin force, 86
Leaky cells, 200
Leukocidin, 180
Leukotriene hydrolase, 118
Leukotrienes, 120, 123, 124, see also specific types
  functional responses to, 125—126
  hydrolysis of, 117, 118
  modulation of synthesis of, 120—121
  phospholipase C and, 128
  synthesis of, 116
Leukotriene synthetase, 124
Lifespan, 91, 142
Ligands, 29, see also specific types
Linoleic acid, 124
Lipids, 34—36, 39, see also specific types
  inositol, 168—173
  metabolism of, 168—170
5-Lipoxygenase pathway, 116—119
12-Lipoxygenase pathway, 119
15-Lipoxygenase pathway, 119
Lipoxygenase pathways, 123—124, see also specific types
Lipoxygenase products, 35—36, see also specific types
  metabolism of, 121—123
  profiles of, 119—120
Lipoxygenases, 125—126, see also specific types
Localization, 127
Locomotion, see Cellular movement

M

Macrophages, 247, 249—252
McCutcheon index, 73, 76
Mean displacements, 68, 69, 73
Measles, 93
Membrane potential
  changes in, 235
  in phagocyte function, 248
  resting, 247
Membrane properties of neutrophils, 9—10
Membranes, 78, see also specific types
  GTPase activity in, 187
  plasma, 180, 181
  PPI-pole and, 181—183
Memory, 78, 87, 89
Memory time, 78
Metabolism, see also specific types
  arachidonic acid, 35—36, 116—124
    5-lipoxygenase pathway of, 116—119
    12-lipoxygenase pathway of, 119
    15-lipoxygenase pathway of, 119
    lipoxygenase product profiles and, 119—120
  inositide, 178—180
  inositol lipid, 168, 170—173
  lipoxygenase product, 121—123
  oxidative, 126
  phosphatidylinositol, 30
Michaelis-Menten equation, 80
Microfilaments, 36—37, 85, 87
Microtubules, 37, 78, 85, 87
Monohydroxy acids, 123
MPO, see Myeloperoxidase
Multiple intracellular signaling, 235—236
Muscle, 143, see also specific types
Myeloperoxidase (MPO), 4, 25, 41
Myosin, 148, 152, 153
Myosin light chain kinase, 148

N

NADPH-oxidase, 208—210
Necrotactic assay, 74
Necrotaxis, 73, 76, 77
Nematic contact guidance, 69—71
Neutrophil oxidase, see Oxidase
Noise, 86, 92
Non-calcium intracellular messengers, 224, 231—235, see also specific types
Nonmitochondrial oxidase, 4
Nonmuscle cells, 148, 152
Nucleotides, 37—38, 234, see also specific types

O

Optical methods of ion transport, 246—247
Order-disorder transition, 91—93
Order parameter, 70, 72—74, 90, 91
Order parameter, 78—82
Orientation, 78
Osmolarity, 39
Osmotic swelling, 151
Oxidase, see also specific types
  dye reductase activities and, 101—102
  electron transfer reactions within, 102—104
  inhibition of, 104
  NADPH, 208—210
  purification of, 104—106
  reactions of, 101—106
  regulation of, 104
  superoxide generating, 105—106
Oxidation, 123
Oxidation-reduction potential, 99, 100

Oxidative burst, 172—173
Oxidative metabolism, 126
Oxygen reduction, 97—106
  neutrophil oxidase and, 101—106
  superoxide generating system and, 98—101
Oxylase, 4

P

PA, see Phosphatidate
Particulate stimuli, 222—223
Patch-clamp technique, 244—246
Pathology, 5—18
  cell response and, 16
  cell size and, 14—15
  cellular responders and, 12
  cellular unpredictability and, 17—18
  cellular unreliability and, 12—14
  inappropriate activation of neutrophils and, 5—11
  probability and, 16
Peptides, 118, 121, see also specific types
pH, 33, 253—255
Phagocytes, 248
Phagocytic stimuli, 172—173
Phagocytosis, 3, 5, 24, 143, 148
  actin filament behavior during, 152—156
  calcium and, 211—212
  cellular movement and, 147
  as continuous process, 152
  contractile proteins during, 153
  cytochalasins in, 154
Phagosome, 4, 5
Phalloidin, 144
Phenomenology, 76—84
Phorbol esters, 159, 180, 206
Phosphates, 171, 233—234, see also specific types
Phosphatidate (PA), 169, 173—176
Phosphatidylinositol (PI), 169, 173—176, 178
Phosphatidylinositol 4,5-bisphosphate (PIP), 168
Phosphatidylinositol metabolism, 30
Phospholipase A, 176—178
Phospholipase C, 128
Phospholipases, 34—35, see also specific types
Phospholipid-dependent protein kinase (protein kinase C), 168
Phosphorylation
  cell activation and, 225—226
  protein, 33
  protein kinase and, 226
Photoproteins, 201, see also specific types
Physical properties of neutrophils, 6—10
PI, see Phosphatidylinositol
PIP, see Phosphatidylinositol 4,5-bisphosphate
Plague forming units, 92
Plasma membranes, 180, 181
Platelet activating factor, 36
Platelets, 119, 121
Polar field, 73, 74, 76, 78
Polar-order parameter, 78—82, 90, 91
Polymerized actin, 145
Polyphosphoinositides, 206, see also specific types
Potassium channels, 249—251
Potassium transport, 253
PPI, 173—176
Primary granules, 25
Probability, 14, 16—18
Profilin, 144
Proteases, 25, see also specific types
Protein kinase C, 130, 206
Protein kinases, 33—34, see also specific types
  phospholipid-dependent (protein kinase C), 168
  phosphorylation and, 226
Protein phosphorylation, 33
Proteins, see also specific types
  actin filament-binding, 146
  band III-like, 146
  contractile, 148, 153
  G-, 29—30, 128
  guanine nucleotide binding, 29—30, 128
  guanine nucleotide regulatory, 181—186
Protrusion, 148, 153
Pumps, 253—255, see also specific types

Q

Quin2, 201—202

R

Radioactivity, 91
Random walk, 76, 86
  anisotropic, 68—72
  isotropic, 66—68
Receptors, 127—128, see also specific types
  agonists of, 221—222
  ligand interactions with, 29
  surface, 126
Refractory time, 89
Respiratory burst, 98
Response triggering, 220—223
  calcium-dependent, 224—231
  calcium-independent, 224—231
Resting membrane potential, 247

S

Secondary granules, 25, 40—41
Secretagogues, 26—27
Secretion, 23—41
  degranulation and, 26—28
  disorders of, 40—41, see also specific types
  fusion and, 38—39
  granule descriptions and, 25—26
  guanine nucleotide regulation of, 187
  soluble agonists for, 172—173
  stimulus-response coupling and, see Stimulus-response coupling
Secretory granules, 24
Sensory machinery, 77, 87
Signal transduction, 126—130
Skeletal muscle, 143
Smectic contact guidance, 71—72

Sodium/hydrogen antiport, 129—130
Sodium transport, 253
Soluble agonists, 172—173
Source of neutrophils, 171—172
Specific granules, see Secondary granules
Stationary cellular motion, 64
Stimulated arachidonic acid release, 114—116
Stimulated inositol lipid metabolism, 168—170
Stimulated neutrophils, 175—176
Stimuli, see also specific types
  for neutrophil function, 172—173, see also specific types
  particulate, 222—223
Stimulus-response coupling, 24, 28—30
  cyclic nucleotides and, 37—38
  cytoskeleton and, 36—37
  G-proteins and, 29—30
  ions and, 30—33
  lipids and, 34—35
  phosphatidylinositol metabolism and, 30
  protein kinases and, 33—34
  receptor-ligand interactions and, 29
Stoichiometry, 144
Subcellular fractions, 200
Superoxide generating oxidase inhibitors, 105—106
Superoxide generating system, 98—101
Surface antigens, 126
Surface receptors, 126
Surface tension, 78, 82
Swelling, 151, see also specific types
Symmetry, 66, 69, 73, 76, 78
Synexins, 39, see also specific types

T

Tactic locomotion, 73
Tactic movement, 77
Tension, surface, 78, 82
Tertiary granules, 25—26
Tetracarboxylates, 201—202, see also specific types
Three-dimensional movement, 147—152
Threshold viral dose, 91, 92
Torque, 82—84
Toxins, 17, see also specific types
Track velocity, 66, 73, 82
Transduction mechanisms, 124—130, see also specific types
Tropomyosin, 148

U

Ubiquinone, 101

V

Velocity, 82, 85, see also specific types
  drift, 76
  track, 66, 73, 82
Virtual force, 82—84
Viruses, 90—91, see also specific types
  cell interactions with, 91
  order-disorder transition and, 91—93
  threshold dose of, 91, 92
Voltage-activated potassium channel, 250—251

W

Weber-Fechner law, 91
White noise, 86

Z

Zigmond chamber, 74
Zymosan, 118